Determined

ALSO BY ROBERT M. SAPOLSKY

Behave: The Biology of
Humans at Our Best and Worst

Monkeyluv: And Other
Essays on Our Lives as Animals

A Primate's Memoir

The Trouble with Testosterone and Other
Essays on the Biology of the Human Predicament

Why Zebras Don't Get Ulcers: A Guide
to Stress, Stress-Related Diseases, and Coping

Stress, the Aging Brain, and
the Mechanisms of Neuron Death

Determined

A Science of Life without Free Will

Robert M. Sapolsky

PENGUIN PRESS
NEW YORK
2023

PENGUIN PRESS
An imprint of Penguin Random House LLC
penguinrandomhouse.com

Library of Congress Cataloging-in-Publication Data

Names: Sapolsky, Robert M., author.
Title: Determined : a science of life without free will / Robert M. Sapolsky.
Description: New York : Penguin Press, 2023. |
Includes bibliographical references and index.
Identifiers: LCCN 2023023790 (print) | LCCN 2023023791 (ebook) |
ISBN 9780525560975 (hardcover) | ISBN 9780525560982 (ebook)
Subjects: LCSH: Free will and determinism.
Classification: LCC BJ1461 .S325 2023 (print) |
LCC BJ1461 (ebook) | DDC 123/.5—dc23/eng/20230705
LC record available at https://lccn.loc.gov/2023023790
LC ebook record available at https://lccn.loc.gov/2023023791

ISBN 9780593656723 (international edition)

Printed in the United States of America
2nd Printing

Designed by Alexis Farabaugh

To L, and to B & R,

Who make it all seem worth it.

Who make it worth it.

CONTENTS

Determined

my brain: click them

me: why?

my brain: you gotta

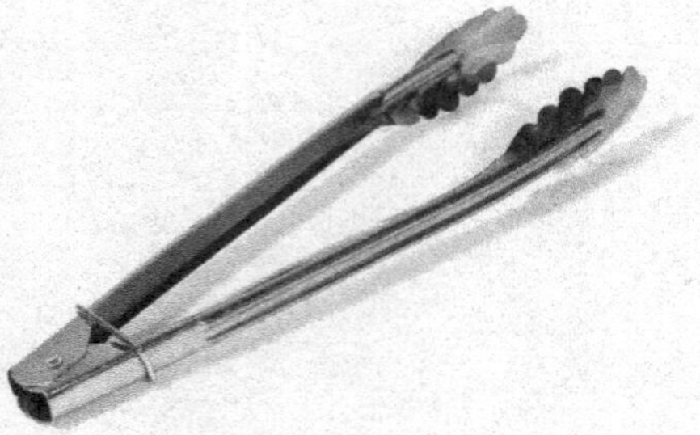

1

Turtles All the Way Down

When I was in college, my friends and I had an anecdote that we retold frequently; it went like this (and our retelling was so ritualistic that I suspect this is close to verbatim, forty-five years later):

> So, it seems that William James was giving a lecture about the nature of life and the universe. Afterward, an old woman came up and said, "Professor James, you have it all wrong."
>
> To which James asked, "How so, madam?"
>
> "Things aren't at all like you said," she replied. "The world is on the back of a gigantic turtle."
>
> "Hmm." said James, bemused. "That may be so, but where does that turtle stand?"
>
> "On the back of another turtle," she answered.
>
> "But madam," said James indulgently, "where does that turtle stand?"
>
> To which the old woman responded triumphantly. "It's no use, Professor James. It's turtles all the way down!"*

*The "turtles all the way down" story has versions featuring other celebrated thinkers as the fall guy, rather than William James. We told our version because we liked James's beard, and there was a building on campus named for him. The "turtles all the way down" punch line has been referenced in numerous cultural contexts, including a great book with that title by John Green

Oh, how we loved that story, always told it with the same intonation. We thought it made us seem droll and pithy and attractive.

We used the anecdote as mockery, a pejorative critique of someone clinging unshakably to illogic. We'd be in the dinner hall, and someone had said something nonsensical, where their response to being challenged had made things worse. Inevitably, one of us would smugly say, "It's no use, Professor James!" to which the person, who had heard our stupid anecdote repeatedly, would inevitably respond, "Screw you, just listen. This actually makes sense."

Here is the point of this book: While it may seem ridiculous and nonsensical to explain something by resorting to an infinity of turtles all the way down, *it actually is much more ridiculous and nonsensical to believe that somewhere down there, there's a turtle floating in the air*. The science of human behavior shows that turtles can't float; instead, it is indeed turtles all the way down.

Someone behaves in a particular way. Maybe it's wonderful and inspiring, maybe it's appalling, maybe it's in the eye of the beholder, or maybe just trivial. And we frequently ask the same basic question: Why did that behavior occur?

If you believe that turtles can float in the air, the answer is that it just happened, that there was no cause besides that person having simply decided to create that behavior. Science has recently provided a much more accurate answer, and when I say "recently," I mean in the last few centuries. The answer is that the behavior happened because something that preceded it caused it to happen. And why did that prior circumstance occur? Because something that preceded it caused *it* to happen. It's antecedent causes all the way down, not a floating turtle or causeless cause to be found. Or as Maria sings in *The Sound of Music*, "Nothing comes from nothing, nothing ever could."*

(Dutton Books, 2017). All the versions of the story have a male Philosopher King Whoever being challenged by an absurd old woman, which now seems kind of sexist and ageist. That didn't particularly register with us then, we adolescent males of our time and place.

*My wife is a musical theater director, and I'm her rusty rehearsal pianist/generalized gofer; as a result, this book is riddled with allusions to musicals. If my college self, being ostensibly cool by

To reiterate, when you behave in a particular way, which is to say when your brain has generated a particular behavior, it is because of the determinism that came just before, which was caused by the determinism just before that, and before that, all the way down. The approach of this book is to show how that determinism works, to explore how the biology over which you had no control, interacting with environment over which you had no control, made you you. And when people claim that there are causeless causes of your behavior that they call "free will," they have (a) failed to recognize or not learned about the determinism lurking beneath the surface and/or (b) erroneously concluded that the rarefied aspects of the universe that do work indeterministically can explain your character, morals, and behavior.

Once you work with the notion that every aspect of behavior has deterministic, prior causes, you observe a behavior and can answer why it occurred: as just noted, because of the action of neurons in this or that part of your brain in the preceding second.[†] And in the seconds to minutes before, those neurons were activated by a thought, a memory, an emotion, or sensory stimuli. And in the hours to days before that behavior occurred, the hormones in your circulation shaped those thoughts, memories, and emotions and altered how sensitive your brain was to particular environmental stimuli. And in the preceding months to years, experience and environment changed how those neurons function, causing some to sprout new connections and become more excitable, and causing the opposite in others.

And from there, we hurtle back decades in identifying antecedent

---

referring to William James, had been told that my future included my family and me debating who was the greatest Elphaba of all time,[*] I would have been astonished—"Musicals? Broadway MUSICALS?! What about atonalism?" It's not what I asked for; sometimes life just slips in through a back door.

([*]Idina Menzel. Obviously.)

[†]The appendix is an introduction to neuroscience, for readers without a background in this area. Also, anyone who has read an agonizingly long book that I wrote (*Behave: The Biology of Humans at Our Best and Worst*, Penguin Press, 2017) will recognize the book summarized in the next few paragraphs: Why did that behavior occur? Because of events one second before, one minute . . . one century . . . one hundred million years before.

causes. Explaining why that behavior occurred requires recognizing how during your adolescence a key brain region was still being constructed, shaped by socialization and acculturation. Further back, there's childhood experience shaping the construction of your brain, with the same then applying to your fetal environment. Moving further back, we have to factor in the genes you inherited and their effects on behavior.

But we're not done yet. That's because everything in your childhood, starting with how you were mothered within minutes of birth, was influenced by culture, which means as well by the centuries of ecological factors that influenced what kind of culture your ancestors invented, and by the evolutionary pressures that molded the species you belong to. Why did that behavior occur? Because of biological and environmental interactions, all the way down.*

As a central point of this book, those are all variables that you had little or no control over. You cannot decide all the sensory stimuli in your environment, your hormone levels this morning, whether something traumatic happened to you in the past, the socioeconomic status of your parents, your fetal environment, your genes, whether your ancestors were farmers or herders. Let me state this most broadly, probably at this point too broadly for most readers: we are nothing more or less than the cumulative biological and environmental luck, over which we had no control, that has brought us to any moment. You're going to be able to recite this sentence in your irritated sleep by the time we're done.

There are all sorts of aspects about behavior that, while true, are not relevant to where we're heading. For example, the fact that some criminal behavior can be due to psychiatric or neurological problems. That some kids have "learning differences" because of the way their brains work. That some people have trouble with self-restraint, because they grew up without any decent role models or because they're still a teenager with a

*"Interactions" implies that those biological influences are meaningless outside the context of social environment (as well as the reverse). They're inseparable. My orientation happens to be biological, and analyzing the inseparability from that angle is clearest in my mind. But at times, framing the inseparability from a biological rather than a social science perspective makes things clunky; I've tried to avoid that to the best of my biologist abilities.

teenager's brain. That someone has said something hurtful merely because they're tired and stressed, or even because of a medication they're taking.

All of these are circumstances where we recognize that sometimes, biology can *impinge on* our behavior. This is essentially a nice humane agenda that endorses society's general views about agency and personal responsibility but reminds you to make exceptions for edge cases: judges should consider mitigating factors in criminals' upbringing during sentencing; juvenile murderers shouldn't be executed; the teacher handing out gold stars to the kids who are soaring in learning to read should do something special too for that kid with dyslexia; college admissions officers should consider more than just SAT cutoffs for applicants who have overcome unique challenges.

These are good, sensible ideas that should be instituted if you decide that *some* people have much less self-control and capacity to freely choose their actions than average, and that at times, we *all* have much less than we imagine.

We can all agree on that; however, we're heading into very different terrain, one that I suspect most readers will not agree with, which is deciding that we have *no* free will at all. Here would be some of the logical implications of that being the case: That there can be no such thing as blame, and that punishment as retribution is indefensible—sure, keep dangerous people from damaging others, but do so as straightforwardly and nonjudgmentally as keeping a car with faulty brakes off the road. That it can be okay to praise someone or express gratitude toward them as an instrumental intervention, to make it likely that they will repeat that behavior in the future, or as an inspiration to others, but never because they *deserve* it. And that this applies to you when you've been smart or self-disciplined or kind. Oh, as long as we're at it, that you recognize that the experience of love is made of the same building blocks that constitute wildebeests or asteroids. That no one has *earned* or is *entitled* to being treated better or worse than anyone else. And that it makes as little sense to hate someone as to hate a tornado because it supposedly decided to level your house, or to love a lilac because it supposedly decided to make a wonderful fragrance.

That's what it means to conclude that there is no free will. This is what I've concluded, for a long, long time. And even I think that taking that seriously sounds absolutely nutty.

Moreover, most people agree that it sounds that way. People's beliefs and values, their behavior, their answers to survey questions, their actions as study subjects in the nascent field of "experimental philosophy," show that people believe in free will when it matters—philosophers (about 90 percent), lawyers, judges, jurors, educators, parents, and candlestick makers. As well as scientists, even biologists, even many neurobiologists, when push comes to shove. Work by psychologists Alison Gopnik at UC Berkeley and Tamar Kushnir at Cornell shows that preschool kids already have a robust belief in a recognizable version of free will. And such a belief is widespread (but not universal) among a wide variety of cultures. We are not machines in most people's view; as a clear demonstration, when a driver or an automated car makes the same mistake, the former is blamed more.[1] And we are not alone in our faith in free will—research that we'll look at in a later chapter suggests that other primates even believe that there is free will.[2]

This book has two goals. The first is to convince you that there is no free will,* or at least that there is much *less* free will than generally assumed when it really matters. To accomplish that, we'll look at the way smart, nuanced thinkers argue *for* free will, from the perspectives of philosophy, legal thought, psychology, and neuroscience. I'll be trying to present their views to the best of my ability, and to then explain why I think they are all mistaken. Some of these mistakes arise from the myopia

*Some of the most extreme "there's NO free will" fellow travelers include philosophers such as Gregg Caruso, Derk Pereboom, Neil Levy, and Galen Strawson; I'll often be discussing their thinking in the pages to come; as an important point, while all reject free will in the everyday sense we understand it when justifying punishment and reward, their rejection is not particularly along biological grounds. In terms of rejecting free will almost entirely on biological grounds, my views are closest to those of Sam Harris, who, appropriately, is not only a philosopher, but a neuroscientist as well.

(used in a descriptive rather than judgmental sense) of focusing solely on just one sliver of the biology of behavior. Sometimes this is because of faulty logic, such as concluding that if it's not possible to ever tell what caused X, maybe nothing caused it. Sometimes the mistakes reflect unawareness or misinterpretation of the science underlying behavior. Most interestingly, I sense that mistakes arise for emotional reasons that reflect that there being no free will is pretty damn unsettling; we'll consider this at the end of the book. So one of my two goals is to explain why I think all these folks are wrong, and how life would improve if people stopped thinking like them.[3]

Right around here, one might ask of me, Where do you get off? As will be seen, free-will debates often revolve around narrow issues—"Does a particular hormone actually cause a behavior or just make it more likely?" or "Is there a difference between wanting to do something and wanting to want something?"—that are usually debated by specialized authorities. My intellectual makeup happens to be that of a generalist. I'm a "neurobiologist" with a lab that does things like manipulate genes in a rat's brain to change behavior. At the same time, I spent part of each year for more than three decades studying the social behavior and physiology of wild baboons in a national park in Kenya. Some of my research turned out to be relevant to understanding how adult brains are influenced by the stress of childhood poverty, and as a result, I've wound up spending time around the likes of sociologists; another facet of my work has been relevant to mood disorders, leading me to hang with psychiatrists. And for the last decade, I've had a hobby of working with public defender offices on murder trials, teaching juries about the brain. As a result, I've been carpetbagging in a number of different fields related to behavior. Which I think has made me particularly prone toward deciding that free will doesn't exist.

Why? Crucially, if you focus on any single field like these—neuroscience, endocrinology, behavioral economics, genetics, criminology, ecology, child development, or evolutionary biology—you are left with plenty of wiggle room for deciding that biology and free will can coexist. In the words of UC San Diego philosopher Manuel Vargas, "Claiming that some scientific result shows the falsity of 'free will' . . . is either bad scholarship or academic hucksterism."[4] He is right, if in-your-face. As we will see in the next chapter, most experimental neurobiology research about free will is narrowly anchored by the result of one study that examined events that happen in the brain a few seconds before a behavior occurs. And Vargas would correctly conclude that this "scientific result" (plus the spin-offs it has generated in the subsequent forty years) doesn't prove there's no free will. Similarly, you can't disprove free will with a "scientific result" from genetics—genes in general are not about inevitability but, rather, about vulnerability and potential, and no single gene, gene variant, or gene mutation has ever been identified that falsifies free will;* you can't even do it when considering *all* our genes at once. And you can't disprove free will from a developmental/sociological perspective by emphasizing the scientific result that a childhood filled with abuse, deprivation, neglect, and trauma astronomically increases the odds of producing a deeply damaged and damaging adult—because there are exceptions. Yeah, no single result or scientific discipline can do that. But—and this is the incredibly important point—*put all the scientific result*s *together*, *from all the relevant scientific discipline*s, and there's no room for free will.†

Why is that? Something deeper than the idea that if you examine enough different disciplines, one -ology after another, you're bound to

---

*That said, there are a few rare diseases that are guaranteed to alter behavior because of a mutation in a single gene (e.g., Tay-Sachs, Huntington's, and Gaucher diseases). Nonetheless, this isn't remotely related to issues of our everyday sense of free will, as these diseases cause massive damage in the brain.

†I'd like to note something in preparation for my spending the first half of the book repeatedly saying, "They're all wrong," about a lot of scholars thinking about this subject. I can be intensely emotional about ideas, with some evoking the closest I can ever feel to religious awe and others seeming so appallingly wrong that I can be bristly, acerbic, arrogantly judgmental, hostile, and unfair in how I critique them. But despite that, I am majorly averse to interpersonal conflict. In other words, with a few exceptions that will be clear, none of my criticisms are meant to be personal. And as a "some of my best friends" cliché, I like being around people with a particular type of belief in free will, because they're generally nicer people than those on "my side" and because I hope some of their peace will rub off on me. What I'm trying to say is that I hope I won't be sounding like a jerk at times, because I very much don't want to.

eventually find one that provides a slam dunk, falsifying free will all by itself. It is also deeper than the idea that even though each discipline has a hole that precludes it from falsifying free will, at least one of the other disciplines compensates for it.

Crucially, all these disciplines collectively negate free will because they are all interlinked, constituting the same ultimate body of knowledge. If you talk about the effects of neurotransmitters on behavior, you are also implicitly talking about the genes that specify the construction of those chemical messengers, and the evolution of those genes—the fields of "neurochemistry," "genetics," and "evolutionary biology" can't be separated. If you examine how events in fetal life influence adult behavior, you are also automatically considering things like lifelong changes in patterns of hormone secretion or in gene regulation. If you discuss the effects of mothering style on a kid's eventual adult behavior, by definition you are also automatically discussing the nature of the culture that the mother passes on through her actions. There's not a single crack of daylight to shoehorn in free will.

As such, the first half of the book's point is to rely on this biological framework in rejecting free will. Which brings us to the second half of the book. As noted, I haven't believed in free will since adolescence, and it's been a moral imperative for me to view humans without judgment or the belief that anyone deserves anything special, to live without a capacity for hatred or entitlement. And I just can't do it. Sure, sometimes I can sort of get there, but it is rare that my immediate response to events aligns with what I think is the only acceptable way to understand human behavior; instead, I usually fail dismally.

As I said, even I think it's crazy to take seriously all the implications of there being no free will. And despite that, the goal of the second half of the book is to do precisely that, both individually and societally. Some chapters consider scientific insights about how we might go about dispensing with free-will belief. Others examine how some of the implications of rejecting free will are not disastrous, despite initially seeming that way.

Some review historical circumstances that demonstrate something crucial about the radical changes we'd need to make in our thinking and feeling: *we've done it before.*

The book's intentionally ambiguous title reflects these two halves—it is both about the science of why there is no free will and the science of how we might best live once we accept that.

STYLES OF VIEWS: WHOM I WILL BE DISAGREEING WITH

I'm going to be discussing some of the common attitudes held by people writing about free will. These come in four basic flavors:*

The world is deterministic and there's no free will. In this view, if the former is the case, the latter has to be as well; determinism and free will are not compatible. I am coming from this perspective of "hard incompatibilism."†

The world is deterministic and there is free will. These folks are emphatic that the world is made of stuff like atoms, and life, in the elegant words of psychologist Roy Baumeister (currently at the University of Queensland in

*Note: I won't be considering any theologically based Judeo-Christian views about these subjects beyond this broad summary here. As far as I can tell, most of the theological discussions center around omniscience—if God's all-knowingness includes knowing the future, how can we ever freely, willingly choose between two options (let alone be judged for our choice)? Amid the numerous takes on this, one answer is that God is outside of time, such that past, present, and future are meaningless concepts (implying, among other things, that God could never relax by going to a movie and being pleasantly surprised by a plot turn—He always knows that the butler didn't do it). Another answer is one of the limited God, something explored by Aquinas—God cannot sin, cannot make a boulder too heavy for Him to lift, cannot make a square circle (or, as another example that I've seen offered by a surprising number of male but not female theologians, even God cannot make a married bachelor). In other words, God cannot do *everything,* He can just do whatever is possible, and foreseeing whether someone will choose good or evil is not knowable, even for Him. Related to this all, Sam Harris mordantly notes that even if we each have a soul, we sure didn't get to pick it.

†Which I'm viewing as synonymous with "hard determinism"; all sorts of philosophers, however, make fine distinctions between the two.

Australia), "is based on the immutability and relentlessness of the laws of nature."[5] No magic or fairy dust involved, no substance dualism, the view where brain and mind are separate entities.* Instead, this deterministic world is viewed as compatible with free will. This is roughly 90 percent of philosophers and legal scholars, and the book will most often be taking on these "compatibilists."

The world is not deterministic; there's no free will. This is an oddball view that everything important in the world runs on randomness, a supposed basis of free will. We'll get to this in chapters 9 and 10.

The world is not deterministic; there is free will. These are folks who believe, like I do, that a deterministic world is not compatible with free will—however, no problem, the world isn't deterministic in their view, opening a door for free-will belief. These "libertarian incompatibilists" are a rarity, and I'll only occasionally touch on their views.

There's a related quartet of views concerning the relationship between free will and moral responsibility. The last word obviously carries a lot of baggage with it, and the sense in which it is used by people debating free will typically calls forth the concept of *basic desert,* where someone can *deserve* to be treated in a particular way, where the world is a morally acceptable place in its recognition that one person can deserve a particular reward, another a particular punishment. As such, these views are:

There's no free will, and thus holding people morally responsible for their actions is wrong. Where I sit. (And as will be covered in chapter 14, this is completely separate from forward-looking issues of punishment for deterrent value.)

There's no free will, but it is okay to hold people morally responsible for their

*Compatibilists make that clear. For example, one paper in the field is entitled "Free Will and Substance Dualism: The Real Scientific Threat to Free Will?" For the author, there's actually no threat to free will; there's a threat, though, of irksome scientists thinking they've scored points against compatibilists by labeling them as substance dualists. Because, to paraphrase a number of compatibilist philosophers, saying that free will doesn't exist because substance dualism is mythical is like saying that love doesn't exist because Cupid is mythical.

actions. This is another type of compatibilism—an absence of free will and moral responsibility coexist without invoking the supernatural.

There's free will, and people should be held morally responsible. This is probably the most common stance out there.

There's free will, but moral responsibility isn't justified. This is a minority view; typically, when you look closely, the supposed free will exists in a very narrow sense and is certainly not worth executing people about.

Obviously, imposing these classifications on determinism, free will, and moral responsibility is wildly simplified. A key simplification is pretending that most people have clean "yes" or "no" answers as to whether these states exist; the absence of clear dichotomies leads to frothy philosophical concepts like partial free will, situational free will, free will in only a subset of us, free will only when it matters or only when it doesn't. This raises the question of whether the edifice of free-will belief is crumbled by one flagrant, highly consequential exception and, conversely, whether free-will skepticism collapses when the opposite occurs. Focusing on gradations between yes and no is important, since interesting things in the biology of behavior are often on continua. As such, my fairly absolutist stance on these issues puts me way out in left field. Again, my goal isn't to convince you that there's no free will; it will suffice if you merely conclude that there's so much less free will than you thought that you have to change your thinking about some truly important things.

Despite starting by separating determinism / free will and free will / moral responsibility, I follow the frequent convention of merging them into one. Thus, my stance is that because the world is deterministic, there can't be free will, and thus holding people morally responsible for their actions is not okay (a conclusion described as "deplorable" by one leading philosopher whose thinking we're going to dissect big time). This incompatibilism will be most frequently contrasted with the compatibilist view that while the world is deterministic, there is still free will, and thus holding people morally responsible for their actions is just.

This version of compatibilism has produced numerous papers by philosophers and legal scholars concerning the relevance of neuroscience to free will. After reading lots of them, I've concluded that they usually boil down to three sentences:

a. Wow, there've been all these cool advances in neuroscience, all reinforcing the conclusion that ours is a deterministic world.

b. Some of those neuroscience findings challenge our notions of agency, moral responsibility, and deservedness so deeply that one must conclude that there is no free will.

c. Nah, it still exists.

Naturally, a lot of time will be spent examining the "nah" part. In doing so, I'll consider only a subset of such compatibilists. Here's a thought experiment for identifying them: In 1848 at a construction site in Vermont, an accident with dynamite hurled a metal rod at high speed into the brain of a worker, Phineas Gage, and out the other side. This destroyed much of Gage's frontal cortex, an area central to executive function, long-term planning, and impulse control. In the aftermath, "Gage was no longer Gage," as stated by one friend. Formerly sober, reliable, and the foreman of his work crew, Gage was now "fitful, irreverent, indulging at times in the grossest profanity (which was not previously his custom) . . . obstinate, yet capricious and vacillating," as described by his doctor. Phineas Gage is the textbook case that we are the end products of our material brains. Now, 170 years later, we understand how the unique function of your frontal cortex is the result of your genes, prenatal environment, childhood, and so on (wait for chapter 4).

Now the thought experiment: Raise a compatibilist philosopher from birth in a sealed room where they never learn anything about the brain. Then tell them about Phineas Gage and summarize our current knowledge about the frontal cortex. If their immediate response is "Whatever, there's still free will," I'm not interested in their views. The compatibilist I have in mind is one who then wonders, "OMG, what if I'm completely wrong about free will?," ponders hard for hours or decades, and concludes that there's still free will, here's why, and it's okay for society to hold people morally responsible for their actions. If a compatibilist has not wrestled through being challenged by knowledge of the biology of who we are, it's not worth the time trying to counter their free-will belief.

GROUND RULES AND DEFINITIONS

What is free will? Groan, we have to start with that, so here comes something totally predictable along the lines of "Different things to different types of thinkers, which gets confusing." Totally uninviting. Nevertheless, we have to start there, followed by "What is determinism?" I'll do my best to mitigate the drag of this.

What Do I Mean by Free Will*?*

People define *free will* differently. Many focus on agency, whether a person can control their actions, act with intent. Other definitions concern whether, when a behavior occurs, the person knows that there are alternatives available. Others are less concerned with what you do than with vetoing what you don't want to do. Here's my take.

Suppose that a man pulls the trigger of a gun. Mechanistically, the muscles in his index finger contracted because they were stimulated by a neuron having an action potential (i.e., being in a particularly excited state). That neuron in turn had its action potential because it was stimulated by the neuron just upstream. Which had its own action potential because of the next neuron upstream. And so on.

Here's the challenge to a free willer: Find me the neuron that started this process in this man's brain, the neuron that had an action potential for no reason, where no neuron spoke to it just before. Then show me that this neuron's actions were not influenced by whether the man was tired, hungry, stressed, or in pain at the time. That nothing about this neuron's function was altered by the sights, sounds, smells, and so on, experienced by the man in the previous minutes, nor by the levels of any hormones marinating his brain in the previous hours to days, nor whether he had experienced a life-changing event in recent months or years. And show me that this neuron's supposedly freely willed functioning wasn't affected by the man's genes, or by the lifelong changes in regulation of those genes caused by experiences during his childhood. Nor by levels of hormones he was exposed to as a fetus, when that brain was being constructed. Nor by the centuries of history and ecology that shaped the invention of the culture in which he was raised. Show me a neuron being a causeless cause in this total sense. The prominent compatibilist philosopher Alfred Mele of Florida State University emphatically feels that requiring something like that of free will is setting the bar "absurdly high."[6] But this bar is neither absurd nor too high. Show me a neuron (or brain) whose generation of a behavior is independent of the sum of its biological past, and for the purposes of this book, you've demonstrated free will. The point of the first half of this book is to establish that this can't be shown.

What Do I Mean by Determinism*?*

It's virtually required to start this topic with the dead White male Pierre Simon Laplace, the eighteenth-/nineteenth-century French polymath (it's also required that you call him a polymath, as he contributed to mathematics, physics, engineering, astronomy, and philosophy). Laplace provided the canonical claim for all of determinism: If you had a superhuman who knew the location of every particle in the universe at this moment, they'd be able to accurately predict every moment in the future. Moreover, if this superhuman (eventually termed "Laplace's demon") could re-create the exact location of every particle at any point in the past, it would lead to a present identical to our current one. The past and future of the universe are already determined.

Science since Laplace's time shows that he wasn't completely right (proving that Laplace was not a Laplacian demon), but the spirit of his demon lives on. Contemporary views of determinism have to incorporate the fact that certain types of predictability turn out to be impossible (the subject of chapters 5 and 6) and certain aspects of the universe are actually nondeterministic (chapters 9 and 10).

Moreover, contemporary models of determinism must also accommodate the role played by meta-level consciousness. What do I mean by this? Consider a classic psychology demonstration of people having less freedom in their choices than they assumed.[7] Ask someone to name their favorite detergent, and if you have unconsciously cued them earlier with the word *ocean*, they become more likely to answer, "Tide." As an important measure of where meta-level consciousness comes in, suppose the person realizes what the researcher is up to and, wanting to show that they can't be manipulated, decides that they won't say "Tide," even if it is their favorite. Their freedom has been just as constrained, a point in many of the coming chapters. Similarly, wind up as an adult exactly like your parents or the exact opposite of them, and you are equally unfree—in the latter case, the pull toward adopting their behavior, the ability to consciously recognize that tendency to do that, the mindset to recoil from that with horror and thus do the opposite, are all manifestations of the ways that you became you outside your control.

Finally, any contemporary view of determinism must accommodate a profoundly important point, one that dominates the second half of the book—despite the world being deterministic, things can change. Brains change, behaviors change. We change. And that doesn't counter this being a deterministic world without free will. In fact, the science of change *strengthens* the conclusion; this will come in chapter 12.

With those issues in mind, time to see the version of determinism that this book builds on.

Imagine a university graduation ceremony. Almost always moving, despite the platitudes, the boilerplate, the kitsch. The happiness, the pride. The families whose sacrifices now all seem worth it. The graduates who were the first in their family to finish high school. The ones whose immigrant parents sit there glowing, their saris, dashikis, barongs broadcasting that their pride in the present isn't at the cost of pride in their past.

And then you notice someone. Amid the family clusters postceremony, the new graduates posing for pictures with Grandma in her wheelchair, the bursts of hugs and laughter, you see the person way in the back, the person who is part of the grounds crew, collecting the garbage from the cans on the perimeter of the event.

Randomly pick any of the graduates. Do some magic so that this garbage collector started life with the graduate's genes. Likewise for getting the womb in which nine months were spent and the lifelong epigenetic consequences of that. Get the graduate's childhood as well—one filled with, say, piano lessons and family game nights, instead of, say, threats of going to bed hungry, becoming homeless, or being deported for lack of papers. Let's go all the way so that, in addition to the garbage collector having gotten all that of the graduate's past, the graduate would have gotten the garbage collector's past. Trade every factor over which they had no control, and you will switch who would be in the graduation robe and who would be hauling garbage cans. This is what I mean by determinism.

AND WHY DOES THIS MATTER?

Because we all know that the graduate and the garbage collector would switch places. And because, nevertheless, we rarely reflect on that sort of fact; we congratulate the graduate on all she's accomplished and move out of the way of the garbage guy without glancing at him.

2

The Final Three Minutes of a Movie

Two men stand by a hangar in a small airfield at night. One is in a police officer's uniform, the other dressed as a civilian. They talk tensely while, in the background, a small plane is taxiing to the runway. Suddenly, a vehicle pulls up and a man in a military uniform gets out. He and the police officer talk tensely; the military man begins to make a phone call; the civilian shoots him, killing him. A vehicle full of police pulls up abruptly, the police emerging rapidly. The police officer speaks to them as they retrieve the body. They depart as abruptly, with the body but not the shooter. The police officer and the civilian watch the plane take off and then walk off together.

What's going on? A criminal act obviously occurred—from the care with which the civilian aimed, he clearly intended to shoot the man. A terrible act, compounded further by the man's remorseless air—this was cold-blooded murder, depraved indifference. It is puzzling, though, that the police officer made no attempt to apprehend him. Possibilities come to mind, none good. Perhaps the officer has been blackmailed by the civilian to look the other way. Maybe all the police who appeared on the scene are corrupt, in the pocket of some drug cartel. Or perhaps the police officer is actually an impostor. One can't be certain, but it's clear that this was a scene of intent-filled corruption and lawless violence, the police officer and the civilian exemplars of humans at their worst. That's for sure.

Intent features heavily in issues about moral responsibility: Did the person intend to act as she did? When exactly was the intent formed? Did she know that she could have done otherwise? Did she feel a sense of ownership of her intent? These are pivotal issues to philosophers, legal scholars, psychologists, and neurobiologists. In fact, a huge percentage of the research done concerning the free-will debate revolves around intent, often microscopically examining the role of intent in the seconds before a behavior happens. Entire conferences, edited volumes, careers, have been spent on those few seconds, and in many ways, this focus is at the heart of arguments supporting compatibilism; this is because all the careful, nuanced, clever experiments done on the subject collectively fail to falsify free will. After reviewing these findings, the purpose of this chapter is to show how, nevertheless, all this is ultimately irrelevant to deciding that there's no free will. This is because this approach misses 99 percent of the story by not asking the key question: *And where did that intent come from in the first place?* This is so important because, as we will see, while it sure may seem at times that we are free do as we intend, we are never free to intend what we intend. Maintaining belief in free will by failing to ask that question can be heartless and immoral and is as myopic as believing that all you need to know to assess a movie is to watch its final three minutes. Without that larger perspective, understanding the features and consequences of intent doesn't amount to a hill of beans.

THREE HUNDRED MILLISECONDS

Let's start off with William Henry Harrison, ninth president of the United States, remembered only for idiotically insisting on giving a record-long two-hour inauguration speech in the freezing cold in January 1841, without coat or hat; he caught pneumonia and died a month later, the first president to die in office and the shortest presidential term.*[1]

With that in place, think about William Henry Harrison. But first, we're going to stick electrodes all over your scalp for an electroencephalogram (EEG), to observe the waves of neuronal excitation generated by your cortex when you're thinking of Bill.

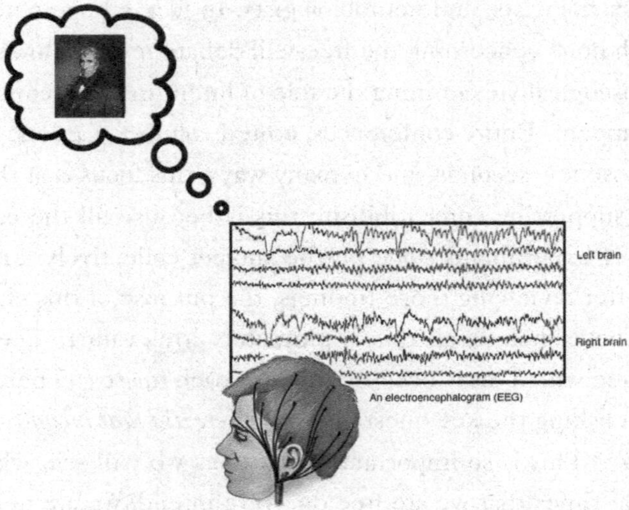

Now *don't* think of Harrison—think about anything else—as we continue recording your EEG. Good, well done. Now don't think about Harrison, but *plan* to think about him whenever you want a little while later,

*Revisionism suggests that rather than at the inauguration, he caught his pneumonia a few weeks later when, again coatless, he went out to buy a cow. But then even more radical revisionism suggests that he didn't die of pneumonia at all but instead from typhoid fever, contracted from the vile, contaminated water available in the White House. This was concluded by writer Jane McHugh and physician Philip Mackowiak, based on the symptoms detailed by Harrison's doctor and the fact that the White House's water supply was just downstream from where "night soil" was dumped. At the time, Washington, DC, was a malarial swamp, its selection having been advocated by powerful Virginians who wanted the capital close to home; this was decided in behind-closed-doors horse-trading between Alexander Hamilton and Virginians Thomas Jefferson and James Madison. "No one really knows how the game is played, the art of the trade, how the sausage gets made," writes noted historian Lin-Manuel Miranda, referring to the mystery of what transpired in those negotiations.

and push this button the instant you do. Oh, also, keep an eye on the second hand on this clock and note when you chose to think about Harrison. We're also going to wire up your hand with recording electrodes to detect precisely when you start the pushing; meanwhile, the EEG will detect when neurons that command those muscles to push the button start to activate. And this is what we find out: those neurons had already activated *before* you thought you were first freely choosing to start pushing the button.

But the experimental design of this study isn't perfect, because of its nonspecificity—we may have just learned what's happening in your brain when it is generically doing something, as opposed to doing this particular something. Let's switch instead to your choosing between doing A and doing B. William Henry Harrison sits down to some typhoid-riddled burgers and fries, and he asks for ketchup. If you decide he would have pronounced it "ketch-up," immediately push this button with your left hand; if it was "cats-up," push this other button with your right. Don't think about his pronunciation of *ketchup* right now; just look at the clock and tell us the instant you chose which button to push. And you get the same answer—the neurons responsible for whichever hand pushes the button activate before you consciously formed your choice.

Let's do something fancier now than looking at brain waves, since EEG reflects the activity of hundreds of millions of neurons at a time, making it hard to know what's happening in particular brain regions. Thanks to a grant from the WHH Foundation, we've bought a neuroimaging system and will do functional magnetic resonance imaging (fMRI) of your brain while you do the task—this will tell us about activity in each individual brain region at the same time. The results show clearly, once again, that particular regions have "decided" which button to push before you believe you consciously and freely chose. Up to ten seconds before, in fact.

Eh, forget about fMRI and the images it produces, where a single pixel's signal reflects the activity of about half a million neurons. Instead, we're going to drill holes in your head and then stick electrodes into your brain to monitor the activity of individual neurons; using this approach, once again, we can tell if you'll go for "ketch-up" or "cats-up" from the activity of neurons *before* you believe you decided.

These are the basic approaches and findings in a monumental series of studies that have produced a monumental shitstorm as to whether they demonstrate that free will is a myth. These are the core findings in virtually every debate about what neuroscience can tell us on the subject. And I think that at the end of the day, these studies are irrelevant.

It began with Benjamin Libet, a neuroscientist at the University of California at San Francisco, in a 1983 study so provocative that at least one philosopher refers to it as "infamous," there are conferences held about it, and scientists are described as doing "Libet-style studies."[*,2]

We know the experimental setup. Here's a button. Push it whenever you want. Don't think about it beforehand; look at this fancy clock that makes it easy to detect fractions of a second and tell us when you decided to push the button, that moment of conscious awareness when you freely made your decision.[†] Meanwhile, we'll be collecting EEG data from you and monitoring exactly when your finger starts moving.

Out of this came the basic findings: people reported that they decided to push the button about two hundred milliseconds—two tenths of a second—before their finger started moving. There was also a distinctive EEG pattern, called a readiness potential, when people prepared to move; this emanated from a part of the brain called the SMA (supplementary motor area), which sends projections down the spine, stimulating muscle movement. But here's the crazy thing: the readiness potential, the evidence that the brain had committed to pushing the button, occurred about three hundred milliseconds *before* people believed they had decided to

---

[*] As a point applying to virtually every scientific finding that I'll discuss in the rest of the book, when I say, "work done by John/Jane Doe," it actually means work done by that lead scientist along with a team of collaborators. As an equally important point (that I'll reiterate in various places, because it can't be mentioned too often), when I say, "Scientists showed that when they'd do this or that, people would do X," I mean that *on average*, people responded this way. There are always exceptions, who are often the most interesting.

[†] In the Libet literature, this point where people thought that they had decided came to be called "W," for the point where they first consciously *w*ished to do something. I'm avoiding using that term, to minimize jargon.

push the button. That sense of freely choosing is just a post hoc illusion, a false sense of agency.

This is the observation that started it all. Read technical papers on biology and free will, and in 99.9 percent of them, *Libet* will appear, usually by the second paragraph. Ditto for articles in the lay press—"Scientist Proves There Is No Free Will; Your Brain Decides Before You Think You Did."* It inspired scads of follow-up research and theorizing; people are still doing studies directly inspired by Libet nearly forty years after his 1983 publication. For example, there's a 2020 paper entitled "Libet's Intention Reports Are Invalid."[3] Having your work be important enough that decades later, people are still trash-talking it is immortality for a scientist.

The basic Libet finding that you're kidding yourself if you think you made a decision when it feels like you did has been replicated. Neuroscientist Patrick Haggard of University College London had subjects choose between two buttons—choosing to do A versus B, rather than choosing to do something versus not. This suggested the same conclusion that the brain has seemingly decided before you think you did.[4]

These findings ushered in Libet 2.0, the work of John-Dylan Haynes and colleagues at Humboldt University in Germany. It was twenty-five years later, with fMRIs available; everything else was the same. Once again, people's sense of conscious choice came about two hundred milliseconds before the muscles started moving. Most important, the study replicated the conclusion from Libet, fleshing it out further.† With fMRI, Haynes was able to spot the which-button decision even farther up in the brain's chain of command, in the prefrontal cortex (PFC). This made sense, as the PFC is where executive decisions are made. (When the PFC, along with the rest of the frontal cortex, is destroyed, à la Gage, one makes

*One paper analyzes the reporting of Libet in the lay press. Eleven percent of the headlines said free will had been disproved; 11 percent said the opposite; many articles were wildly inaccurate in describing how the experiment was done (e.g., saying that it was the researcher who would push the button). And on other fronts, there's even a piece of music called "Libet's Delay." It's moody and so repetitive that I felt a conscious sense of wishing to scream; I can only conclude that it was composed by a deeply depressed AI.

†I'm using "the conclusion from Libet" rather than "Libet's conclusion" in that the latter suggests what Libet himself was thinking about his finding. We'll get to what he thought.

terrible, disinhibited decisions.) To simplify a bit, once having decided, the PFC passes the decision on to the rest of the frontal cortex, which passes it to the premotor cortex, then to the SMA and, a few steps later, on to your muscles.* Supporting the view of Haynes having spotted decision-making farther upstream, the PFC was making its decision up to *ten seconds* before subjects felt they were consciously deciding.[†,5]

Then Libet 3.0 explored free-will-is-an-illusion down to monitoring the activity of individual neurons. Neuroscientist Itzhak Fried of UCLA worked with patients with intractable epilepsy, unresponsive to antiseizure medications. As a last-ditch effort, neurosurgeons remove the part of the brain where these seizures initiate; with Fried's patients, it was the frontal cortex. One obviously wants to minimize the amount of tissue removed, and in preparation for that, electrodes are implanted in the targeted area prior to the surgery, allowing for monitoring activity there. This provides a fine-grained map of function, telling you what subparts you should avoid removing, if there's any leeway.

So Fried would have the subjects do a Libet-style task while electrodes in their frontal cortex detected when particular neurons there activated. Same punch line: some neurons activated in preparation for a particular movement decision seconds before subjects claimed they had consciously decided. In fascinating related studies, he has shown that neurons in the hippocampus that code for a specific episodic memory activate one to two seconds before the person becomes aware of freely recalling that memory.[6]

Thus, three different techniques, monitoring the activity of hundreds of millions of neurons down to single neurons, all show that at the moment when we believe that we are consciously and freely choosing to do something, the neurobiological die has already been cast. That sense of conscious intent is an irrelevant afterthought.

This conclusion is reinforced by studies showing how malleable the

*One neuroscientist aptly describes the SMA as the "gateway" by which the PFC talks to your muscles.

†Haynes and colleagues have since identified the exact subregion of the PFC involved. They also implicated an additional brain region, the parietal cortex, as part of the decision-making process.

sense of intent and agency is. Back to the basic Libet paradigm; this time, pushing a button caused a bell to ring, and the researchers would vary how long of a fraction-of-a-second time delay there'd be between the pushing and the ringing. When the bell ringing was delayed, subjects reported their intent to push the button coming a bit later than usual—without the readiness potential or actual movement changing. Another study showed that if you feel happy, you perceive that conscious sense of choice sooner than if you're unhappy, showing how our conscious sense of choosing can be fickle and subjective.[7]

Other studies of people undergoing neurosurgery for intractable epilepsy, meanwhile, showed that the sense of intentional movement and actual movement can be separated. Stimulate an additional brain region relevant to decision-making,* and people would claim they had just moved voluntarily—without so much as having tensed a muscle. Stimulate the pre-SMA instead, and people would move their finger while claiming that they hadn't.[8]

One neurological disorder reinforces these findings. Stroke damage to part of the SMA produces "anarchic hand syndrome," where the hand controlled by that side of the SMA† acts against the person's will (e.g., grabbing food from someone else's plate); sufferers even restrain their anarchic hand with their other one.‡ This suggests that the SMA keeps volition on task, binding "intention to action," all before the person believes they've formed that intention.[9]

Psychology studies also show how the sense of agency can be illusory.

*The parietal cortex, mentioned a few footnotes back.

†As a technical detail completely unrelated to any of this, the right half (hemisphere) of the brain regulates movements in the left half of the body; the left hemisphere the reverse.

‡Anarchic hand syndrome, and the closely related "alien hand syndrome," is sometimes called "Dr. Strangelove syndrome"—for the titular character in the 1964 Stanley Kubrick movie. Strangelove was mostly modeled after rocket scientist Wernher von Braun, who went from faithfully serving his Nazi masters during World War II to serving his American ones after; turns out he was a patriotic American all along, that whole Nazi thing just a misunderstanding. Strangelove, wheelchair bound after a stroke, has anarchic hand syndrome, his hand constantly trying to give a Nazi salute to his American overlords. Stanley Kubrick, the famed director of the movie, also incorporated elements of John von Neumann, Herman Kahn, and Edward Teller into Strangelove (but not, despite urban legends, Henry Kissinger).

In one study, pushing a button would be followed immediately by a light going on . . . some of the time. The percentage of time the light would go on was varied; subjects were then asked how much control they felt they had over the light. People consistently overestimate how reliably the light occurs, feeling that they control it.* In another study, subjects believed they were voluntarily choosing which hand to use in pushing a button. Unbeknownst to them, hand choice was being controlled by transcranial magnetic stimulation† of their motor cortex; nonetheless, subjects perceived themselves as controlling their decisions. Meanwhile, other studies used manipulations straight out of the playbook of magicians and mentalists, with subjects claiming agency over events that were actually foregone and out of their control.[10]

If you do X and this is followed by Y, what increases the odds of your feeling like you caused Y? Psychologist Daniel Wegner of Harvard, a key contributor in this area, identified three logical variables. One is *priority*—the shorter the delay between X and Y, the more readily we have an illusory sense of will. There are also *consistency* and *exclusivity*—how consistently Y happens after you've done X, and how often Y happens in the absence of X. The more of the former and the less of the latter, the stronger the illusion.[11]

Collectively, what does this Libetian literature, starting with Libet, show? That we can have an illusory sense of agency, where our sense of freely, consciously choosing to act can be disconnected from reality;‡ we can be manipulated as to when we first feel a sense of conscious control;

---

*Interestingly, people with depression are resistant to being tricked into this sense of "illusory will." This will be returned to in the final chapter.

†In TMS, an electromagnetic coil is placed on the scalp and used to activate or inactivate the patch of cortex just beneath (I had that done to me once, with the colleague controlling when I bent my index finger; it felt beyond creepy). How's this for a finding whose implications resonate through this book? TMS can be used to alter people's judgments of the moral appropriateness of a behavior.

‡Although, in response to this, philosopher Peter Tse of Dartmouth writes, "Just as the existence of visual illusions does not prove that all vision is illusory, the existence of illusions of conscious agency does not prove that conscious operations cannot be causal of action in certain cases."

most of all, this sense of agency comes after the brain has already committed to an action. Free will is a myth.[12]

Surprise!, people have been screaming at each other about these conclusions ever since, incompatibilists perpetually citing Libet and his descendants, and compatibilists being scornful shade throwers about the entire literature. It didn't take long to start. Two years after his landmark paper, Libet published a review in a peer-commentary journal (where someone presents a theoretical paper on a controversial topic, followed by short commentaries by the scientist's friends and enemies); commentators beating on Libet accused him of "egregious errors," overlooking "fundamental measurement concepts," conceptual unsophistication ("Pardon, your dualism is showing," accused one critic), and having an unscientific faith in the accuracy of his timing measurements (sarcastically proclaiming Libet as practicing "chronotheology").[13]

The criticisms of the work of Libet, Haynes, Fried, Wegner, and friends continue unabated. Some focus on minutiae like the limitations of using EEGs, fMRI, and single-neuron recordings, or the pitfalls inherent in subjects self-reporting most anything. But most criticisms are more conceptual and collectively show that rumors of Libetianism killing free will are exaggerated. These are worth detailing.

YOU GUYS PROCLAIM THE DEATH OF FREE WILL, BASED ON SPONTANEOUS FINGER MOVEMENTS?

The Libetian literature is built around people spontaneously deciding to do something. In the view of Manuel Vargas, free will revolves around being future oriented, enduring an immediate cost for a long-term goal, and thus "Libet's experiment insisted on a purely immediate, impulsive action—which is precisely not what free will is for."[14]

Moreover, what was being spontaneously decided was to push a button, and this bears little resemblance to whether we have free will concerning our beliefs and values or our most consequential actions. In the words of psychologist Uri Maoz of Chapman University, this is a contrast between "picking" and "choosing"—Libet is about picking which box of Cheerios to take off the supermarket shelf, not about choosing something major. Dartmouth philosopher Adina Roskies, for example, views Libet-world picking as a caricature of real choice, dwarfed even by the complexity of deciding between tea and coffee.[*,15]

Does the Libet finding apply to something more interesting than button pushing? Fried replicated the Libet effect when subjects in a driving simulator chose between turning left and turning right. Another study merged neuroscience with getting out of the lab on a sunny day, checking for the Libet phenomenon in subjects just before they bungee-jumped. Did the neuroscientists, clutching their equipment, jump too? No, a wireless EEG device was strapped to the jumpers' heads, making them look like Martians persuaded to bungee-jump by frat bros after some beer pong. Results? Replication of Libet, where a readiness potential preceded the subjects' believing they had decided to jump.[16]

To which the compatibilists replied, This is still totally artificial—choosing when to leap into an abyss or whether to turn left or right in a driving simulator tells us nothing about our free will in choosing between, say, becoming a nudist versus a Buddhist, or becoming an algologist versus an allergologist. This criticism was backed by a particularly elegant study. In the first situation, subjects would be presented with two buttons and told that each represented a particular charity; press one of the buttons and that charity will be sent a thousand dollars. Second version: two buttons, two charities, push whichever button you feel like, each charity is getting five hundred dollars. The brain was commanding the same movement in both scenarios, but the choice in the first one was highly consequential, while that in the second was as arbitrary as the one in the Libet

*While usually classified as a philosopher, Roskies leaves the rest of us pikers in the dust by having a PhD in neuroscience, in addition to her philosophy PhD.

study. The boring, arbitrary situation evoked the usual readiness potential before there was a sense of conscious decision; the consequential one didn't. In other words, Libet doesn't tell us anything about free will worth wanting. In the wonderfully sarcastic words of one leading compatibilist, the take-home message of this entire literature is "Don't play rock paper scissors for money [with one of these free will skeptic researchers] if your head is in an fMRI machine."[17]

But then, the revenge of the free will skeptics. Haynes's group brain-imaged subjects participating in a nonmotoric task, choosing whether to add or subtract one number from another; they found a neural signature of decision coming before conscious awareness, but coming from a *different* brain region than the SMA (called the posterior cingulate / precuneus cortex). So maybe the pick-your-charity scientists were just looking in the wrong part of the brain—simple brain regions decide things before you think you've consciously made a simple decision, more complicated regions before you think you've made a complicated choice.[18]

The jury is still out, because the Libetian literature remains almost entirely about spontaneous decisions regarding some fairly simple things. On to the next broad criticism.

60 PERCENT? REALLY?

What does it mean to become aware of a conscious decision? What do "deciding" and "intending" really mean? Again with semantics that aren't just semantic. The philosophers run wild here in subtle ways that leave many neuroscientists (e.g., me) gasping in defanged awe. How long does it take to focus on focusing on the second hand on a clock? In her writing, Roskies emphasizes the difference between conscious intention and consciousness of intention. Alfred Mele speculates that the readiness potential is the time when, in fact, you have legitimately freely chosen, and it then takes a bit of time for you to be consciously aware of your freely willed choice. Arguing against this, one study showed that at the time of the onset of the readiness potential, rather than thinking about when they were going to move, many subjects were thinking about things like dinner.[19]

Can you decide to decide? Are intending and having an intent the same thing? Libet instructed subjects to note the time when they first became aware of "the subjective experience of 'wanting' or intending to act"—but are "wanting" and "intending" the same? Is it possible to be spontaneous when you've been told to be spontaneous?

As long as we're at it, what actually is a readiness potential? Remarkably, nearly forty years after Libet, a paper can still be entitled "What Is the Readiness Potential?" Could it be deciding-to-do, actual "intention," while the conscious sense of decision is deciding-to-do-now, an "implementation of intention"? Maybe the readiness potential doesn't mean anything—some models suggest that it is just the point where random activity in the SMA passes a detectable threshold. Mele forcefully suggests that the readiness potential is not a decision but an urge, and physicist Susan Pockett and psychologist Suzanne Purdy, both of the University of Auckland, have shown that the readiness potential is less consistent and shorter when subjects are planning to identify when they made a decision, versus when they felt an urge. For others, the readiness potential is the process leading to deciding, not the decision itself. One clever experiment supports this interpretation. In it, subjects were presented four random letters and then instructed to choose one in their minds; sometimes they were then signaled to press a button corresponding to that letter, sometimes not—thus, the same decision-making process occurred in both scenarios, but only one actually produced movement. Crucially, a similar readiness potential occurred in both cases, suggesting, in the words of compatibilist neuroscientist Michael Gazzaniga, that rather than the SMA deciding to enact a movement, it's "warming up for its participation in the dynamic events."[20]

So are readiness potentials and their precursors decisions or urges? A decision is a decision, but an urge is just an increased likelihood of a decision. Does a preconscious signal like a readiness potential ever occur and despite that, the movement *doesn't* then happen? Does a movement ever occur *without* a preconscious signal preceding it? Combining these two questions, how accurately do these preconscious signals predict actual behavior? Something close to 100 percent accuracy would be a major blow to free-will belief. In contrast, the closer accuracy is to chance (i.e., 50 percent), the less likely it is that the brain "decides" anything before we feel a sense of choosing.

As it turns out, predictability isn't all that great. The original Libet study was done in such a way that it wasn't possible to generate a number for this. However, in the Haynes studies, fMRI images predicted which behavior occurred with only about 60 percent accuracy, almost at the chance level. For Mele, a "60-percent accuracy rate in predicting which button a participant will press next doesn't seem to be much of a threat to free will." In Roskies's words, "All it suggests is that there are some physical factors that influence decision-making." The Fried studies recording from individual neurons pushed accuracy up into the 80 percent range; while certainly better than chance, this sure doesn't constitute a nail in free will's coffin.[21]

Now for the next criticisms.

WHAT IS CONSCIOUSNESS?

Giving this section this ridiculous heading reflects how unenthused I am about having to write this next stretch. I don't understand what consciousness is, can't define it. I can't understand philosophers' writing about it. Or neuroscientists', for that matter, unless it's "consciousness" in the boring neurological sense, like not experiencing consciousness because you're in a coma.[*,22]

Nevertheless, consciousness is central to Libet debates, sometimes, in a fairly heavy-handed way. For example, take Mele, in a book whose title

---

*Naturally, it turns out that the neurological distinction between consciousness and unconscious is not boring, simple, or dichotomous, but that's another can of worms.

trumpets that he's not pulling any punches—*Free: Why Science Hasn't Disproved Free Will.* In its first paragraph, he writes, "There are two main scientific arguments today against the existence of free will." One arises from social psychologists showing that behavior can be manipulated by factors that we're not aware of—we've seen examples of these. The other is neuroscientists whose "basic claim is that *all our decisions* are made unconsciously and therefore not freely" (my italics). In other words, that consciousness is just an epiphenomenon, an illusory, reconstructive sense of control irrelevant to our actual behavior. This strikes me as an overly dogmatic way of representing just one of many styles of neuroscientific thought on the subject.

The "ooh, you neuroscientists not only eat your dead but also believe all our decisions are unconscious" nyah-nyah matters, because we shouldn't be held morally responsible for our unconscious behaviors (although neuroscientist Michael Shadlen of Columbia University, whose excellent research has informed free-will debates, makes a spirited argument along with Roskies that we should be held morally responsible for even our unconscious acts).[23]

Compatibilists trying to fend off the Libetians often make a last stand with consciousness: Okay, okay, suppose that Libet, Haynes, Fried, and so on really have shown that the brain decides something before we have a sense of having consciously and freely done so. Let's grant the incompatibilists that. But does turning that preconscious decision into actual behavior require that conscious sense of agency? Because if it does, rather than bypassing consciousness as an irrelevancy, free will can't be ruled out.*

As we saw, knowing what a brain's preconscious decision was moderately predicts whether the behavior will actually occur. But what about the relationship between the preconscious brain's decision and the sense of

*Note that while related, this is subtly different from the issue of whether the sense of conscious decision-making always occurs with the same time lag after the readiness potential; as we saw, the timing of that sense of agency can be manipulated by other factors.

conscious agency—is there ever a readiness potential followed by a behavior without a conscious sense of agency coming in between? One cool study done by Dartmouth neuroscientist Thalia Wheatley and collaborators* shows precisely this—subjects were hypnotized and implanted with a posthypnotic suggestibility that they make a spontaneous Libet-like movement. In this case, when triggered by the cued suggestion, there'd be a readiness potential and the subsequent movement, without conscious awareness in between. Consciousness is an irrelevant hiccup.[24]

Sure, retort compatibilists, this doesn't mean that intentional behavior *always* bypasses consciousness—rejecting free will based on what happens in the posthypnotic brain is kind of flimsy. And there is a higher-order level to this issue, something emphasized by incompatibilist philosopher Gregg Caruso of the State University of New York—you're playing soccer, you have the ball, and you consciously decide that you are going to try to get past this defender, rather than pass the ball off. In the process of then trying to do this, you make a variety of procedural movements that you're not consciously choosing; what does it mean that you have made the explicit choice to let a particular implicit process take over? The debate continues, not just over whether the preconscious requires consciousness as a mediating factor but also over whether both can simultaneously cause a behavior.[25]

Amid these arcana, it's hugely important if the preconscious decision requires consciousness as a mediator. Why? Because during that moment of conscious mediation we should then be expected to be able to *veto* a decision, prevent it from happening. And you can hang moral responsibility on that.[26]

*The study was a collaboration not just between philosophers and neuroscientists but also between people with decidedly incompatibilist stances (Wheatley) and the notable compatibilists Roskies, Tse, and Duke philosopher Walter Sinnott-Armstrong. This is the process of questing for knowledge at its objective best.

FREE WON'T: THE POWER TO VETO

Even if we don't have free will, do we have free won't, the ability to slam our foot on the brake between the moment of that conscious sense of freely choosing to do something and the behavior itself? This is what Libet concluded from his studies. Clearly we have that veto power. Writ small, you're about to reach for more M&M's but stop an instant before. Writ larger, you're about to say something hugely inappropriate and disinhibited but, thank God, you stop yourself as your larynx warms up to doom you.

The basic Libetian findings gave rise to a variety of studies looking at where vetoing actions fits in. Do it or not: once that conscious sense of intent occurs, subjects have the option to stop. Do it now or in a bit: once that conscious sense of intent occurs, immediately push the button or first count to ten. Impose an external veto: In a brain-computer interface study, researchers used a machine learning algorithm that monitored a subject's readiness potential, predicting in real time when the person was about to move; some of the time, the computer would signal the subject to stop the movement in time. Of course, people could generally stop themselves up until a point of no return, which roughly corresponded to when the neurons that send a command directly to muscles were about to fire. As such, a readiness potential doesn't constitute an unstoppable decision, and one would generally look the same whether the subject was definitely going to push a button or there was the possibility of a veto.[*,27]

How does the vetoing work, neurobiologically? Slamming a foot on the brake involved activating neurons just upstream of the SMA.[†] Libet may have spotted this in a follow-up study examining free won't. Once subjects

[*]As a fascinating finding in these studies, failing to stop in time activates the anterior cingulate cortex, a brain region associated with subjective feelings of pain; in other words, a few dozen milliseconds is enough time for you to feel like a loser because a computer has gotten a faster draw on you.

[†]Depending on the study, the "*pre*-SMA," anterior frontomedial cortex, and/or right inferior frontal gyrus. Note that the last two places, logically, implicate the frontal cortex in executive vetoing.

had that conscious sense of intent, they were supposed to veto the action; at that point, the tail end of the readiness potential would lose steam, flatten out.[*,28]

Meanwhile, other studies explored interesting spin-offs of free won't–ness. What's the neurobiology of a gambler on a losing streak who manages to stop gambling, versus one who doesn't?[†] What happens to free won't when there's alcohol on board? How about kids versus adults? It turns out that kids need to activate more of their frontal cortex than do adults to get the same effectiveness at inhibiting an action.[29]

So what do all these versions of vetoing a behavior in a fraction of a second say about free will? Depends on whom you talk to, naturally. Findings like these have supported a two-stage model about how we are supposedly the captains of our fate, one espoused by the likes of everyone from William James to many contemporary compatibilists. Stage one, the "free" part: your brain spontaneously chooses, amid alternative possibilities, to generate the proclivity toward some action. Stage two, the "will" part, is where you consciously consider this proclivity and either green-light it or free-won't it. As one proponent writes, "Freedom arises from the creative and indeterministic generation of alternative possibilities, which present themselves to the will for evaluation and selection." Or in Mele's words, "even if urges to press are determined by unconscious brain activity, it may be up to the participants whether they act on those urges or not."[30] Thus, "our brains" generate a suggestion, and "we" then judge it; this dualism sets our thinking back centuries.

---

*The original Libet publication didn't mention anything about flattening out; it was only in a later review that he decided that it occurred. And to be a bit of a killjoy, after looking at the original paper, which had only four subjects, I just don't see it in the shapes of the readiness potentials displayed, and there's no real way to rigorously analyze the shape of each curve, given the data available in the paper; this study happened during a less quantitative, more innocent time for analyzing data.

†Continuing to gamble activated brain regions associated with incentives and reward; in contrast, quitting activated regions related to subjective pain, anxiety, and conflict. This is amazing—continuing to gamble with the possibility of losing is less neurobiologically aversive than quitting and contemplating the possibility that you *would have* won if you hadn't stopped. We're a really screwed-up species.

The alternative conclusion is that free won't is just as suspect as free will, and for the same reasons. Inhibiting a behavior doesn't have fancier neurobiological properties than activating a behavior, and brain circuitry even uses their components interchangeably. For example, sometimes brains do something by activating neuron X, sometimes by inhibiting the neuron that is inhibiting neuron X. Calling the former "free will" and calling the latter "free won't" are equally untenable. This recalls chapter 1's challenge to find a neuron that initiated some act without being influenced by any other neuron or by any prior biological event. Now the challenge is to find a neuron that was equally autonomous in preventing an act. Neither free-will nor free-won't neurons exist.

Having now reviewed these debates, what can we conclude? For Libetians, these studies show that our brains decide to carry out a behavior before we think that we've freely and consciously done so. But given the criticisms that have been raised, I think all that can be concluded is that in some fairly artificial circumstances, certain measures of brain function are moderately predictive of a subsequent behavior. Free will, I believe, survives Libetianism. And yet I think that is irrelevant.

JUST IN CASE YOU THOUGHT THIS WAS ALL ACADEMIC

The debates over Libet and his descendants can be boiled down to a question of intent: When we consciously decide that we intend to do something, has the nervous system already started to act upon that intent, and what does it mean if it has?

A related question is screamingly important in one of the areas where this free-will hubbub is profoundly consequential—in the courtroom. When someone acts in a criminal manner, did they intend to?

By this I'm not suggesting bewigged judges arguing about some low-life's readiness potentials. Instead, the questions that define "intent" are whether a defendant could foresee, without substantial doubt, what was going to happen as a result of their action or inaction, and whether they were okay with that outcome. From that perspective, unless there was intent in that sense, a person shouldn't be convicted of a crime.

Naturally, this generates complex questions. For example, should intending to shoot someone but missing count as a lesser crime than shooting successfully? Should driving with a blood alcohol level in the range that impairs control of a car count as less of a transgression if you lucked out and happened not to kill a pedestrian than if you did (an issue that Oxford philosopher Neil Levy has explored with the concept of "moral luck")?[31]

As another wrinkle, the legal field distinguishes between *general* and *specific* intent. The former is about intending to commit a crime, whereas the latter is intending to commit a crime as well as intending a specific consequence; the charge of the latter is definitely more serious than the former.

Another issue that can come up is deciding whether someone acted intentionally out of fear or anger, with fear (especially when reasonable) seen as more mitigating; trust me, if the jury consisted of neuroscientists, they'd deliberate for eternity trying to decide which emotion was going on. How about if someone intended to do something criminal but instead unintentionally did something else criminal?

An issue that we all recognize is how long before a behavior the intent was formed. This is the world of premeditation, the difference between, say, a crime of passion with a few milliseconds of intent versus an action long planned. It is pretty unclear legally exactly how long one needs to meditate upon an intended act for it to count as premeditated. As an example of this lack of clarity, I once was a teaching witness in a trial where a pivotal issue was whether eight seconds (as recorded by a CCTV camera) is enough time for someone in a life-threatening circumstance to *pre*meditate a murder. (My two cents was that under the circumstances involved, eight seconds not only wasn't enough time for a brain to do premeditated thinking, it wasn't enough time for it to do any thinking, and free won't–ness was an irrelevant concept; the jury heartily disagreed.)

Then there are questions that can be at the core of war crime trials. What kind of threat is needed for someone's criminality to count as coerced? What about agreeing to do something with criminal intent while knowing that if you refused, someone else would do it immediately and more brutally? Taking things even further, what should be done with someone who intentionally chose to commit a crime, not knowing that they would have been forced to commit that act if they had tried to do otherwise?[*,32]

At this juncture, we appear to have two wildly different realms of thinking about agency and responsibility—people arguing about the supplementary motor area in neurophilosophy conferences and prosecutors and public defenders jousting in courtrooms. Yet they share something that potentially strikes a blow against free-will skepticism:

> Suppose it turns out that our sense of conscious decision-making doesn't actually come after things like readiness potentials, that activity in the SMA, the prefrontal cortex, the parietal cortex, wherever, is

*It seems intuitive that someone should be punished if they thought they had willingly chosen to do something illegal without knowing that they actually didn't have a choice. The late Princeton philosopher Harry Frankfurt has taken the implications of this intuition in a particular compatibilist direction. Step 1: Incompatibilists say that if the world is deterministic, there shouldn't be moral responsibility. Step 2: Consider someone choosing to do something, not knowing that they would have been coerced if they hadn't. Step 3: Therefore, this would be a deterministic world, in that the person didn't actually have the option of doing otherwise . . . yet our intuitions are to hold him morally responsible, perceiving him as having had free will. Huzzah, we've thus just proven that free will and moral responsibility are compatible with determinism. I feel bad saying this because Frankfurt looks cherubic in his pictures, but this seems like more than a bit of sophistry and sure doesn't represent the Demise of Incompatibilism. Moreover, I get the sense from friends in the know that while Frankfurt is enormously influential in some corners of legal philosophy, millennia go by without these "Frankfurt counterexamples" being relevant in an actual courtroom; it is unlikely for there to be scenarios where "the defendant chose to slap the Oscar host across the face, not aware that if he had not chosen to do so, he would nevertheless have been forced to."

> never better than only moderately predicting behavior, and only for the likes of pushing buttons. You sure can't say free will is dead based on that.
>
> Likewise, suppose a defendant says, "I did it. I knew there were other things I could do, but I intended to do it, planned it in advance. I not only knew that X could have been the outcome, I wanted that to happen." Good luck convincing someone that the defendant lacked free will.

But the point of this chapter is that even if either or both of these are the case, I still think that free will doesn't exist. To appreciate why, time for a Libet-style thought experiment.

THE DEATH OF FREE WILL IN THE SHADOW OF INTENT

You have a friend doing research for her doctorate in neurophilosophy, and she asks you to be a test subject. Sure. She's upbeat because she's figured out how to both get another data point for her study *and* simultaneously accomplish something else that she's keen on—win-win. It involves ambulatory EEG, out of the lab, like in the bungee jumping study. You're out there now, wired up with the leads, electromyography being done on your hand, a clock in view.

As with the classic Libet, the motoric action involved is to move your index finger. Hey, aren't we decades past that sort of really artificial scenario? Fortunately, the study is more sophisticated than that, thanks to your friend's careful experimental design—you'll be making a simple movement, but with a nonsimple consequence. Don't plan ahead to make this movement, you're told, do it spontaneously, and note on the clock what time it is when you first consciously intend to. All set? Now, when you feel like it, pull a trigger and kill this person.

Maybe the person is an enemy of the Fatherland, a terrorist blowing up bridges in one of the gloriously occupied colonies. Maybe it's the person behind the cash register in the liquor store you're robbing. Maybe they're a terminally ill loved one in unspeakable pain, begging you to do this. Maybe it's someone who is about to harm a child; maybe it is the infant Hitler, cooing in his crib.

You are free to choose not to shoot. You're disillusioned with the regime's brutality and refuse; you think killing the clerk ups the ante too much if you're caught; despite your loved one begging, you just can't do it. Or maybe you're Humphrey Bogart, your friend is Claude Rains, you're confusing reality with story line and figure that if you let Major Strasser escape, the story doesn't end and you'll get to star in a sequel to *Casablanca*.*

But suppose you have to pull the trigger or else there'll be no readiness potential to detect and your friend's research will be slowed down. Nonetheless, you still have options. You can shoot the person. You can shoot but intentionally miss. You can shoot yourself rather than comply.† As a major plot twist, you can shoot your friend.

It makes intuitive sense that if you want to understand what you wind up doing with your index finger on that trigger, that you should explore Libetian concerns, studying particular neurons and particular milliseconds in order to understand the instant you feel you have chosen to do something, the instant your brain has committed to that action, and whether those two things are the same. But here's why these Libetian debates, as well as a criminal justice system that cares only about whether someone's actions are intentional, are irrelevant to thinking about free will. As first aired at the beginning of this chapter, that is because neither asks a question central to every page of this book: *Where did that intent come from in the first place?*

If you don't ask that question, you've restricted yourself to a domain of

*Aha!

†The Dalai Lama was once asked what he would do in the "runaway trolley" problem (a trolley whose brakes have failed is hurtling down the tracks, about to kill five people; is it okay to push someone in front of the trolley, intentionally killing them but preventing the deaths of five?); he said he would throw himself in front of the trolley.

a few seconds. Which is fine by many people. Frankfurt writes, "The questions of how the actions and his identifications with their springs are caused are irrelevant to the questions of whether he performs the actions freely or is morally responsible for performing them." Or in the words of Shadlen and Roskies, Libetian-ish neuroscience "can provide a basis for accountability and responsibility that focuses on the agent, *rather than on prior causes*" (my emphasis).

Where does intent come from? Yes, from biology interacting with environment one second before your SMA warmed up. But also from one minute before, one hour, one millennium—this book's main song and dance. Debating free will can't start and end with readiness potentials or with what someone was thinking when they committed a crime.* Why have I spent page after page going over the minutiae of the debates about what Libet means before blithely dismissing all of it with "And yet I think that is irrelevant"? Because Libet is viewed as *the* most important study ever done exploring the neurobiology of whether we have free will. Because virtually every scientific paper on free will trots out Libet early on. Because maybe you were born at the precise moment that Libet published his first study and now, all these years later, you're old enough that your music is called "classic" rock and you have started to make little middle-aged grunting sounds when you get up from a chair . . . *and they're still debating Libet.* And as noted before, this is like trying to understand a movie solely by watching its final three minutes.[33]

This charge of myopia is not meant to sound pejorative. Myopia is central to how we scientists go about finding out new things—by learning

*This contrast between proximal versus distal explanations of behavior (i.e., causes in proximity to a behavior versus those at a distance) is caught perfectly by neurosurgeon Rickard Sjöberg of Umeå University, Sweden. He imagines walking down a hall of his hospital and someone asking him why he just put his left foot in front of his right foot. Yes, one type of reply plunges us into the world of readiness potentials and milliseconds. But equally valid replies would be "Because when I woke up this morning, I decided not to call in sick" or "Because I decided to pursue a neurosurgery residency despite knowing about the long on-call hours." Sjöberg has done important work on the effects of removing the SMA on issues of volition, and in an extremely judicious review concludes that whatever resolution there is to free will debates, it isn't going to be found in the milliseconds of SMA activity.

more and more about less and less. I once spent nine years on a single experiment; this can become the center of a very small universe. And I'm not accusing the criminal justice system of myopically focusing solely on whether there was intent—after all, where intent came from, someone's history and potential mitigating factors, are considered when it comes to sentencing.

Where I am definitely trying to sound pejorative and worse is when this ahistorical view of judging people's behavior is moralistic. Why would you ignore what came before the present in analyzing someone's behavior? Because you don't care why someone else turned out to be different from you.

As one of the few times in this book where I will knowingly be personal, this brings me to the thinking of Daniel Dennett of Tufts University. Dennett is one of the best-known and most influential philosophers out there, a leading compatibilist who has made his case both in technical work within his field and in witty, engaging popular books.

He implicitly takes this ahistorical stance and justifies it with a metaphor that comes up frequently in his writing and debates. For example, in *Elbow Room: The Varieties of Free Will Worth Wanting*, he asks us to imagine a footrace where one person starts off way behind the rest at the starting line. Would this be unfair? "Yes, if the race is a hundred-yard dash." But it is fair if this is a marathon, because "in a marathon, such a relatively small initial advantage would count for nothing, since one can reliably expect other fortuitous breaks to have even greater effects." As a succinct summary of this view, he writes, "After all, luck averages out in the long run."[34]

No, it doesn't.* Suppose you're born a crack baby. In order to counterbalance this bad luck, does society rush in to ensure that you'll be raised in relative affluence and with various therapies to overcome your neurodevelopmental problems? No, you are overwhelmingly likely to be born into poverty and stay there. Well then, says society, at least let's make sure your mother is loving, is stable, has lots of free time to nurture you with

*A point elegantly argued by philosopher Gregg Caruso in some stirring debates with Dennett.

books and museum visits. Yeah, right; as we know, your mother is likely to be drowning in the pathological consequences of her own miserable luck in life, with a good chance of leaving you neglected, abused, shuttled through foster homes. Well, does society at least mobilize then to counterbalance that additional bad luck, ensuring that you live in a safe neighborhood with excellent schools? Nope, your neighborhood is likely to be gang-riddled and your school underfunded.

You start out a marathon a few steps back from the rest of the pack in this world of ours. And counter to what Dennett says, a quarter mile in, because you're still lagging conspicuously at the back of the pack, it's your ankles that some rogue hyena nips. At the five-mile mark, the rehydration tent is almost out of water and you can get only a few sips of the dregs. By ten miles, you've got stomach cramps from the bad water. By twenty miles, your way is blocked by the people who assume the race is done and are sweeping the street. And all the while, you watch the receding backsides of the rest of the runners, each thinking that they've earned, they're entitled to, a decent shot at winning. Luck does not average out over time and, in the words of Levy, "we cannot undo the effects of luck with more luck"; instead our world virtually guarantees that bad and good luck are each amplified further.

In the same paragraph, Dennett writes that "a good runner who starts at the back of the pack, if he is really good enough to *DESERVE* winning, will probably have plenty of opportunity to overcome the initial disadvantage" (my emphasis). This is one step above believing that God invented poverty to punish sinners.

Dennett has one more thing to say that summarizes this moral stance. Switching sports metaphors to baseball and the possibility that you think there's something unfair about how home runs work, he writes, "If you don't like the home run rule, don't play baseball; play some other game." Yeah, I want another game, says our now-adult crack baby from a few paragraphs ago. This time, I want to be born into a well-off, educated family of tech-sector overachievers in Silicon Valley who, once I decide that, say, ice-skating seems fun, will get me lessons and cheer me on from my first wobbly efforts on the ice. Fuck this life I got dumped into; I want to change games to *that* one.

Thinking that it is sufficient to merely know about intent in the present is far worse than just intellectual blindness, far worse than believing that it is the very first turtle on the way down that is floating in the air. In a world such as we have, it is deeply ethically flawed as well.

Time to see where intent comes from, and how the biology of luck doesn't remotely average out in the long run.[35]

3

Where Does Intent Come From?

Because of our fondness for all things Libetian, we sit you in front of two buttons; you must push one of them. You're given only hazy information about the consequences of pushing each button, beyond being told that if you pick the wrong button, thousands of people will die. Now pick.

No free will skeptic insists that sometimes you form your intent, lean way over to push the appropriate button, and suddenly, the molecules comprising your body deterministically fling you the other way and make you push the other button.

Instead, the last chapter showed how the Libetian debate concerns when exactly you formed that intent, when you became conscious of having formed it, whether neurons commanding your muscles had already activated by then, when it was that you could still veto that intention. Plus, questions about your SMA, frontal cortex, amygdala, basal ganglia—what they knew and when they knew it. Meanwhile, in parallel in the courtroom next door, lawyers argue over the nature of your intent.

The last chapter concluded by claiming that all these minutiae of milliseconds are completely irrelevant to why there is no free will. Which is why we didn't bother sticking electrodes into your brain just before seating you. They wouldn't reveal anything useful.

This is because the Libetian Wars don't ask the most fundamental question: Why did you form the intent that you did?

This chapter shows how you don't ultimately control the intent you form. You wish to do something, intend to do it, and then successfully do so. But no matter how fervent, even desperate, you are, *you can't successfully wish to wish for a different intent.* And you can't meta your way out—you can't successfully wish for the tools (say, more self-discipline) that will make you better at successfully wishing what you wish for. None of us can.

Which is why it would tell us nothing to stick electrodes in your head to monitor what neurons are doing in the milliseconds when you form your intent. To understand where your intent came from, all that needs to be known is what happened to you in the seconds to minutes before you formed the intention to push whichever button you choose. As well as what happened to you in the hours to days before. And years to decades before. And during your adolescence, childhood, and fetal life. And what happened when the sperm and egg destined to become you merged, forming your genome. And what happened to your ancestors centuries ago when they were forming the culture you were raised in, and to your species millions of years ago. Yeah, all that.

Understanding this turtleism shows how the intent you form, the person you are, is the result of all the interactions between biology and environment that came before. All things out of your control. Each prior influence flows without a break from the effects of the influences before. As such, there's no point in the sequence where you can insert a freedom of will that will be in that biological world but not of it.

Thus, we'll now see how who we are is the outcome of the prior seconds, minutes, decades, geological periods before, over which we had no control. And how bad and good luck sure as hell don't balance out in the end.

SECONDS TO MINUTES BEFORE

We ask our first version of the question of where that intent came from: What sensory information flowing into your brain (including some you're not even conscious of) in the preceding seconds to minutes helped form that intent?* This can be obvious—"I formed the intent to push that button because I heard the harsh demand that I do so, and saw the gun pointed in my face."

But things can be subtler. You view a picture of someone holding an object, for a fraction of a second; you must decide whether it was a cell phone or a handgun. And your decision in that second can be influenced by the pictured person's gender, race, age, and facial expression. We all know real-life versions of this experiment resulting in police mistakenly shooting an unarmed person, and about the implicit bias that contributed to that mistake.[1]

Some examples of intent being influenced by seemingly irrelevant stimuli have been particularly well studied.† One domain concerns how sensory disgust shapes behavior and attitudes. In one highly cited study, subjects rated their opinions about various sociopolitical topics (e.g., "On a scale of 1 to 10, how much do you agree with this statement?"). And if subjects were sitting in a room with a disgusting smell (versus a neutral one), the average level of warmth both conservatives and liberals reported for gay men decreased. Sure, you think—you'd feel less warmth for anyone if you're gagging. However, the effect was specific to gay men, with no change in warmth toward lesbians, the elderly, or African Americans. Another study showed that disgusting smells make subjects less accepting of gay marriage (as well as about other politicized aspects of sexual behavior).

---

*If you have read my book *Behave*, you'll recognize that the rest of this chapter is a summary of its first four hundred or so pages. Good luck . . .

†I'm being diplomatic. Many readers will know of the "replication crisis" in psychology, where an alarming percentage of published findings, even some in textbooks, turn out to be hard or impossible for other scientists to independently replicate (including some findings, I admit ruefully, that wound up being cited in my 2017 book, where I should have been more discerning). Thus, this section considers only findings whose broad conclusions have been independently replicated.

Moreover, just thinking about something disgusting (eating maggots) makes conservatives less willing to come into contact with gay men.[2]

Then there's a fun study where subjects were either made uncomfortable (by placing their hand in ice water) or disgusted (by placing their thinly gloved hand in imitation vomit).* Subjects then recommended punishment for norm violations that were purity related (e.g., "John rubbed someone's toothbrush on the floor of a public restroom" or the supremely distinctive "John pushed someone into a dumpster which was swarming with cockroaches") or violations unrelated to purity (e.g., "John scratched someone's car with a key"). Being disgusted by fake puke, but not being icily uncomfortable, made subjects more selectively punitive about purity violations.[3]

How can a disgusting smell or tactile sensation change unrelated moral assessments? The phenomenon involves a brain region called the insula (aka the insular cortex). In mammals, it is activated by the smell or taste of rancid food, automatically triggering spitting out the food and the species's version of barfing. Thus, the insula mediates olfactory and gustatory disgust and protects from food poisoning, an evolutionarily useful thing.

But the versatile human insula also responds to stimuli we deem *morally* disgusting. The insula's "this food's gone bad" function in mammals is probably a hundred million years old. Then, a few tens of thousands of years ago, humans invented constructs like morality and disgust at moral norm violations. That's way too little time to have evolved a new brain region to "do" moral disgust. Instead, moral disgust was added to the insula's portfolio; as it's said, rather than inventing, evolution tinkers, improvising (elegantly or otherwise) with what's on hand. Our insula neurons don't distinguish between disgusting smells and disgusting behaviors, explaining metaphors about moral disgust leaving a bad taste in your mouth, making you queasy, making you want to puke. You sense something disgusting,

*For DIYers, the paper contained the imitation vomit recipe: cream of mushroom soup, cream of chicken soup, black beans, pieces of fried gluten; quantities were unspecified, suggesting that you just have to get a feel for this sort of thing—a pinch of this, a smidgen of that. The study also noted that this recipe was *partially* based on one in a prior study—i.e., plucky innovation is advancing imitation-vomit science.

yech . . . and unconsciously, it occurs to you that it's disgusting and wrong when *those people* do X. And once activated this way, the insula then activates the amygdala, a brain region central to fear and aggression.[4]

Naturally, there is the flip side to the sensory disgust phenomenon—sugary (versus salty) snacks make subjects rate themselves as more agreeable and helpful individuals and rate faces and artwork as more attractive.[5]

Ask a subject, Hey, in last week's questionnaire you were fine with behavior A, but now (in this smelly room) you're not. Why? They won't explain how a smell confused their insula and made them less of a moral relativist. They'll claim some recent insight caused them, bogus free will and conscious intent ablaze, to decide that behavior A isn't okay after all.

It's not just sensory disgust that can shape intent in seconds to minutes; beauty can as well. For millennia, sages have proclaimed how outer beauty reflects inner goodness. While we may no longer openly claim that, beauty-is-good still holds sway unconsciously; attractive people are judged to be more honest, intelligent, and competent, are more likely to be elected or hired, and with higher salaries; are less likely to be convicted of crimes, then getting shorter sentences. Jeez, can't the brain distinguish beauty from goodness? Not especially. In three different studies, subjects in brain scanners alternated between rating the beauty of something (e.g., faces) or the goodness of some behavior. Both types of assessments activated the same region (the orbitofrontal cortex, or OFC); the more beautiful or good, the more OFC activation (and the less insula activation). It's as if irrelevant emotions about beauty gum up cerebral contemplation of the scales of justice. Which was shown in another study—moral judgments were no longer colored by aesthetics after temporary inhibition of a part of the PFC that funnels information about emotions into the frontal cortex.* "Interesting," the subject is told. "Last week, you sent that other person to prison for life. But just now, when looking at this other person who had done the same thing, you voted for them for Congress—how come?" And

*The region was the dorsomedial PFC, as shown with transcranial magnetic stimulation. As a control, no effect was seen when inhibiting the more "cerebral" dorsolateral PFC. Lots more on these brain regions in the next chapter.

the answer isn't "Murder is definitely bad, but OMG, those eyes are like deep, limpid pools." Where did the intent behind the decision come from? The fact that the brain hasn't had enough time yet to evolve separate circuits for evaluating morality and aesthetics.[6]

Next, want to make someone more likely to choose to clean their hands? Have them describe something crummy and unethical they've done. Afterward, they're more likely to wash their hands or reach for hand sanitizer than if they'd been recounting something ethically neutral they'd done. Subjects *instructed* to lie about something rate cleansing (but not noncleansing) products as more desirable than do those instructed to be honest. Another study showed remarkable somatic specificity, where lying orally (via voice mail) increased the desire for mouthwash, while lying by hand (via email) made hand sanitizers more desirable. One neuroimaging study showed that when lying by voice mail boosts preference for mouthwash, a different part of the sensory cortex activates than when lying by email boosts the appeal of hand sanitizers. Neurons believing, literally, that your mouth or hand, respectively, is dirty.

Thus, feeling morally soiled makes us want to cleanse. I don't believe there's a soul for such moral taint to weigh on, but it sure weighs on your frontal cortex; after disclosing an unethical act, subjects are less effective at cognitive tasks that tap into frontal function . . . unless they got to wash their hands in between. The scientists who first reported this general phenomenon poetically named it the "Macbeth effect," after Lady Macbeth, washing her hands of that imaginary damned spot caused by her murderousness.* Reflecting that, induce disgust in subjects, and if they can then wash their hands, they judge purity-related norm violations less harshly.[7]

Our judgments, decisions, and intentions are also shaped by sensory information coming from our bodies (i.e., interoceptive sensation). Consider one study concerning the insula confusing moral and visceral disgust. If you're ever on a ship in rough waters and are heaving over the rail, it's guaranteed that someone will sidle over and smugly tell you that

*And don't forget Pontius Pilate being reported to "wash his hands" of that crucifixion bother.

they're feeling great because they ate some ginger, which settles the stomach. In the study, subjects judged the wrongness of norm violations (e.g., a morgue worker touching the eye of a corpse when no one is looking; drinking out of a new toilet); consuming ginger beforehand lessened disapproval. Interpretation? First, hearing about that illicit eyeball touching pushes your stomach toward lurching, thanks to your weird human insula. Your brain then decides your feelings about that behavior based in part on lurching severity—less lurching, thanks to ginger, and funeral home shenanigans don't seem as bad.[*,8]

Particularly interesting findings regarding interoception concern hunger. One much-noted study suggested that hunger makes us less forgiving. Specifically, across more than a thousand judicial decisions, the longer it had been since judges had eaten, the less likely they were to grant a prisoner parole. Other studies also show that hunger changes prosocial behavior. "Changes"—decreasing prosociality, as with the judges, or increasing it? It depends. Hunger seems to have different effects on how charitable subjects say they are going to be, versus how charitable they actually are,[†] or where subjects have either only one or multiple chances to be naughty or nice in an economic game. But as the key point, people don't cite blood glucose levels when explaining why, say, they were nice just now and not earlier.[9]

In other words, as we sit there, deciding which button to push with supposed freely chosen intent, we are being influenced by our sensory environment—a foul smell, a beautiful face, the feel of vomit goulash, a gurgling stomach, a racing heart. Does this disprove free will? Nah—the effects are typically mild and only occur in the average subject, with

---

[*]Psychology fans will recognize how this study supports the James-Lange theory of emotion (yes, William James!). In its modern incarnation, it posits that our brain "decides" how strongly we feel about something, in part, by canvassing interoceptive info from the body; for example, if your heart is racing (thanks to unknowingly being given an adrenaline-like drug), you perceive your feelings as being more intense.

[†]With at least one paper inevitably making reference in its title to "hunger games." By the way, in chapter 11, we'll be looking at a really key circumstance where there is a major discrepancy between how charitable people say they are and how much they actually are.

plenty of individuals who are exceptions. This is just the first step in understanding where intentions come from.[10]

MINUTES TO DAYS BEFORE

The choice you'd seemingly freely make about the life-or-death button-pressing task can also be powerfully influenced by events in the preceding minutes to days. As one of the most important routes, consider the scads of different types of hormones in our circulation—each secreted at a different rate and effecting the brain in varied ways from one individual to the next, all without our control or awareness. Let's start with one of the usual suspects when it comes to hormones altering behavior, namely testosterone.

How does testosterone (T) in the preceding minutes to days play a role in determining whether you kill that person? Well, testosterone causes aggression, so the higher the T level, the more likely you'll be to make the more aggressive decision.* Simple. But as a first complication, T doesn't actually cause aggression.

For starters, T rarely generates new patterns of aggression; instead, it makes preexisting patterns more likely to happen. Boost a monkey's T levels, and he becomes more aggressive to monkeys *already* lower-ranking than him in the dominance hierarchy, while brown-nosing his social betters as per usual. Testosterone makes the amygdala more reactive, but only if neurons there are already being stimulated by looking at, say, the face of a stranger. Moreover, T lowers the threshold for aggression most dramatically in individuals already prone toward aggression.[11]

The hormone also distorts judgment, making you more likely to interpret a neutral facial expression as threatening. Boosting your T levels makes you more likely to be overly confident in an economic game, re-

*Regardless of your sex, since both secrete T (albeit in differing amounts) and have T receptors in the brain. The hormone has broadly similar effects in both sexes, just typically more strongly in males.

sulting in being less cooperative—who needs anyone else when you're convinced you're fine on your own?* Moreover, T tilts you toward more risk-taking and impulsivity by strengthening the ability of the amygdala to directly activate behavior (and weakening the ability of the frontal cortex to rein it in—stay tuned for the next chapter).† Finally, T makes you less generous and more self-centered in, for example, economic games, as well as less empathic toward and trusting of strangers.[12]

A pretty crummy picture. Back to your deciding which button to press. If T is having particularly strong effects in your brain at the time, you become more likely to perceive threat, real or otherwise, less caring about others' pain, and more likely to fall into aggressive tendencies that you already have.

What factors determine whether T has strong effects in your brain? Time of day matters, as T levels are nearly twice as high during the daily circadian peak as during the trough. Whether you're sick, are injured, just had a fight, or just had sex all influence T secretion. It also depends on how high your average T levels are; they can vary fivefold among healthy individuals of the same sex, even more so in adolescents. Moreover, the brain's sensitivity to T also varies, with T receptor numbers in some brain regions varying up to tenfold among individuals. And why do individuals differ in how much T their gonads make or how many receptors there are in particular brain regions? Genes and fetal and postnatal environment matter. And why do individuals differ in the extent of their preexisting tendencies toward aggression (i.e., how the amygdala, frontal cortex, and

*These are almost always "double-blind" studies, in which half the subjects get the hormone, the rest get saline, and neither the subjects nor the researchers testing them know who got which.

†What do I mean by T "strengthening" a projection from the amygdala to another part of the brain (the basal ganglia, in this case)? The amygdala is particularly sensitive to T, has lots of receptors for it; T lowers the threshold for amygdaloid neurons to have action potentials, making it more likely—"strengthening"—that a signal would propagate from one neuron to the next, down the line. Meanwhile, T is having the opposite effect when "weakening" projections. Dotting i's and crossing t's—T receptors are technically called androgen receptors, reflecting there being an array of "androgenic" hormones, with T as the most powerful. We're going to ignore that for all-around sanity.

so on differ)? Above all, because of how much life has taught them at a young age that the world is a menacing place.*[13]

Testosterone is not the only hormone that can influence your button-pressing intentions. There's oxytocin, acclaimed for having prosocial effects among mammals. Oxytocin enhances mother-infant bonding in mammals (and enhances human-dog bonding). The related hormone vasopressin makes males more paternal in the rare species where males help parent. These species also tend to form monogamous pair bonds; oxytocin and vasopressin strengthen the bond in females and males, respectively. What's the nuts-and-bolts biology of why males in some rodent species are monogamous and others not? Monogamous species are genetically prone toward higher concentrations of vasopressin receptors in the dopaminergic "reward" part of the brain (the nucleus accumbens). The hormone is released during sex, the experience with that female feels really *really* pleasurable because of the higher receptor number, and the male sticks around. Amazingly, boost vasopressin receptor levels in that part of the brain in males from polygamous rodent species, and they become monogamous (wham, bam, thank . . . weird, I don't know what just came over me, but I'm going to spend the rest of my life helping this female raise our kids).[14]

Oxytocin and vasopressin have effects that are the polar opposite of T's. They decrease excitability in the amygdala, making rodents less aggressive and people calmer. Boost your oxytocin levels experimentally, and you're more likely to be charitable and trusting in a competitive game. And showing how this is the endocrinology of sociality, you wouldn't have the response to oxytocin if you thought you were playing against a computer.[15]

As an immensely cool wrinkle, oxytocin doesn't make us warm and fuzzy and prosocial to everyone. Only to in-group members, people who count as an Us. In one study in the Netherlands, subjects had to decide if

*Just as an important complication, testosterone can make people more prosocial under circumstances where doing so gains them status (for example, in an economic game where status is gained by making more generous offers). In other words, testosterone is about aggression only under circumstances where the right type of aggression gets you high status.

it was okay to kill one person to save five; oxytocin had no effects when the potential victim had a Dutch name but made subjects more likely to sacrifice someone with a German or Middle Eastern name (two groups that evoke negative connotations among the Dutch) and increased implicit bias against those two groups. In another study, while oxytocin made team members more cooperative in a competitive game, as expected, it made them more preemptively aggressive to opponents. The hormone even enhances gloating over strangers' bad luck.[16]

Thus, the hormone makes us nicer, more generous, empathic, trusting, loving . . . to people who count as an Us. But if it is a Them, who looks, speaks, eats, prays, loves differently than we do, forget singing "Kumbaya."*

On to individual differences related to oxytocin. The hormone's levels vary manyfold among different individuals, as do levels of receptors for oxytocin in the brain. Those differences arise from the effects of everything from genes and fetal environment to whether you woke up this morning next to someone who makes you feel safe and loved. Moreover, oxytocin receptors and vasopressin receptors each come in different versions in different people. Which flavor you were handed at conception influences parenting style, stability of romantic relationships, aggressiveness, sensitivity to threat, and charitableness.[17]

Thus, the decisions you supposedly make freely in moments that test your character—generosity, empathy, honesty—are influenced by the levels of these hormones in your bloodstream and the levels and variants of their receptors in your brain.

One last class of hormones. When an organism is stressed, whether mammal, fish, bird, reptile, or amphibian, it secretes from the adrenal gland hormones called glucocorticoids, which do roughly the same things to the body in all these cases.† They mobilize energy from storage sites in the body, like the liver or fat cells, to fuel exercising muscle—very helpful

*Note before how testosterone can have opposite effects on neurons in two different parts of the brain. Here we have oxytocin having opposite effects on behavior in two different social contexts.

†Minor detail: Glucocorticoids, coming from the adrenal gland during stress, are different from adrenaline, also coming from the adrenal during stress. Different hormone classes but broadly

if you are stressed because, say, a lion is trying to eat you, or if you're that lion and will starve unless you predate something. Following the same logic, glucocorticoids increase blood pressure and heart rate, delivering oxygen and energy to those life-saving muscles that much faster. They suppress reproductive physiology—don't waste energy, say, ovulating, if you're running for your life.[18]

As might be expected, during stress, glucocorticoids alter the brain. Amygdala neurons become more excitable, more potently activating the basal ganglia and disrupting the frontal cortex—all making for fast, habitual responses with low accuracy in assessing what's happening. Meanwhile, as we'll see in the next chapter, frontal cortical neurons become less excitable, limiting their ability to make the amygdala act sensibly.[19]

Based on these particular effects in the brain, glucocorticoids have predictable effects on behavior during stress. Your judgments become more impulsive. If you're reactively aggressive, you become more so, if anxious, more so, if depressive, ditto. You become less empathic, more egoistic, more selfish in moral decision-making.[20]

The workings of every bit of this endocrine system will reflect whether you've been stressed recently by, say, a mean boss, a miserable morning's commute, or surviving your village being pillaged. Your gene variants will influence the production and degradation of glucocorticoids, as well as the number and function of glucocorticoid receptors in different parts of your brain. And the system would have developed differently in you depending on things like the amount of inflammation you experienced as a fetus, your parents' socioeconomic status, and your mother's parenting style.*

Thus, three different classes of hormones work over the course of minutes to hours to alter the decision you make. This just scratches the surface; Google "list of human hormones," and you'll find more than seventy-five, most effecting behavior. All rumbling below the surface,

---

similar effects. The major glucocorticoid in humans and other primates is cortisol, aka hydrocortisone.

*For what it's worth, and as a demonstration of how narrow the focus of science can be, I spent more than three decades of my life obsessing over issues related to the last four paragraphs.

influencing your brain without your awareness. Do these endocrine effects over the course of minutes to hours disprove free will? Certainly not on their own, because they typically alter the likelihood of certain behaviors, rather than cause them. On to our next turtle heading all the way down.[21]

WEEKS TO YEARS BEFORE

So hormones can change the brain over the course of minutes to hours. In those cases, "change the brain" isn't some abstraction. As a result of a hormone's actions, neurons might release packets of neurotransmitter when they otherwise wouldn't; particular ion channels might open or close; the number of receptors for some messenger might change in a specific brain region. The brain is structurally and functionally malleable, and your pattern of hormone exposure this morning will have altered your brain now, as you contemplate the two buttons.

The point of this section is that such "neuroplasticity" is small potatoes compared with how the brain can change in response to experience over longer periods. Synapses might permanently become more excitable, more likely to send a message from one neuron to the next. Pairs of neurons can form entirely new synapses, or disconnect existing ones. Branchings of dendrites and axons might expand or contract. Neurons can die; others are born.* Particular brain regions might expand or atrophy so dramatically that you can see the changes on a brain scan.[22]

*Time to step into a minefield. Since humans first learned to make fire, introductory neuroscience classes taught that the adult brain doesn't make new neurons. Then, starting in the 1960s, doughty pioneers found hints that there actually is "adult neurogenesis" after all. They were ignored for decades until the evidence became incontrovertible, and adult neurogenesis became the sexiest, most revolutionary topic in neuroscience. There have been reams of findings about how/when/why it occurs in animals, what sort of things promote it (e.g., voluntary exercise, estrogen, an enriched environment), and what inhibits it (e.g., stress, inflammation). What are the new neurons good for? Various rodent studies indicate that they contribute to stress resilience, anticipating a new reward, and something called pattern separation—once you've learned the general features of something, the new neurons help you learn distinctions among different examples of

Some of this neuroplasticity is immensely cool but tangential to free-will squabbles. If someone goes blind and learns to read braille, her brain remaps—i.e., the distribution and excitability of synapses to particular brain regions change. Result? Reading braille with her fingertips, a tactile experience, stimulates neurons in the visual cortex, as if she were reading printed text. Blindfold a volunteer for a week and his auditory projections start colonizing the snoozing visual cortex, enhancing his hearing. Learn a musical instrument and the auditory cortex remaps to devote more space to the instrument's sound. Persuade some wildly invested volunteers to practice a five-finger exercise on the piano two hours a day for weeks, and their motor cortex remaps to devote more space to controlling finger movements in that hand; get this—the same thing happens if the volunteer spends that time *imagining* the finger exercise.[23]

But then there's neuroplasticity relevant to free will–lessness. Developing post-traumatic stress disorder after trauma transforms the amygdala. Synapse number increases along with the extent of the circuitry by which the amygdala influences the rest of the brain. The overall size of the amygdala increases, and it becomes more excitable, with a lower threshold for triggering fear, anxiety, and aggression.[24]

Then there's the hippocampus, a brain region central to learning and memory. Suffer from major depression for decades and the hippocampus shrinks, disrupting learning and memory. In contrast, experience two weeks of rising estrogen levels (i.e., be in the follicular stage of your

it—say, once you've learned to recognize a performance of *Next to Normal*, you rely on pattern separation in the hippocampus to teach you the difference between a performance of it on Broadway and one in a high school (the distinctions can be minimal and subtle, if the latter is in the hands of a superb director).❤

As this neurogenesis literature matured, there was evidence that the adult human brain could make new neurons also. Then an extremely thorough 2018 paper in *Nature*, using the largest number of human brains studied to date, suggested that maybe there wasn't much/any neurogenesis in the adult human brain after all (amid there being plenty in other species). Massive controversy ensued, still raging. I find that study to be convincing (but, full disclosure, I'm not really objective, since the lead author on the paper, Shawn Sorrells, now of the University of Pittsburgh, was one of my star grad students).

ovulatory cycle), and the hippocampus beefs up. Likewise, if you enjoy exercising regularly or are stimulated by an enriching environment.[25]

Moreover, experience-induced changes aren't limited to the brain. Chronic stress expands the adrenal glands, which then pump out more glucocorticoids, even when you're not stressed. Becoming a father reduces testosterone levels; the more nurturing you are, the bigger the drop.[26]

How's this for how unlikely the subterranean biological forces on your behavior can be over weeks to months—your gut is filled with bacteria, most of which help you digest your food. "Filled with" is an understatement—there are more bacteria in your gut than cells in your own body,* of hundreds of different types, collectively weighing more than your brain. As a burgeoning new field, the makeup of the different species of bacteria in your gut over the previous weeks will influence things like appetite and food cravings . . . and gene expression patterns in your neurons . . . and proclivity toward anxiety and the ferocity with which some neurological diseases spread through your brain. Clear out all of a mammal's gut bacteria (with antibiotics) and transfer in the bacteria from another individual, and you'll have transferred those behavioral effects. These are mostly subtle effects, but who would have thought that bacteria in your gut were influencing what you mistake for free agency?

The implications of all these findings are obvious. How will your brain function as you contemplate the two buttons? It depends in part on events during previous weeks to years. Have you been barely managing to pay the rent each month? Experiencing the emotional swell of finding love or of parenting? Suffering from deadening depression? Working successfully at a stimulating job? Rebuilding yourself after combat trauma or sexual assault? Having had a dramatic change in diet? All will change your brain and behavior, beyond your control, often beyond your awareness. Moreover, there will be a metalevel of differences outside your control, in that your genes and childhood will have regulated how *easily* your brain changes

*Meaning, among other things, that if someone centrifuged you and then extracted your DNA, if they were not careful, they'd mostly be inadvertently studying the DNA of your gut bacteria.

in response to particular adult experiences—there is plasticity as to how much and what kind of neuroplasticity each person's brain can manage.[27]

Does neuroplasticity show that free will is a myth? Not by itself. Next turtle.[28]

BACK TO ADOLESCENCE

As will be familiar to any reader who is, was, or will be an adolescent, this is one complex time of life. Emotional gyrations, impulsive risk-taking and sensation seeking, the peak time of life for extremes of both pro- and antisocial behavior, for individuated creativity and for peer-driven conformity; behaviorally, it is a beast unto itself.

Neurobiologically as well. Most research examines why adolescents behave in adolescent ways; in contrast, our purpose is to understand how features of the adolescent brain help explain button-pushing intentions in adulthood. Conveniently, the same hugely interesting bit of neurobiology is relevant to both. By early adolescence, the brain is a fairly close approximation of the adult version, with adult densities of neurons and synapses, and the process of myelinating the brain already achieved. Except for one brain region which, amazingly, won't fully mature for another decade. The region? The frontal cortex, of course. Maturation of this region lags way behind the rest of the cortex—to some degree in all mammals, and dramatically so in primates.[29]

Some of that delayed maturation is straightforward. Starting with fetal brain building, there's a steady increase in myelination up to adult levels, including in the frontal cortex, just with a huge delay. But the picture is majorly different when it comes to neurons and synapses. At the start of adolescence, the frontal cortex has *more* synapses than in the adult. Adolescence and early adulthood consist of the frontal cortex pruning synapses that turn out to be superfluous, poky, or plain wrong, as the region gets progressively leaner and meaner. As a great demonstration of this, while a thirteen-year-old and a twenty-year-old may perform equally on some test of frontal function, the former needs to mobilize more of the region to accomplish this.

So the frontal cortex—with its roles in executive function, long-term planning, gratification postponement, impulse control, and emotion regulation—isn't fully functional in adolescents. Hmm, what do you suppose that explains? Just about everything in adolescence, especially when adding the tsunamis of estrogen, progesterone, and testosterone flooding the brain then. A juggernaut of appetites and activation, constrained by the flimsiest of frontal cortical brakes.[30]

For our purposes, the main point about delayed frontal maturation isn't that it produces kids who got really bad tattoos but the fact that adolescence and early adulthood involve a massive construction project in the brain's most interesting part. The implications are obvious. If you're an adult, your adolescent experiences of trauma, stimulation, love, failure, rejection, happiness, despair, acne—the whole shebang—will have played an outsize role in constructing the frontal cortex you're working with as you contemplate those buttons. Of course, the enormous varieties of adolescence experiences will help produce enormously varied frontal cortexes in adulthood.

A fascinating implication of the delayed maturation is important to remember when we get to the section on genes. By definition, if the frontal cortex is the last part of the brain to develop, it is the brain region least shaped by genes and most shaped by environment. This raises the question of why the frontal cortex matures so slowly. Is it intrinsically a tougher building project than the rest of the cortex? Are there specialized neurons, neurotransmitters unique to the region that are tough to synthesize, distinctive synapses that are so fancy that they require thick construction manuals? No, virtually nothing unique like that.[*,31]

*There is a neuron type called the von Economo neuron (VEN) that is found pretty much only in two brain regions tightly linked to the frontal cortex—the insular cortex and the anterior cingulate cortex. For a while, there was massive excitement in that it appeared to be a neuron type unique to humans, a first. But things were actually even more interesting—VENs also occur in the brains of some of the most socially complex species on earth, such as other apes, cetaceans, and elephants. No one is quite sure what they are for, but there's been some progress. But despite

Thus, delayed maturation isn't inevitable, given the complexity of frontal construction, where the frontal cortex would develop faster, if only it could. Instead, the delay actively evolved, was selected for. If this is the brain region central to doing the right thing when it's the harder thing to do, no genes can specify what counts as the right thing. It has to be learned the long, hard way, by experience. This is true for any primate, navigating social complexities as to whether you hassle or kowtow to someone, align with them or stab them in the back.

If that's the case for some baboon, just imagine humans. We have to learn our culture's rationalizations and hypocrisies—thou shalt not kill, unless it's one of them, in which case here's a medal. Don't lie, except if there's a huge payoff, or it's a profoundly good act ("Nope, no refugees hiding in my attic, no siree"). Laws to be followed strictly, laws to be ignored, laws to be resisted. Reconciling acting as if each day is your last with today being the first day of the rest of your life. On and on. Reflecting that, while frontocortical maturation finally tops out around puberty in other primates, we need another dozen years. This suggests something remarkable—the genetic program of the human brain evolved to free the frontal cortex from genes as much as possible. Much more to come about the frontal cortex in the next chapter.

Next turtle.[32]

AND CHILDHOOD

So adolescence is the final phase of frontal cortical construction, with the process heavily shaped by environment and experience. Moving further back into childhood, there are massive amounts of construction of *everything* in the brain,* a process of a smooth increase in the complexity or

VENs' existence, the similarities between the building blocks of the frontal cortex and the rest of the cortex are much greater than the differences.

*Note: "Everything in the brain" includes the frontal cortex; amid the drama of delayed maturation, a substantial percentage of its construction occurs during childhood.

neuron neuronal circuitry and of myelination. Naturally, this is paralleled by growing behavioral complexity. There's maturation of reasoning skills and of cognition and affect relevant to moral decision-making (e.g., transitioning from obeying laws to avoid punishment to obeying because where would society be without people obeying them?). There's maturation of empathy (with growing capacities to empathize with someone's emotional rather than physical state, about abstract pain, about pains you've never experienced, about pain for people totally different from you). Impulse control is also maturing (from successfully restraining yourself for a few minutes from eating a marshmallow in order to then be rewarded with two marshmallows, to staying focused on your eighty-year project to get into the nursing home of your choice).

In other words, simpler things precede more complicated things. Child-development researchers have typically framed these trajectories of maturation as coming in "stages" (for example, Harvard psychologist Lawrence Kohlberg's canonical stages of moral development). Predictably, there are huge differences as to what particular maturational stage different kids are at, the speed of stage transitions, and the stage carried stably into adulthood.[*,33]

Speaking to our interests, you have to ask where individual differences in maturation come from, how much control we have over that process, and how it helps generate the you that is you, contemplating the buttons. What sorts of influences effect maturation? An overlapping list of the most usual suspects, with incredibly brief summaries:

1. Parenting, of course. Differences in parenting styles were the focus of highly influential work originating with Berkeley psychologist Diana Baumrind. There's authoritative parenting, where high levels of demands and expectation are placed on the child, coupled with lots of

*Naturally, there are problems with an overly literal reliance on stage thinking—the transitions from one stage to the next can be smooth continua, rather than crossings of distinctive borders; a child's stage of, say, moral reasoning may differ with differing emotional states; insights have mostly come from studies of boys in Western cultures. Nonetheless, the basic idea is really useful.

flexibility in responding to the child's needs; this is usually the style aspired to by neurotic middle-class parents. Then there's authoritarian parenting (high demand, low responsiveness—"Do this because I said so"), permissive parenting (low demand, high responsiveness), and negligent parenting (low demand, low responsiveness). And each tends to produce a different sort of adult. As we'll see in the next chapter, parental socioeconomic status (SES) is also enormously important; for example, low familial SES predicts stunted maturation of the frontal cortex in *kindergarteners*.[34]

2. Peer socialization, with different peers modeling different behaviors with varying allure. The importance of peers has often been underappreciated by developmental psychologists but is no surprise to any primatologists. Humans invented a novel way to transmit information across generations, where an adult expert intentionally directs information at young'uns—i.e., a teacher. In contrast, the usual among primates is kids learning by watching their somewhat older peers.[35]

3. Environmental influences. Is the neighborhood park safe? Are there more bookstores or liquor stores? Is it easy to buy healthy food? What's the crime rate? All the usual.

4. Cultural beliefs and values, which influence these other categories. As we'll see, culture dramatically influences parenting style, the behaviors modeled by peers, the sorts of physical and social communities that are constructed. Cultural variability in overt and covert rites of passage, the brands of places of worship, whether kids aspire to earn lots of merit badges versus getting skilled at harassing out-group members.

A pretty straightforward list. And, of course, there are loads of individual differences in childhood patterns of hormone exposure, nutrition, pathogen load, and so on. All converging to produce a brain that, as we'll see in chapter 5, has to be unique.

The huge question then becomes, How do different childhoods produce different adults? Sometimes, the most likely pathway seems pretty clear without having to get all neurosciencey. For example, a study examining more than a million people across China and the U.S. showed the effects of growing up in clement weather (i.e., mild fluctuations around an average of seventy degrees). Such individuals are, on the average, more individualistic, extroverted, and open to novel experience. Likely explanation: the world is a safer, easier place to explore growing up when you don't have to spend significant chunks of each year worrying about dying of hypothermia and/or heatstroke when you go outside, where average income is higher and food stability greater. And the magnitude of the effect isn't trivial, being equal to or greater than that of age, gender, the country's GDP, population density, and means of production.[36]

The link between weather clemency in childhood and adult personality can be framed biologically in the most informative way—the former influences *the type of brain you're constructing* that you will carry into adulthood. As is almost always the case. For example, lots of childhood stress, by way of glucocorticoids, impairs construction of the frontal cortex, producing an adult less adept at helpful things like impulse control. Lots of exposure to testosterone early in life makes for the construction of a highly reactive amygdala, producing an adult more likely to respond aggressively to provocation.

The nuts and bolts of how this happens revolves around the massively trendy field of "epigenetics," revealing how early life experience causes long-lasting changes in gene expression in particular brain regions. Now, this is not experience changing genes themselves (i.e., changing DNA sequences), but instead changing their regulation—whether some gene is always active, never active, or active in one context but not another; a lot is known by now about how this works. As one celebrated example, if you're a baby rat growing up with an atypically inattentive mother,* epigenetic changes in the regulation of one gene in your hippocampus will make it harder for you to recover from stress as an adult.[37]

Where do differences in rodential mothering style come from? Obvi-

*Whoa, different rat mothers mother differently? Sure, with variation as to how often they groom or licks their pups, respond to their vocalizations, and so on. This is landmark work pioneered by neuroscientist Michael Meaney of McGill University.

ously, from one second, one minute, one hour, before in that rat mom's biological history. Knowledge about epigenetic bases of this has grown at breakneck speed, showing, for example, how some epigenetic changes in the brain can have multigenerational consequences (e.g., helping to explain why being a rat, monkey, or human abused in childhood increases the odds of being an abusive parent). Just to show the scale of epigenetic complexity, differences in mothering styles in monkeys cause epigenetic changes in more than a thousand genes expressed in the offspring's frontal cortex.[38]

If you had to compress the variability in all those facets of childhood influences into a single axis, it would be easy—how lucky was the childhood you were handed? This massively important fact has been formalized into an Adverse Childhood Experience (ACE) score. What count as adverse experiences in this measure? A logical list:

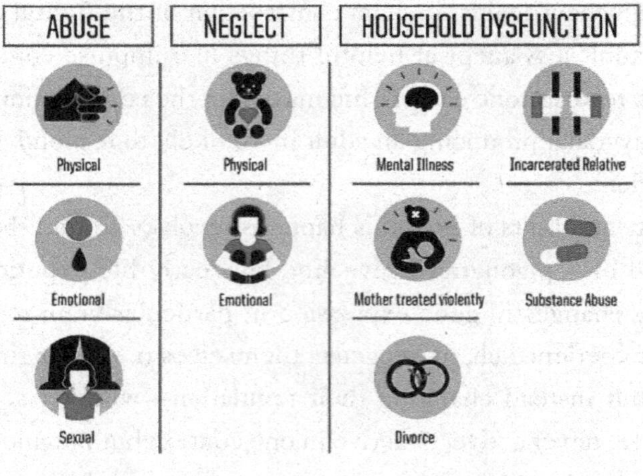

Source: Centers for Disease Control and Prevention

For each of these experienced, you get a point on the checklist, where the unluckiest have scores approaching an unimaginable ten and the luckiest luxuriating around zero.

This field has produced a finding that should floor anyone holding out for free will. For every step higher in one's ACE score, there is roughly a 35 percent increase in the likelihood of adult antisocial behavior, including violence; poor frontocortical-dependent cognition; problems with impulse control; substance abuse; teen pregnancy and unsafe sex and other risky behaviors; and increased vulnerability to depression and anxiety disorders. Oh, and also poorer health and earlier death.[39]

You'd get the same story if you flipped the approach 180 degrees. As a child, did you feel loved and safe in your family? Was there good modeling about sexuality? Was your neighborhood crime-free, your family mentally healthy, your socioeconomic status reliable and good? Well then, you'd be heading toward a high RLCE score (Ridiculously Lucky Childhood Experiences), predictive of all sorts of important good outcomes.

Thus, essentially every aspect of your childhood—good, bad, or in between—factors over which you had no control, sculpted the adult brain you have while contemplating those buttons. How's this for an example outside of someone's control—because of the randomness of month of birth, some kids can be as much as six months older or younger than the average of their peer group. Older kindergarteners, for example, are typically more cognitively advanced. Result—they get more one-on-one attention and praise from teachers, so that by first grade their advantage is even greater, so that by second grade . . . And in the UK, which has an August 31 cutoff for kindergarten, this "relative age effect" produces a major skew in educational attainment. For example:

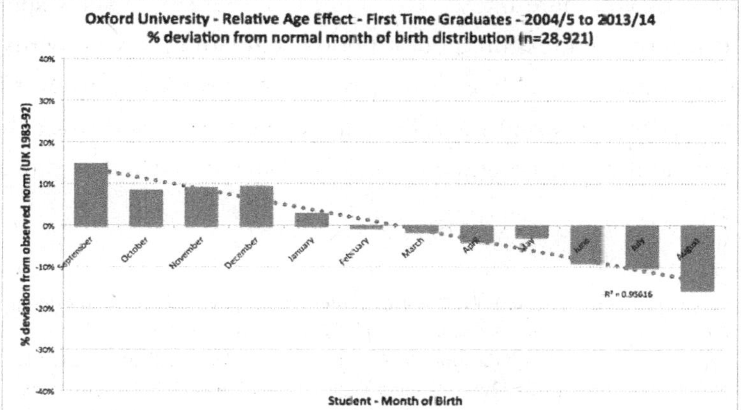

Luck evens out over time, my ass.*[,40]

Does the role of childhood invalidate free will? Nope—the likes of ACE scores are about adult potential and vulnerability, not inevitable destiny, and there are plenty of people whose adulthoods are radically different from what you'd expect, given their childhoods. This is just another piece of the sequence of influences.[41]

BACK TO THE WOMB

If you couldn't control what family you landed in at birth, you sure had no control over which womb you hung out in for nine influential months. Environmental influences begin long before birth. The biggest source of these influences is what's in the maternal circulation, which will help determine what's in the fetus—levels of a huge array of different hormones, immune factors, inflammatory molecules, pathogens, nutrients, environmental toxins, illicit substances, all which regulate brain function in adulthood. Not surprising, the general themes echo those of childhood. Lots of glucocorticoids from Mom marinating your fetal brain, thanks to maternal stress, and there's increased vulnerability to depression and anxiety in your adulthood. Lots of androgens in your fetal circulation (coming from Mom; females secrete androgens, though to a lesser extent than do males) makes you more likely as an adult of either sex to show spontaneous and reactive aggression, poor emotion regulation, low empathy, alcoholism, criminality, even lousy handwriting. A shortage of nutrients for the fetus, caused by maternal starvation, and there's increased risk of schizophrenia in adulthood, along with a variety of metabolic and cardiovascular diseases.†[,42]

---

*The same effect holds for sports. Professional athletic teams are way disproportionately filled with players who were older than average in their childhood sports cohort.

†These effects on fetuses were first identified in humans in two horrifically unnatural "natural experiments of starvation"—the Dutch Hunger Winter of 1944, when Holland was starved by the occupying Nazis, and the Great Leap Forward famine in China in the late 1950s.

The implications of fetal environmental effects? Another route toward how lucky or unlucky you're likely to be in the world that awaits you.[43]

BACK TO YOUR VERY BEGINNING: GENES

Down to the next turtle. If you didn't choose the womb you grew in, you certainly didn't choose the unique mixture of genes you inherited from your parents. Genes have plenty to do with decision-making crossroads, and in more interesting ways than commonly believed.

We start with an unbelievably superficial primer on genes, to position us to appreciate things when we get to genes and free will.

First, what are genes, and what do they do? Our bodies are filled with thousands of different types of proteins doing dizzyingly varied jobs. Some are "cytoskeletal" proteins that give different cell types their distinctive shapes. Some are messengers—many neurotransmitters, hormones, and immune messengers are proteins. It's proteins that make up enzymes that construct those messengers and that tear them apart when they're obsolete; virtually all receptors for messengers throughout the body are made of protein.

Where does all this proteinaceous versatility come from? Each type of protein is constructed from a distinctive sequence of different types of amino acid building blocks; the sequence determines the shape of the protein; the shape determines function. A "gene" is the stretch of DNA that specifies the sequence/shape/function of a particular protein. Each of our approximately twenty thousand genes codes for the production of a unique protein.*

How does a gene "decide" when to initiate the construction of the protein it codes for, and whether there will be one or ten thousand copies made? Implicit in this question is the popular view of genes as the be-all

*For those with a background in this, it's worth noting a few of the things that I ignored in this paragraph: the intronic/exonic structure of genes, gene splicing, multiple conformations in prion proteins, transposons, genes coding for small interfering RNA, and RNA enzymes . . .

and end-all, the code of codes in regulating what goes on in your body. As it turns out, genes decide nothing, are out at sea. Saying that a gene decides when to generate its associated protein is like saying that the recipe decides when to bake the cake that it codes for.

Instead, genes are turned on and off by environment. What is meant here by *environment*? It can be the environment within a single cell—a cell is running low on energy, which generates a messenger molecule that activates the genes that code for proteins that boost energy production. Environment can encompass the entire body—a hormone is secreted and is carried in the circulation to target cells at the other end of the body, where it binds to its distinctive receptors; as a result, particular genes are turned on or off. Or environment can take the form of our everyday usage, namely events happening in the world around us. These different versions of environment are linked. For example, living in a stressful, dangerous city will produce chronically elevated levels of glucocorticoids secreted by your adrenal glands, which will activate particular genes in neurons in the amygdala, making those cells more excitable.*

How do different environmentally activated messengers turn on different genes? Not every stretch of DNA contributes to the code in a gene; instead, long stretches don't code for anything. Instead, they are the on/off switches for activating nearby genes. Now for a wild fact—only about 5 percent of DNA constitutes genes. The remaining 95 percent? The dizzyingly complex on/off switches, the means by which various environmental influences regulate unique networks of genes, with multiple types of switches on a single gene and multiple genes being regulated by the same type of switch. In other words, most DNA is devoted to gene regulation rather than to genes themselves. Moreover, evolutionary changes in DNA are usually more consequential when they alter on/off switches rather than the gene. As another measure of the importance of the regulation, the

*Things left out of this paragraph: transcription factors, signal transduction pathways, the fact that it is only steroid hormones, in contrast to peptidergic hormones, which directly regulate transcription . . .

more complex the organism, the greater the percentage of its DNA is devoted to gene regulation.*

Where have we gotten in this primer? Genes code for workhorse proteins; genes don't decide when they are active but are, instead, regulated by environmental signals; the evolution of DNA is disproportionately about gene regulation rather than about genes.

So environmental signals have activated some gene, leading to the production of its protein; the newly made proteins then do their usual thing. As a next key point, the same protein can work differently in different environments. Such "gene/environment interactions" are less important in species that inhabit only one type of environment. But they're plenty relevant in species that inhabit multiple types of environments—species like, say, us. We can live in tundra, desert, or rain forest; in an urban megalopolis of millions or in small hunter-gatherer bands; in capitalist or socialist societies, polygamous or monogamous cultures. When it comes to humans, it can be silly to ask what a particular gene does—only what it does in a particular environment.

What might gene/environment interactions look like? Suppose someone has a gene variant related to aggression; depending on the environment, that can result in an increased likelihood of street brawling or of playing chess really aggressively. Or a gene related to risk-taking that, depending on environment, will influence whether you rob a store or gamble on founding a start-up. Or a gene related to addiction that, depending on environment, produces a Brahmin drinking too much Scotch in his club or someone desperately stealing to get money for heroin.†

Final bit of the primer. Most genes come in more than one flavor, with people inheriting their particular variants from their parents. Such gene

*Some of the things left out here: promoters and other regulatory elements in DNA, transcriptional cofactors imparting tissue specificity of gene transcription, selfish DNA derived from self-replicating retroviruses . . .

†Things left out include how simplistic it is to focus on a single gene and its singular effect, even after accounting for environment. This is because of pleiotropic and polygenic genetic effects; startling evidence for the importance of the latter comes from genome-wide survey studies, indicating that even the most boringly straightforward human traits, such as height, are coded for by hundreds of different genes.

variants code for slightly different versions of their protein, with some being better at their job than others.*

Where have we gotten? People differing in the flavors of genes they possess, those genes being regulated differently in different environments, producing proteins whose effects vary in different environments. We now consider how genes relate to this free-will obsession of ours.

It's button time; how will your brain be influenced in that moment by the flavors of particular genes you inherited? Consider the neurotransmitter serotonin—differing profiles of serotonin signaling among people help explain individual differences related to mood, levels of arousal, tendency toward compulsive behavior, ruminative thoughts, and reactive aggression. And how can individual differences in gene variants contribute to differences in serotonin signaling? Easily—different flavors exist for the genes coding for the proteins that synthesize serotonin, that remove it from the synapse, and that degrade it,† plus variants in the genes that code more than a dozen different types of serotonin receptors.[44]

Same story with the neurotransmitter dopamine. To barely scratch the surface, individual differences in dopamine signaling are relevant to reward, anticipation, motivation, addiction, gratification postponement, long-term planning, risk-taking, novelty seeking, salience of cues, and ability to focus—you know, things pertinent to our judging, say, whether someone could have transcended their dire circumstances if only they could have shown some self-discipline. And the genetic sources of dopaminergic differences among people? Genetic variants related to dopamine's synthesis, degradation, and removal from the synapse,‡ as well as in the various dopamine receptors.[45]

We could go on now to the neurotransmitter norepinephrine. Or en-

*Some of the things left out: homozygosity versus heterozygosity, dominant versus recessive traits . . .

†Aficionados: the genes coding for tryptophan hydroxylase and aromatic amino acid decarboxylase, the 5HTT serotonin transporter, monoamine oxidase-alpha, respectively.

‡More details: the genes for tyrosine hydroxylase, the DAT dopamine transporter, catechol-O-methyltransferase.

zymes that synthesize and degrade various hormones and hormone receptors. Or pretty much anything pertinent to brain function. There's usually extensive individual variation in every relevant gene, and you weren't consulted as to which you'd choose to inherit.

What about the flip side—a bunch of people all have the identical gene variant but live in different environments? You get precisely what was discussed above, namely dramatically different effects of the gene variant depending on environment. For example, one variant of the gene whose protein breaks down serotonin will increase your risk of antisocial behavior . . . but only if you were severely abused during childhood. A variant of a dopamine receptor gene makes you either more or less likely to be generous, depending on whether you grew up with or without secure parental attachment. That same variant is associated with poor gratification postponement . . . if you were raised in poverty. One variant of the gene that directs dopamine synthesis is associated with anger . . . but only if you were sexually abused as a kid. One version of the gene for the oxytocin receptor is associated with less sensitive parenting . . . but only when coupled with childhood abuse. On and on (and with many of the same relationships being seen in other primate species as well).[46]

Dang, how can environment cause genes to work so differently, even in diametrically opposite ways? Just to start to put all the pieces together, because different environments will cause different sorts of epigenetic changes in the same gene or genetic switch.

Thus, people have all these different versions of all of these, and these different versions work differently, depending on childhood environment. Just to put some numbers to it, humans have roughly twenty thousand genes in our genome; of those, approximately 80 percent are active in the brain—sixteen thousand. Of those genes, nearly all come in more than one flavor (are "polymorphic"). Does this mean that in each of those genes, the polymorphism consists of one spot in that gene's DNA sequence that can differ among individuals? No—there are actually an average of 250 spots in the DNA sequence of each gene . . . which adds up to there being individual variability in approximately four million spots in the sequence of DNA that codes for genes active in the brain.*[,47]

Does behavior genetics disprove free will? Not on its own—as a familiar theme, genes are about potentials and vulnerabilities, not inevitabilities, and the effects of most of these genes on behavior are relatively mild. Nonetheless, all these effects on behavior arise from genes you didn't choose, interacting with a childhood you didn't choose.[48]

BACK CENTURIES: THE SORT OF PEOPLE YOU COME FROM

The Libetian buttons beckon. What does your culture have to do with the intent you will act upon? Tons. Because from your moment of birth, you were subject to a universal, which is that every culture's values include ways to make their inheritors recapitulate those values, to become "the sort of people you come from." As a result, your brain reflects who your ancestors were and what historical and ecological circumstances led them to invent those values surrounding you. If a fairly tunnel-visioned neurobiologist became dictator of the world, anthropology would be defined as "the study of the ways that different groups of people attempt to shape brain construction in their children."

Cultures produce dramatically different behaviors with consistent patterns. One of the most studied contrasts concerns "individualist" versus "collectivist" cultures. The former emphasize autonomy, personal achievement, uniqueness, and the needs and rights of the individual; it's looking out for number one, where your actions are "yours." Collectivist cultures, in contrast, espouse harmony, interdependence, and conformity, where the needs of the community guide behavior; the priority is that your actions make the community proud, because you are "theirs." Most studies

*If each of those polymorphic spots comes in one of merely two possible versions, the number of different genetic makeups would be two to the four millionth power, a pretty good approximation of infinity—two to the mere fortieth power is something like a trillion.

of these contrasts compare individuals from the poster child of individualist cultures, the United States, with those from the textbook collectivist cultures of East Asia. The differences make sense. People from the U.S. are more likely to use first-person-singular pronouns, to define themselves in personal rather than relational terms ("I'm a lawyer" versus "I'm a parent"), to organize memory around events rather than social relations ("the summer I learned to swim" versus "the summer we became friends"). Ask subjects to draw a sociogram—a diagram with circles representing themselves and the people who matter in their lives, connected by lines—Americans typically place themselves in the biggest circle, in the center. Meanwhile, an East Asian's circle typically is no bigger than the others, and is not front and center. The American goal is to distinguish yourself by getting ahead of everyone else; the East Asian is to avoid being distinguishable.* And from these differences come major differences as to what count as norm violations and what you do about them.[49]

Naturally, this reflects different workings of the brain and body. On average, in East Asian individuals, the dopamine "reward" system activates more when looking at a calm versus excited facial expression; for Americans, it's the opposite. Show subjects a picture of a complex scene. Within *milliseconds*, East Asians typically scan the entire scene as a whole, remembering it; Americans focus on the person in the center of the picture. Force an American to tell you about times that other people influenced them, and they secrete glucocorticoids; someone East Asian will secrete the stress hormone when forced to tell you about times they influenced other people.[50]

Where do these differences come from? The standard explanations for American individualism include (a) not only are we a nation of immigrants (as of 2017, ~37 percent immigrants or children of), but it's not random who emigrates; instead, immigrating is a filtering process selecting for people willing to leave their world and culture behind, sustain an arduous journey

*Just to reiterate a point about every fact in this chapter: These are broad populational differences that differ with statistical significance from chance, not reliable predictors of every individual's behavior. Every statement is tacitly preceded by "*on the average.*"

to a place with barriers impeding their entry, and labor at the most shit jobs when granted admission; and (b) most of American history has been spent with an expanding western border settled by similarly tough, individualist pioneers. Meanwhile, the standard explanation for East Asian collectivism is ecology dictating the means of production—ten millennia of rice farming, which demands massive amounts of collective labor to turn mountains into terraced rice paddies, collective planting and harvesting of each person's crops in sequence, collective construction and maintenance of massive and ancient irrigation systems.*[,51]

A fascinating exception that proves the rule concerns parts of northern China where the ecosystem precludes rice growing, producing millennia of the much more individualistic process of wheat farming. Farmers from this region, and even their university student grandchildren, are as individualistic as Westerners. As one finding that is beyond cool, Chinese from rice regions accommodate and avoid obstacles (in this case, walking around two chairs experimentally placed to block the way in Starbucks); people from wheat regions remove obstacles (i.e., moving the chairs apart).[52]

Thus, cultural differences arising centuries, millennia, ago, influence behaviors from the most subtle and minuscule to dramatic.† Another literature compares cultures of rain forest versus desert dwellers, where the former tend toward inventing polytheistic religions, the latter, monotheistic ones. This probably reflects ecological influences as well—life in the desert is a furnace-blasted, desiccated singular struggle for survival; rain

---

*As an example that floored me, an irrigation system near Djiuangyan City in China irrigates five thousand square kilometers of rice fields and has been collectively used and maintained for *two thousand* years.

†To introduce a hot potato, are there genetic differences between individualist and collective cultures? Whatever there are can't be too important; after a generation or two, descendants of Asian American immigrants are as individualistic as European Americans. Nonetheless, genetics differences have been found that are *so* interesting. Consider the gene DRD4, which codes for a dopamine receptor. You know, dopamine—that's about motivation, anticipation, and reward. One DRD4 variant makes a receptor that is less responsive to dopamine and increases the likelihood of novelty seeking, extroversion, and impulsivity in people. Europeans and European Americans: a 23 percent incidence of that variant. East Asians: 1 percent, a difference way above chance, suggesting the variant being selected *against* in East Asia for thousands of years.

forests teem with a multitude of species, biasing toward the invention of a multitude of gods. Moreover, monotheistic desert dwellers are more warlike and more effective conquerors than rain forest polytheists, explaining why roughly 55 percent of humans proclaim religions invented by Middle Eastern monotheistic shepherds.[53]

Shepherding raises another cultural difference. Traditionally, humans make livings as agriculturalists, hunter-gatherers, or pastoralists. The last are folks in deserts, grasslands, or plains of tundra, with their herds of goats, camels, sheep, cows, llamas, yaks, or reindeer. Such pastoralists are uniquely vulnerable. It's hard to sneak in at night and steal someone's rice field or rain forest. But you can be a sneaky varmint and rustle someone's herd, stealing the milk and meat they survive on.* This pastoralist vulnerability has generated "cultures of honor" with the following features: (a) extreme but temporary hospitality to the stranger passing through—after all, most pastoralists are wanderers themselves with their animals at some point; (b) adherence to strict codes of behavior, where norm violations are typically interpreted as insulting someone; (c) such insults demanding retributive violence—the world of feuds and vendettas lasting generations; (d) the existence of warrior classes and values where valor in battle produces high status and a glorious afterlife. Much has been made of the hospitality, conservatism (as in strictly conserving cultural norms), and violence of the traditional culture of honor of the American South. The pattern of violence tells a ton: murders in the South, which typically has the highest rates in the country, are not about stickups gone wrong in a city; they're about murdering someone who has seriously tarnished your honor (by conspicuously bad-mouthing you, failing to repay a debt, coming on to your significant other . . .), particularly if living in a

*Among the pastoralist Maasai I've lived near in Africa, group violence increasingly revolves around clashes with neighboring agricultural folks, Sharks-versus-Jets moments in market areas visited by both. But the historical enemies of my Maasai are the Kuria people of Tanzania, pastoralists prone to cattle rustling from Maasai at night; this leads to spear-laden retributive raids that can kill dozens. As a measure of the combativeness of the Kuria after independence, Tanzania's army was 50 percent Kuria, despite their being only 1 percent of the population.

rural area.* Where does the Southern culture of honor come from? A widely accepted theory among historians makes this paragraph's point perfectly—while colonial New England filled with Pilgrims, and the mid-Atlantic with mercantile folks like Quakers, the South was disproportionately peopled by wild-assed pastoralists from northern England, Scotland, and Ireland.[54]

One last cultural comparison, between "tight" cultures (with numerous and strictly enforced norms of behavior) and "loose" ones. What are some predictors of a society being tight? A history of lots of cultural crises, droughts, famines, and earthquakes, and high rates of infectious diseases.† And I mean it with "history"—in one study of thirty-three countries, tightness was more likely in cultures that had high population densities back in *1500*.[‡, 55]

Five hundred years ago!? How can that be? Because generation after generation, ancestral culture influenced the likes of how much physical contact mothers had with their children; whether kids were subject to scarification, genital mutilation, and life-threatening rites of passage; whether myths and songs were about vengeance or turning the other cheek.

Does the influence of culture disprove free will? Obviously not. As usual, these are tendencies, amid lots of individual variation. Just consider Gandhi, Anwar Sadat, Yitzhak Rabin, and Michael Collins, atypically

---

*As a great experimental example, stage things so that your male subject is insulted by someone; if they come from the South, there is a huge increase in circulating cortisol and testosterone levels and an increased likelihood of advocating a violent response to a hypothetical honor violation (relative to uninsulted Southern subjects). Northerners? No such changes.

†The infectious-disease link may help explain the additional finding that cultures originating in the tropics tend toward more extreme in-group/out-group differentiation than cultures from regions farther from the equator. Temperate ecosystems make for cultures that are more temperate about outsiders.

‡And as a possible neurobiological underpinning of this, consider people from cities, suburbs, and rural areas. The larger the population someone grew up in, the more reactive their amygdala is likely to be during stress. This has prompted various articles with titles revolving around "Stress and the City."

inclined toward peacemaking, assassinated by coreligionists atypically inclined toward extremism and violence.[*,56]

OH, WHY NOT? EVOLUTION

For various reasons, humans were sculpted by evolution over millions of years to be, on the average, more aggressive than bonobos but less so than chimps, more social than orangutans but less so than baboons, more monogamous than mouse lemurs but more polygamous than marmosets. 'Nuff said.[57]

SEAMLESS

Where does intent come from? What makes us who we are at any given minute? What came before.[†] This raises an immensely important point first brought up in chapter 1, which is that the biology/environment

*As a final vote for the power of ecological influences underlying many of these cultural patterns, humans and other animals living in the same ecosystem tend to share numerous traits. For example, high levels of biodiversity in a particular ecosystem predict high levels of linguistic diversity among the humans living there (and places where large numbers of species are in danger of extinction are also where languages and cultures are most at risk of extinction). A study of 339 hunter-gatherer cultures from around the world showed even more dramatic convergence between humans and other animals—human cultures with high degrees of polygamy tend (at higher-than-chance level) to be surrounded by other animals with high rates of the same. There is also human/animal covariance in likelihood of males helping to take care of kids, of storing food, and of subsisting predominantly on a fish diet. And statistically, the human/animal resemblances are explained by ecological features like latitude, altitude, rainfall, and extreme versus temperate weather. Once again, we're just another animal, if a weird one

†It's worth noting that similar, if not identical, types of turtles all the way down also explain why, say, some chimp is the most gifted member of her generation at making tools: good social and observational skills allowing her to hang out closely and learn the trade from an older master, impulse control allowing patience with trial and error, attention to detail, the combination of innovativeness and the confidence to ignore how the cool kids are doing it—all arising from events one minute before, one hour before, and so on. Not a smidgen of "when the going gets tough, tough chimps *choose to* get going."

interactions of, say, a minute ago and a decade ago are not separate entities. Suppose we are considering the genes someone inherited, back when they were a fertilized egg, and what those genes have to do with that person's behavior. Well then, we are being geneticists thinking about genetics. We could even make our club more exclusive and be "behavior geneticists," publishing our research only in a journal called, well, *Behavior Genetics.* But if we are talking about the genes inherited that are relevant to the person's behavior, we're automatically also talking about how the person's brain was constructed—because brain construction is primarily carried out by the proteins coded for by "genes implicated in neurodevelopment." Similarly, if we are studying the effects of childhood adversity on adult behavior, often best understood on the psychological or sociological level, we're implicitly also considering how the molecular biology of childhood epigenetics helps explain adult personality and temperament. If we are evolutionary biologists thinking about human behavior, by definition we're also being behavior geneticists, developmental neurobiologists, and neuroplasticians (spell-check just went crazy). This is because *evolving* means changes in what variants of genes you find in organisms and thus the ways in which they shape brain construction. Study hormones and behavior, and we're also studying what fetal life had to do with the development of the glands that secrete those hormones. So on and so on. Each moment flowing from all that came before. And whether it's the smell of a room, what happened to you when you were a fetus, or what was up with your ancestors in the year 1500, all are things that you couldn't control.* A seamless stream of influences that, as said at the beginning, precludes being able to shoehorn in this thing called free will that is supposedly in the brain but not of it. In the words of legal scholar Pete Alces, there is "no remaining gap between

*This approach is implicit in the thinking of Cornell philosopher Derk Pereboom; he posits four scenarios: you do something awful because (1) scientists manipulated your brain a second ago; (2) they manipulated your childhood experiences; (3) they manipulated the culture you were raised in; (4) they manipulated the physical nature of the universe. These are ultimately equally deterministic scenarios, though most people's intuitions solidly view the first as far more so than the other three, because of its close proximity to the behavior itself.

nature and nurture for moral responsibility to fill." Philosopher Peter Tse hits the nail on the head when referring to the biological turtles all the way down as a "responsibility destroying regress."*[,58]

This seamless stream shows why bad luck doesn't get evened out, why it amplifies instead. Have some particular unlucky gene variant, and you'll be unluckily sensitive to the effects of adversity during childhood. Suffering from early-life adversity is a predictor that you'll be spending the rest of your life in environments that present you with fewer opportunities than most, and that enhanced developmental sensitivity will unluckily make you less able to benefit from those rare opportunities—you may not understand them, may not recognize them as opportunities, may not have the tools to make use of them or to keep you from impulsively blowing the opportunity. Fewer of those benefits make for a more stressful adult life, which will change your brain into one that is unluckily bad at resilience, emotional control, reflection, cognition . . . Bad luck doesn't get evened out by good. It is usually amplified until you're not even on the playing field that needs to be leveled.

This is the view forcefully argued by philosopher Neil Levy in his 2011 book, *Hard Luck: How Luck Undermines Free Will and Moral Responsibility* (Oxford University Press). He focuses on two categories of luck. One, present luck, examines its role in the difference between driving while so drunk that, when coupled with events in the seconds to minutes before, you would have killed someone if they had happened to be crossing the street, and the bad luck of being in that state and actually killing someone. As we saw, whether this distinction is meaningful is often the domain of legal scholars. More meaningful to Levy is what he calls constitutive luck, the fortune, good or bad, that sculpted you up to this moment. In other words, our world of one second before, one minute before . . . (although he only passingly frames the idea biologically). And when you recognize that that is all there is to explain who we are, he concludes, "it is not ontology

*Mind you, the compatibilist Tse isn't pleased with this, writing somewhere between how this regress can't exist and how it shouldn't exist—a contrast that anchors parts of chapter 15.

that rules out free will, it is *luck* (his emphasis)."* In his view, not only does it make no sense to hold us responsible for our actions; we also had no control over the formation of our *beliefs* about the rightness and consequences of that action or about the availability of alternatives. You can't successfully believe something different from what you believe.†

In the first chapter, I wrote about what is needed to prove free will, and this chapter has added details to that demand: show me that the thing a neuron just did in someone's brain was unaffected by any of these preceding factors—by the goings-on in the eighty billion neurons surrounding it, by any of the infinite number of combinations of hormone levels percolated that morning, by any of the countless types of childhoods and fetal environments were experienced, by any of the two to the four millionth power different genomes that neuron contains, multiplied by the nearly as large range of epigenetic orchestrations possible. Et cetera. All out of your control.

"Turtles all the way down" is a joke because the confident claim presented to William James is not just absurd but immune to every challenge he raises. It's a highbrow version of the insult battles that would go on in schoolyards in my youth: "You're a sucky baseball player." "I know you are, but what am I?" "Now you're being annoying." "I know you are, but what am I?" "Now you're indulging in lazy sophistry." "I know you are . . ." If the old woman going at James were, at some point, to report that the next turtle down floats in the air, the anecdote wouldn't be funny; while the answer is still absurd, the rhythm of the infinite regress has been broken.

Why did that moment just occur? "Because of what came before it." Then why did *that* moment just occur? "Because of what came before

*As a small clarification, Levy doesn't necessarily believe that we have *no* control over our actions, just that we have no relevant control.

†Levy has an interesting analysis that focuses on a file-this-away-for-future-use word, *akrasia*, which is when an agent acts against her expressed judgment. When certain akrasias become common enough, we have seemingly insoluble inconsistencies . . . until we generate a view of ourselves that consistently accommodates the akrasia. "I'm normally a very self-disciplined person . . . except when it comes to chocolate."

that," forever,* isn't absurd and is, instead, how the universe works. The absurdity amid this seamlessness is to think that we have free will and that it exists because at some point, the state of the world (or of the frontal cortex or neuron or molecule of serotonin . . .) that "came before that" happened out of thin air.

In order to prove there's free will, you have to show that some behavior just happened out of thin air in the sense of considering all these biological precursors. It may be possible to sidestep that with some subtle philosophical arguments, but you can't with anything known to science.

As noted in the first chapter, the prominent compatibilist philosopher Alfred Mele judged this requirement of free will as setting the bar "absurdly high." Some subtle semantics come into play; what Levy calls "constitutive" luck is luck that is "remote" to Mele, "remote" as in so detached in time—a whole million years before you decide, a whole minute before you decide—that it doesn't preclude free will and responsibility. This is supposedly because the remoteness is so remote as to not be remotely relevant, or because the consequences of that remote biological and environmental luck are still filtered through some sort of immaterial "you" at the end picking and choosing among the influences, or because remote bad luck, á la Dennett, will be balanced out by good luck in the long run and can thus be ignored. This is how some compatibilists arrive at the conclusion that someone's history is irrelevant. Levy's wording of "constitutive" luck suggests something very different, namely that not only is history relevant but, in his words, "the problem of history *is* a problem of

*"Forever" may not really be the case because, at some point in this regression, you get to the big bang and whatever came before that, about which I understand precisely zero. Regardless of whether things go back infinitely, as a key point, the further back you go, the smaller the influence is likely to be—how you respond to this stranger who may have just insulted you is more influenced by your circulating stress hormone levels at the time than by the infectious-disease load experienced by your distant ancestors. When trying to explain our behavior, I'm perfectly happy to call a time-out on "what came before that" when it's going far enough back to explain, say, why we're a carbon-based rather than silicon-based life form. But we have ample evidence for the relevance of what-came-befores that people used to feel justified in ignoring—the trauma that occurred a few months before a person behaved as they did, the ideal level of stimulation experienced in their childhood, the alcohol levels their fetal brain was pickled in . . .

luck." It is why it is anything but an absurdly high bar or straw man to say that free will can exist only if neurons' actions are completely uninfluenced by all the uncontrollable factors that came before. It's the only requirement there can be, because all that came before, with its varying flavors of uncontrollable luck, is what came to *constitute* you. This is how you became you.[59]

4

Willing Willpower: The Myth of Grit

The last two chapters were devoted to how you can believe in free will by ignoring history. And you can't—to repeat our emerging mantra, all we are is the history of our biology, over which we had no control, and of its interaction with environments, over which we also had no control, creating who we are in the moment.

However, not all free-will fans deny the importance of history, and this chapter dissects two ways in which it is invoked. The first, which we'll blow over relatively quickly, is a silly effort by some serious scholars to incorporate history into the picture, as part of a larger strategy of saying, "Yes, of course free will exists. Just not where you're looking." It happened in the past. It'll happen in your future. It happens wherever you're not looking in the brain. It happens *outside* you, floating on interactions between people.

We'll look at the second misuse of history more deeply. Those last two chapters were about the damage caused if you decide that punishment and reward are morally justifiable because history doesn't matter when explaining someone's behavior. This chapter is about how it's just as destructive to conclude that history is relevant only to *some* aspects of behavior.

WAS-NESS

Suppose you have some guy in a tough situation—being threatened by a stranger who's coming at him with a knife. Our guy pulls out a gun and shoots once, leaving the assailant on the ground. What does our guy then do? Does he conclude, "It's over, he's incapacitated, I'm safe?" Or does he keep shooting? What if he waits eleven seconds before attacking the assailant further? In the final scenario he is charged with premeditated murder—if he had stopped after the first shot, it would have counted as self-defense; but he had eleven seconds to think about his options, meaning that his second round of shots was freely chosen and premeditated.

Let's consider the guy's history. He was born with fetal alcohol syndrome, due to his mother's drinking. She abandoned him when he was five, resulting in a string of foster homes featuring physical and sexual abuse. A drinking problem by thirteen, homeless at fifteen, multiple head injuries from fights, surviving by panhandling and being a sex worker, robbed numerous times, stabbed a month earlier by a stranger. An outreach psychiatric social worker saw him once and noted that he might well have PTSD. Ya think?

Someone has tried to kill *you* and you have eleven seconds to make a life-or-death decision; there's a well-understood neurobiology as to why you readily make a terrible decision during this monumental stressor. Now, instead, it's our guy with a neurodevelopmental disorder due to fetal neurotoxicity, repeated childhood trauma, substance abuse, repeated brain injuries, and a recent stabbing in a similar situation. His history has resulted in this part of his brain being enlarged, this other part atrophied, this pathway disconnected. And as a result, there's, like, zero chance that he'll make a prudent, self-regulated decision in those eleven seconds. And you'd have done the same thing if life had handed you that brain. In this context, "eleven seconds to premeditate" is a joke.*

*I've testified saying something like this paragraph to about a dozen juries as a teaching witness, in case after case where someone with that sort of life story had a few seconds to make a similar decision and went back to the prone assailant and stabbed him an additional sixty-two times. So

Despite that, the compatibilist philosophers (and most prosecutors . . . and judges . . . and juries) don't think it's a joke. Sure, life has thrown awful things at the guy, but he's had plenty of time in the past to have *chosen* to not be the sort of person who would go back and put another bullet in the assailant's brain.

A great summary of this viewpoint is given by philosopher Neil Levy (one that he does *not* agree with):

> Agents are not responsible as soon as they acquire a set of active dispositions and values; instead, they become responsible by taking responsibility for their dispositions and values. Manipulated agents are not immediately responsible for their actions, because it is only after they have had sufficient time to reflect upon and experience the effects of their new dispositions that they qualify as fully responsible agents. The passing of time (under normal conditions) offers opportunities for deliberation and reflection, thereby enabling agents to become responsible for who they are. Agents become responsible for their dispositions and values in the course of normal life, even when these dispositions and values are the product of awful constitutive luck. At some point bad constitutive luck ceases to excuse, because agents have had time to take responsibility for it.[1]

Sure, maybe no free will just now, but there *was* relevant free will in the past.

As implied in Levy's quote, the process of freely choosing what sort of person you become, despite whatever bad constitutive luck you've had, is usually framed as a gradual, usually maturational process. In a debate with Dennett, incompatibilist Gregg Caruso outlined chapter 3's essence—we have no control over either the biology or the environment thrown at us. Dennett's response was "So what? The point I think you are missing is that autonomy is something one *grows into,* and this is indeed a process that is *initially* entirely beyond one's control, but as one matures, and

far, with one exception that I now view as a fluke, the juries have decided that it's premeditated murder and convicted on all charges.

learns, one begins to be able to control more and more of one's activities, choices, thoughts, attitudes, etc." This is a logical outcome of Dennett's claim that bad and good luck average out over time: Come on, get your act together. You've had enough time to take responsibility, to choose to catch up to everyone else in the marathon.[2]

A similar view comes from the distinguished philosopher Robert Kane, of the University of Texas: "Free will in my view involves more than merely *free of action*. It concerns *self-formation*. The relevant question for free will is this: *How did you get to be the kind of person you now are?*" Roskies and Shadlen write, "It is plausible to think that agents might be held morally responsible even for decisions that are not conscious, if those decisions are due to policy settings which are expressions of the agent [in other words, acts of free will in the past]."[3]

Not all versions of this idea require gradual acquisition of past-tense free will. Kane believes that "choose what sort of person you're going to be" happens at moments of crisis, at major forks in the road, at moments of what he calls "Self-Forming Actions" (and he proposes a mechanism by which this supposedly occurs, which we'll touch on briefly in chapter 10). In contrast, psychiatrist Sean Spence, of the University of Sheffield, believes that those I-*had*-free-will-back-then moments happen when life is at its optimal, rather than in crisis.[4]

Whether that free will was-ness was a slow maturational process or occurred in a flash of crisis or propitiousness, the problem should be obvious. *Was* was once *now*. If the function of a neuron right now is embedded in its neuronal neighborhood, effects of hormones, brain development, genes, and so on, you can't go away for a week and then show that the function a week prior wasn't embedded after all.

A variant on this idea is that you may not have free will now about *now*, you have free will now about who *you are going to be* in the future. Philosopher Peter Tse, who calls this second-order free will, writes how the brain can "cultivate and create new types of options for itself in the future." Not just any brains, however. Tigers, he notes, can't have this sort of free will (e.g., choosing that they're going to become vegans). "Humans, in contrast, bear a degree of responsibility for having chosen to become the kind of chooser who they now are." Combine this with Dennett's retrospective view and we have something akin to the idea that somewhere in the future, you will have had free will in the past—I will freely choosed.[5]

Rather than there being free will, "just not when you're looking," there's free will, "just not *where* you're looking"—you may have shown that free will isn't coming from the area of the brain you're studying; it's coming from the area you *aren't*. Roskies writes, "It is possible that an indeterministic event elsewhere in the larger system affects the firing of [neurons in brain region X], thus making the system as a whole indeterministic, even though the relation between [neuronal activity in brain region X] and behavior is deterministic." And neuroscientist Michael Gazzaniga moves the free will outside the brain entirely: "Responsibility exists at a different level of organization: the social level, not in our determined brains." There are two big problems with this: First, it isn't free will and responsibility just because, on the social level, everyone says it is—that's a central point of this book. Second, sociality, social interactions, organisms being social with each other, are as much an end product of biology interacting with environment as is the shape of your nose.[6]

Throw down the gauntlet from chapter 3—present me with the neuron, right *here*, right *now*, that caused that behavior, independent of any other current or historical biological influence. The answer can't be "Well, we can't, but that happened before." Or "That's going to occur, but not yet." Or "That's occurring right now but not here—instead, over there; no, not *that* there, that *other* there. . . ." It's turtles in every place and time; there are no cracks in the process by which *was* generates *is* in which to squeeze free will.

We move now to probably the most important topic in this half of the book, a way to erroneously see free will that isn't there.

WHAT YOU WERE GIVEN AND WHAT YOU DO WITH IT

Kato and Finn (names changed to protect their identities) have a good thing going, backing each other in a fight and serving as each other's wingman in the sex department. Each has a fairly dominant personality, and working together, they're unstoppable.

I'm watching them racing across a field. Kato got the head start, but Finn is catching up. They're trying to run down a gazelle, which is tearing away from them. Kato and Finn are baboons, intent on a meal. If they do catch the gazelle, which seems increasingly likely, Kato will eat first, as he is number two in the hierarchy, Finn, number three.

Finn is still catching up. I note a subtle shift in his running, something I can't describe, but having observed Finn for a long time, I know what's coming next. "Idiot, you're going to blow it," I think. Finn has seemingly decided, "Screw it with this waiting for the leftovers. I want first dibs on the best parts." He accelerates. "What fools these baboons be," I think. Finn leaps on Kato's back, biting him, knocking him over so that Finn can get the gazelle himself. Naturally, he trips over Kato in the process and sprawls ass over teakettle. They get up, glowering at each other, the gazelle long gone; end of their cooperative coalition. With Kato no longer willing to back him up in a fight, Finn is soon toppled by Bodhi, number four in the hierarchy, followed by being trounced by number five, Chad.

Some baboons are just that way. They're full of potential—big, muscular, with sharp canines—but go nowhere in the hierarchy because they never miss an opportunity to miss an opportunity. They break up their coalition with an impulsive act, like Finn did. They can't keep themselves from challenging the alpha male for a female, and get pummeled. They're in a bad mood and can't stop themselves from displacing aggression by biting the wrong nearby female, then get chased out of the troop by her irate high-ranking relatives. Major underachievers that can resist anything except temptation.

We are replete with human examples, always featuring the word *squan-*

der. Athletes who squander their natural talents by partying. Smart kids squandering their academic potential with drugs* or indolence. Dissipated jet-setters who squander their families' fortunes on crackpot vanity projects—according to one study, 70 percent of family fortunes are lost by the second generation of inheritors. From Finn on, squanderers all.[7]

And then there are the people who overcame bad luck with spectacular tenacity and grit. Oprah, growing up wearing potato sack dresses. Harland Sanders, eventually the Colonel, who failed to sell his fried chicken recipe to 1,009 restaurants before striking gold. Marathoner Eliud Kibet, who collapsed a few meters from the finish line and crawled to the end; fellow Kenyan Hyvon Ngetich, who crawled the final fifty meters of her marathon; Japanese runner Rei Iida, who fell, fracturing her leg, and crawled the final *two hundred meters* to the finish line. Nobel laureate geneticist Mario Capecchi, who was a homeless street kid in World War II Italy. Then, of course, there's Helen Keller and Anne Sullivan with the w-a-t-e-r. Desmond Doss, an unarmed conscientious objector medic, who returned under enemy fire to carry seventy-five injured servicemen to safety in the Battle of Okinawa. Five-foot-three Muggsy Bogues playing in the NBA. Madeleine Albright, future secretary of state, who, as a teenage Czechoslovakian refugee, sold bras in a Denver department store. The Argentinian guy working as a janitor and bouncer who put his nose to the grindstone and became the pope.

Whether considering Finn and the squanderers or Albright selling bras, we are moths pulled to the flame of the most entrenched free-will myth. We've already examined versions of partial free will—not now but in the past; not here but where you're not looking. This is another version of partial free will—yes, there are our attributes, gifts, shortcomings, and deficiencies over which we had no control, but it is us, we agentic, free, captain-of-our-own-fate selves who choose *what we do* with those attributes. Yes, you had no control over that ideal ratio of slow- to fast-twitch fibers in your leg muscles that made you a natural marathoner, but it's you

*To my surprise, some studies have shown that high-IQ kids are more prone than average toward illegal drug use and alcohol abuse in adulthood.

who fought through the pain at the finish line. Yes, you didn't choose the versions of glutamate receptor genes you inherited that gave you a great memory, but you're responsible for being lazy and arrogant. Yes, you may have inherited genes that predispose you to alcoholism, but it's you who commendably resists the temptation to drink.

A stunningly clear statement of this compatibilist dualism concerns Jerry Sandusky, the Penn State football coach who was sentenced to sixty years in prison in 2012 for being a horrific serial child molester. Soon after this, a provocative CNN piece ran under the title "Do Pedophiles Deserve Sympathy?" Psychologist James Cantor of the University of Toronto reviewed the neurobiology of pedophilia. The wrong mix of genes, endocrine abnormalities in fetal life, and childhood head injury all increase the likelihood. Does this raise the possibility that a neurobiological die is cast, that some people are destined to be this way? Precisely. Cantor concludes correctly, "One cannot choose to not be a pedophile."

But then he does an Olympian leap across the Grand Canyon–size false dichotomy of compatibilism. Does any of that biology lessen the condemnation and punishment that Sandusky deserved? No. "One cannot choose to not be a pedophile, *but one can choose to not be a child molester*" (my emphasis).[8]

The following table formalizes this dichotomy. On the left are things that most people accept as outside our control—biological stuff. Sure, sometimes we have trouble remembering that. We praise, single out, the chorus member who is an anchor of reliability because of their perfect pitch (which is a biologically heritable trait).* We praise a basketball player's dunk, ignoring that being seven-foot-two has something to do with it. We smile more at someone attractive, are more likely to vote for them in an election, less likely to convict them of a crime. Yeah, yeah, we agree sheepishly when this is pointed out, they obviously didn't choose the

*Perfect pitch is actually a classic example of genes being about potential, not certainty. Research suggests that you probably need to have inherited the potential for perfect pitch; however, it is not expressed in someone unless they were exposed to a fair amount of music early in life.

shape of their cheekbones. We're usually pretty good at remembering that the biological stuff on the left is out of our control.[9]

"Biological stuff"	*Do you have grit?*
Having destructive sexual urges	*Do you resist acting upon them?*
Being a natural marathoner	*Do you fight through the pain?*
Not being all that bright	*Do you triumph by studying extra hard?*
Having a proclivity toward alcoholism	*Do you order ginger ale instead?*
Having a beautiful face	*Do you resist concluding that you're entitled to people being nice to you because of it?*

And then on the right is the free will you supposedly exercise in choosing what you do with your biological attributes, the *you* who sits in a bunker *in* your brain but not *of* your brain. Your you-ness is made of nanochips, old vacuum tubes, ancient parchments with transcripts of Sunday-morning sermons, stalactites of your mother's admonishing voice, streaks of brimstone, rivets made out of gumption. Whatever that real *you* is composed of, it sure ain't squishy biological brain yuck.

When viewed as evidence of free will, the right side of the chart is a compatibilist playground of blame and praise. It seems so hard, so counterintuitive, to think that willpower is made of neurons, neurotransmitters, receptors, and so on. There seems a much easier answer—willpower is what happens when that nonbiological essence of you is bespangled with fairy dust.

And as one of the most important points of this book, we have as little control over the right side of the chart as over the left. Both sides are equally the outcome of uncontrollable biology interacting with uncontrollable environment.

To understand the biology of the right side of the chart, time to focus on the fanciest part of the brain, the frontal cortex, which was lightly touched on in the last two chapters.

DOING THE RIGHT THING WHEN IT'S THE HARDER THING TO DO

Bragging for the frontal cortex, it's the newest part of the brain; we primates have, proportionately, more of it than other mammals; when you examine gene variants that are unique to primates, a disproportionate percentage of them are expressed in the frontal cortex. Our human frontal cortex is proportionately bigger and/or more complexly wired than that of any other primate. As noted in the last chapter, it's the last part of the brain to fully mature, not being fully constructed until your midtwenties; this is outrageously delayed, given that most of the brain is up and running within a few years of birth. And as a major implication of this delay, a quarter century of environmental influences shape how the frontal cortex is being put together. It's one of the hardest-working parts of the brain, in terms of energy consumption. It has a type of neuron found nowhere else in the brain. And the most interesting part of the frontal cortex—the prefrontal cortex (PFC)—is proportionately even larger than the rest of the frontal cortex, and more recently evolved.[*,10]

As a reminder, the PFC is central to executive function, decision-making. We saw this in chapter 2, where, way up in the chain of Libetian commands, there was the PFC making decisions up to ten seconds before subjects first became aware of that intent. What the PFC is most about is

*Neuroanatomists will turn over in their graves, but from here on out, I'm going to refer to the entire frontal cortex as the PFC, for simplicity's sake.

making *tough* decisions in the face of temptation—gratification postponement, long-term planning, impulse control, emotional regulation. The PFC is essential for getting you to do the right thing when it is the harder thing to do. Which is *so* pertinent to that false dichotomy between what attributes fate hands you and what you do with them.

THE COGNITIVE PFC

As a warm-up, let's examine "doing the right thing" in the cognitive realm. It's the PFC that inhibits you from doing something the habitual way when you're supposed to be doing it in a novel manner. Sit someone in front of a computer and say to them, "Here's the rule—when a blue light flashes on the screen, hit the button on the left as fast as possible; red light, hit the button on the right." Have them do that a bunch of times, get the hang of it. "*Now* reverse that—blue light, button on the right; red, left." Have them do that awhile. "Now switch back again." Each time the rule changes, the PFC is in charge of "Remember, blue now means . . ."

Now, quick, say the months of the year backward. The PFC activates, suppressing the overlearned response—"Remember, September-August this time, not September-October." More frontal activation predicts a better performance here.

One of the best ways to appreciate these frontal functions is to examine people with a damaged PFC (as after certain types of strokes or dementias). There are huge problems with "reversal" tasks like these. It's too hard to do that right thing when it is a change from the usual.

Thus, the PFC is for learning a new rule, or a new variant of a rule. Implied in that is that the functioning of the PFC can change. Once that novel rule persists and has stopped being novel, it becomes the task of other, more automatic brain circuitry. Few of us need to activate the PFC to pee nowhere but in the bathroom; but we sure did when we were three.

"Doing the right thing" requires two different skills from the PFC. There's sending the decisive "do this" signal along the path from the PFC to the frontal cortex to the supplementary motor area (the SMA of chapter 2) to the motor cortex. But even more important, there is the "and don't do that, even if that's the usual" signal. Even more than sending excitatory signals to the motor cortex, the PFC is about inhibiting habitual brain circuits. To hark back again to chapter 2, the PFC is central to showing that we lack both free will and the conscious veto power of free won't.[11]

THE SOCIAL PFC

Obviously, the crowning achievement of millions of years of frontocortical evolution is not reciting months backward. It's social—it's suppressing the emotionally easier thing to do. The PFC is the center of our social brain. The bigger the average size of the social group in a primate species, the greater a percentage of the brain is devoted to the PFC; the bigger the size of some human's texting network, the larger a particular subregion of the PFC and its connectivity with the limbic system. So does sociality enlarge the PFC, or does a large PFC drive sociality? At least partially the former—take individually housed monkeys and put them together in big, complex social groups, and a year later, everyone's PFC will have enlarged; moreover, the individual who emerges at the top of the hierarchy shows the largest increase.[*,12]

Neuroimaging studies show the PFC reining in more emotional brain regions in the name of doing (or thinking) the right thing. Stick a volunteer in a brain scanner and flash up pictures of faces. And in a depressing, well-replicated finding, flash up the face of someone of another race and in about 75 percent of subjects, there is activation of the amygdala, the

*Which tells you something very important about primate dominance. For example, for a male baboon, attaining high rank is all about muscles, sharp canines, and winning the right fight. But *maintaining* high rank is about avoiding fights, having the self-control to ignore provocations, avoiding fighting by being psychologically intimidating, being a sufficiently self-disciplined, stable coalition partner (unlike Finn) to always have someone watching your back. An alpha male who is constantly fighting won't be in the corner office long; successful alphaship is a minimalist art of nonwar.

brain region central to fear, anxiety, and aggression.* In under a *tenth* of a second.† And then the PFC does the harder thing. In most of those subjects, a few seconds after the amygdala activates, the PFC kicks in, turning off the amygdala. It's a delayed frontocortical voice—"Don't think that way. That's not who I am." And who are the folks in which the PFC doesn't muzzle the amygdala? People whose racism is avowedly, unapologetically explicit—"That *is* who I am."[13]

In another experimental paradigm, a subject in a brain scanner plays an online game with two other people—each is represented by a symbol on the screen, forming a triangle. They toss a virtual ball around—the subject presses one of two buttons, determining which of the two symbols the ball is tossed to; the other two toss it to each other, toss it back to the subject. This goes on for a while, everyone having a fine time, and then, oh no, the other two people stop tossing the ball to the subject. It's the middle-school nightmare: "They know I'm a dork." The amygdala rapidly activates, along with the insular cortex, a region associated with disgust and distress. And then, after a delay, the PFC inhibits these other regions—"Get this in perspective; this is just a stupid game." In a subset of individuals, however, the PFC doesn't activate as much, and the amygdala and insular cortex just keep going, as the subject feels more subjective distress. Who are these impaired individuals? Teenagers—the PFC isn't up to the task yet of dismissing social ostracism as meaningless. There you have it.‡,[14]

*There's a world of complexity to this. It depends on whom the picture is of—strapping young guy, and the amygdala roars into activity; frail, grandmotherly type, not so much. More for a stranger than for an other-race beloved celebrity—that person counts as an honorary Us. What about the 25 percent of people who don't have the amygdala response? They were typically raised in multiracial communities, have had intimate relationships with people of that other race, or have been psychologically primed before the experiment to process each face as an individual. In other words, the implicit racism coded in the amygdala is not remotely inevitable.

†These studies have produced another distressing finding. When we look at faces, there is activation of a very primate part of the cortex called the fusiform face area. And in most subjects, the face of an other-race Them activates the fusiform less than usual. Their face doesn't count as being much of a face.

‡Studies like this include a key control, showing that it is social anxiety that is being generated: the other two stop tossing the ball to the subject, who is told that it's because of some problem

More of the PFC reining in the amygdala. Give a volunteer a mild shock now and then; the amygdala majorly wakes up each time. Now condition the volunteer: just before each shock, show them a picture of some object with completely neutral associations—say, a pot, a pan, a broom, or a hat. Soon the mere sight of that previously innocuous object activates the amygdala.* The next day, show the subject a picture of that object that activates a conditioned fear response in them. Amygdala activation. Except today, there's no shock. Do it again, and again. Each time, no shock. And slowly you "extinguish" the fear response; the amygdala stops reacting. Unless the PFC isn't working. Yesterday it was the amygdala that learned "brooms are scary." Today it is the PFC that learns, "but not today," and calms down the amygdala.[†,15]

More insight into the PFC comes from brilliant studies by neuroscientist Josh Greene of Harvard. Subjects in a brain scanner play repeated rounds of a chance guessing game with a 50 percent success rate. Then comes the fiendishly clever manipulation. Tell subjects there's been a computer glitch so that they can't enter their guess; that's okay, they're told, we'll show you the answer and you can just tell us whether you were right. In other words, an opportunity to *cheat.* Throw in enough of those there-goes-that-computer-glitch-again opportunities, and you can tell if someone starts cheating—their success rate averages above 50 percent. What happens in the brains of cheaters when temptation arises? Massive activation of the PFC, the neural equivalent of the person wrestling with whether to cheat.[16]

And then for the profound additional finding. What about the people

---

with the computer. If it's that, rather than social ostracism, there's no equivalent brain response.

*Depressing finding: Instead of conditioning subjects to a neutral, innocuous object, condition them to a picture of an out-group Them. People learn to associate that with a shock faster than if it were an in-group member.

†Is the PFC causing the amygdala to forget that bells are scary? No—the insight is still there but is just being suppressed by frontal cortex. How can you tell this? On day three of the study, go back to the sight of that arbitrary object being followed by a shock. The person relearns the association faster than they learned it in the first place—the amygdala remembers.

who never cheated—how do they do it? Maybe their astonishingly strong PFC pins Satan to the mat each time. Major willpower. But that's not what happens. In those folks, the PFC doesn't stir. At some point after "don't pee in your pants" no longer required the PFC to flex its muscles, an equivalent happened in such individuals, generating an automatic "I don't cheat." As framed by Greene, rather than withstanding the siren call of sin thanks to "will," this instead represents a state of "grace." Doing the right thing isn't the harder thing.

The frontal cortex reins in inappropriate behavior in additional ways. One example involves a brain region called the striatum that has to do with automatic, habitual behaviors, exactly the sort of things that the amygdala can take advantage of by activating. The PFC sends inhibitory projections to the striatum as a backup plan—"I warned the amygdala not to do it, but if that hothead does it anyway, don't listen to it."[17]

What happens to social behavior if the PFC is damaged? A syndrome of "frontal disinhibition." We all have thoughts—hateful, lustful, boastful, petulant—we'd be mortified if anyone knew. Be frontally disinhibited and you say and do exactly those things. When one of those diseases* occurs in an eighty-year-old, it's off to a neurologist. When it's a fifty-year-old, it's usually a psychiatrist. Or the police. As it turns out, a substantial percentage of people incarcerated for violent crime have a history of concussive head trauma to the PFC.[18]

---

*Here are some factoids that emphasize the extent to which social demands sculpt the evolution of the PFC. The PFC contains a neuron type not found elsewhere in the brain. To add to its coolness, for a while people thought that these "von Economo neurons," introduced in the footnote on page 61, occurred only in humans. But as something even cooler, the neurons also occur in the most socially complex species out there—other apes, elephants, cetaceans. A neurological disease called behavioral frontotemporal dementia demonstrates that PFC damage causes inappropriate social behavior. What are the first neurons that die in that disease? The von Economo neurons. So whatever they do (which isn't at all clear), it has "doing the harder thing" written all over it. (Brief screed of interest to only a few readers—despite quasi–New Age neuroscientific claims, von Economo neurons are not mirror neurons responsible for empathy. These aren't mirror neurons. And mirror neurons don't do empathy. Don't get me started.)

COGNITION VERSUS EMOTION, COGNITION AND EMOTION, OR COGNITION VIA EMOTION?

Thus, the frontal cortex isn't just this cerebral, eggheady brain region weighing the pluses and minuses of each decision, sending nice rational Libetian commands to the motor cortex—i.e., an excitatory role. It's also an inhibitory, rule-bound goody-goody telling more emotional parts of the brain not to do something because they're going to regret it. And basically, those other brain regions think of the PFC as this moralizing pain with a stick up its butt, especially when it turns out to be right. This generates a dichotomy (spoiler alert: it's false), that there is a major fault line between thought and emotion, between the cortex, captained by the PFC, and the part of the brain that processes emotions (broadly called the limbic system, containing the amygdala along with other structures* related to sexual arousal, maternal behavior, sadness, pleasure, aggression . . .).

A picture of a war of wills between the PFC and the limbic system certainly makes sense by now. After all, it's the former telling the latter to stop those implicit racist thoughts, to put a stupid game in perspective, to resist cheating. And it's the latter that runs wild with crazy stuff when the PFC is silent—e.g., during REM sleep, when you're dreaming. But it's not always the two regions wrestling.† Sometimes they simply have different purviews. The PFC handles April 15; the limbic system, February 14. The former makes you grudgingly respect *Into the Woods*; the latter makes you tearful during *Les Mis*, despite knowing that you're being manipulated. The former is engaged when juries decide guilt or innocence; the latter, when they decide how much to punish the guilty.[19]

But—and this is a truly key point—rather than the PFC and limbic system either being in opposition or ignoring each other, they are usually

*Such as the hippocampus, septum, habenula, hypothalamus, mammillary bodies, and nucleus accumbens.

†And of considerable importance, we'll be getting to circumstances where the limbic system convinces the PFC to rubber-stamp strongly emotional decisions.

intertwined. In order to do the correct, harder thing, the PFC requires a huge amount of limbic, emotional input.

To appreciate this, we must sink deeper into minutiae, considering two subregions of the PFC.

The first is the dorsolateral PFC (dlPFC), the definitive rational decider in the frontal cortex. Like a Russian nesting doll, the cortex is the newest part of the brain to evolve, the frontal cortex is the newest part of the cortex, the PFC is the newest part of the frontal cortex, and the dlPFC is the newest part of the PFC. The dlPFC is the last part of the PFC to fully mature.

The dlPFC is the essence of the PFC as tight-assed superego. It's the most active part of the PFC during "count the months backward" tasks, or when considering temptation. It is fiercely utilitarian—more dlPFC activity during a moral-judgment task predicts that the subject chooses to kill an innocent person to save five.[20]

What happens when the dlPFC is silenced is really informative. This can be done experimentally with an immensely cool technique called transcranial magnetic stimulation (TMS—introduced on page 26 in the footnote), in which a strong magnetic pulse to the scalp can temporarily activate or inactivate the small patch of cortex just below. Activate the dlPFC this way, and subjects become more utilitarian in deciding to sacrifice one to save many. Inactivate the dlPFC, and subjects become more impulsive—they rate a lousy offer in an economic game as unfair but lack the self-control needed to hold out for a better reward. This is all about sociality—manipulating the dlPFC has no effect if subjects think their opponent is a computer.[*,21]

Then there are people who have sustained selective damage to their dlPFC. The outcome is just what you'd expect—impaired planning or gratification postponement, perseveration on strategies that offer imme-

*Heads up, running-dog capitalists: one study has used TMS to manipulate the projection from the dlPFC to dopaminergic reward pathways in the striatum, thereby transiently changing people's music tastes—enhancing the subjective appreciation of a piece of music and the physiological response to it . . . as well as boosting the monetary value subjects assign to the music.

diate reward, plus poor executive control over socially inappropriate behavior. A brain with no voice saying, "I wouldn't do that if I were you."

The other key subregion of the PFC is called the ventromedial PFC (vmPFC), and to savagely simplify, it's the opposite of the dlPFC. That cerebral dlPFC is mostly getting inputs from other cortical regions, canvassing the outer districts to find out their well-considered thoughts. But the vmPFC carries in information from the limbic system, that brain region that's swoony or overwrought with emotion—the vmPFC is how the PFC finds out what you're feeling.*

What happens if the vmPFC is damaged? Great things, if you're not big on emotion. For that crowd, we are at our best when we are rational, optimizing machines, thinking our way to our best moral decisions. In this view, the limbic system gums up decision-making by being all sentimental, sings too loud, dresses flamboyantly, has unsettling amounts of armpit hair. In this view, if we just could get rid of the vmPFC, we'd be calmer, more rational, and function better.

As a deeply significant finding, someone with vmPFC damage makes terrible decisions, but of a very different type from those with dlPFC damage. For starters, people with vmPFC damage have trouble making decisions, because they're not getting gut feelings about how they should decide. When we are making a decision, the dlPFC is musing philosophically, running thought experiments about what decision to make. What the vmPFC is reporting to the dlPFC are the results of a *feel* experiment. "How will I feel if I do X and Z then happens?" And without that gut-feeling input, it's immensely hard to make decisions.[22]

Moreover, the decisions made can be wrong by anyone's standards. People with vmPFC damage don't shift their behavior based on negative feedback. Suppose subjects are repeatedly choosing between two tasks, one of which is more rewarding. Switch which task is the more rewarding

---

*Starting in the 1960s, the esteemed neuroanatomist Walle Nauta of MIT nearly ruined his career by stating that the vmPFC should be viewed as part of the limbic system. Horror—the cortex is about solving Fermat's theorem, not getting all weepy when Mimi is dying in Roger's arms. And it took years for everyone else to see that the vmPFC is the limbic system's portal to the PFC.

one, and people typically shift their strategy accordingly (even if they're not consciously aware of the change in reward rates). But with vmPFC damage, the person can even say that it's the other task that is now more rewarding . . . while sticking with the previous task. Without a vmPFC, you still know what negative feedback means, but not how it feels.[23]

As we saw, dlPFC damage produces inappropriate, emotionally disinhibited behaviors. But without a vmPFC, you desiccate into heartless detachment. This is the person who, meeting someone, says, "Hello, good to meet you. I see that you're quite overweight." And when castigated later by their mortified partner will ask with calm puzzlement, "What's wrong? It's true." Unlike most people, those with vmPFC damage don't advocate harsher punishment for violent versus nonviolent crimes, don't alter game play if they think they're playing against a computer rather than a human, and don't distinguish between a loved one and a stranger when deciding whether to sacrifice them in order to save five people. The vmPFC is not the vestigial appendix of the PFC, where emotion is like appendicitis, inflaming a sensible brain. Instead, it's essential.

So the PFC does the harder thing when it's the right thing to do. But as a crucial point, *right* is used in a neurobiological and instrumental sense rather than a moral one.

Consider lying, and the obvious role the PFC plays in resisting the temptation to lie. But you also use the PFC to lie competently; pathological liars, for example, have atypically complex wiring in the PFC. Moreover, lying competently is value-free, amoral. A child schooled in situational ethics lies about how she loves the dinner that Grandma made. A Buddhist monk plays liar's dice superbly. A dictator fabricates the occurrence of a massacre as an excuse to invade a country. A spawn of Ponzi defrauds investors. As with much about the frontal cortex, it's context, context, context.

With this tour of the PFC complete, we return to the hideously destructive false dichotomy between your attributes, those natural gifts and weaknesses that you just happen to have, and your supposedly freely chosen choices as to what you do with those attributes.

"Biological stuff"	*Do you have grit?*
Having destructive sexual urges	*Do you resist acting upon them?*
Being a natural marathoner	*Do you fight through the pain?*
Not being all that bright	*Do you triumph by studying extra hard?*
Having a proclivity toward alcoholism	*Do you order ginger ale instead?*
Having a beautiful face	*Do you resist concluding that you're entitled to people being nice to you because of it?*

THE SAME EXACT STUFF

Look once again at the actions in the right column, those crossroads that test our mettle. Do you resist acting on your destructive sexual urges? Do you fight through the pain, work extra hard to overcome your weaknesses? You can see where this is heading. If you want to finish this paragraph and then skip the rest of the chapter, here are the three punch lines: (a) grit, character, backbone, tenacity, strong moral compass, willing spirit winning out over weak flesh, are all produced by the PFC; (b) the PFC is made of biological stuff identical to the rest of your brain; (c) your current PFC is the outcome of all that uncontrollable biology interacting with all that uncontrollable environment.

Chapter 3 explored the biological answer to the question, Why did that behavior just occur?, the answer being, because of what came a second before, and a minute before, and . . . Now we ask the more focused question of why that PFC functioned the way it did just now. And it's the same answer.

THE LEGACY OF THE PRECEDING SECONDS TO AN HOUR

You sit there, alert, on task. Each time the blue light comes on, you rapidly hit the button on the left; red light, button on the right. Then, the rule reverses—blue right, red left. Then it reverses again, and then again . . .

What's going on in your brain during this task? Each time a light flashes, your visual cortex briefly activates. An instant later, there's brief activation of the pathway carrying that information from the visual cortex to the PFC. An instant later, the pathways from there to your motor cortex and then from your motor cortex to your muscles activate your motor cortex to your muscles. What's happening IN the PFC? It's sitting there having to focus, repeating, "Blue left, red right" or "Blue right, red left." It's working hard *the entire time*, chanting which rule is in effect. When you're trying to do the right, harder thing, the PFC becomes the most expensive part of the brain.

Expensive. Nice metaphor. But it's not a metaphor. Any given neuron in the PFC is firing nonstop, each action potential triggering waves of ions flowing across membranes and then having to be corralled and pumped back to where they started. And those action potentials can occur *a hundred times a second* while you're concentrating on the rule that is now in place. Those PFC neurons consume mammoth amounts of energy.

You can demonstrate this with brain-imaging techniques, showing how a working PFC consumes tons of glucose and oxygen from the bloodstream, or by measuring how much biochemical cash is available in each neuron at any given time.* Which leads to the main point of this section—when the PFC doesn't have enough energy on board, *it doesn't work well.*

This is the cellular underpinning of concepts like "cognitive load" or

*Cash = ATP, aka adenosine triphosphate, just to tap into the recesses of your memory, dredging up a factoid from ninth-grade biology.

"cognitive reserve," alluded to in chapter 3.* As your PFC works hard on a task, those reserves are depleted.[24]

For example, place a bowl of M&M's in front of someone dieting. "Here, have all you want." They're trying to resist. And if the person has just done something frontally demanding, even some idiotically irrelevant red light / blue light task, the person snacks on more candy than usual. In the words of part of the charming title of a paper on the subject, "Deplete us not into temptation." Same thing in reverse—deplete frontal reserve by sitting for fifteen minutes resisting those M&M's, and afterward you'll be lousy at red light / blue light.[25]

PFC function and self-regulation go down the tubes if you're terrified or in pain—the PFC is using up energy dealing with the stress. Recall the Macbeth effect, where reflecting on something unethical you once did impairs frontal cognition (unless you've relieved yourself of that burdensome soiling by washing your hands). Frontal competence even declines if it's keeping you from being distracted by something positive—patients are more likely to die as a result of surgery if it is the surgeon's birthday.[26]

Fatigue also depletes frontal resources. As the workday progresses, doctors take the easier way out, ordering up fewer tests, being more likely to prescribe opiates (but not a nonproblematic drug like an anti-inflammatory, or physical therapy). Subjects are more likely to behave unethically and become less morally reflective as the day progresses, or after they've struggled with a cognitively challenging task. In an immensely unsettling study of emergency room doctors, the more cognitively demanding the workday (as measured by patient load), the higher the levels of implicit racial bias by the end of the day.[27]

It's the same with hunger. Here's one study that should stop you in your tracks (and was first referred to in the last chapter). The researchers studied a group of judges overseeing more than a thousand parole board decisions. What best predicted whether a judge granted someone parole versus

*Similar concepts that are invoked include "ego depletion" and "decision fatigue." See notes for how the core concepts of cognitive reserve and ego depletion have been heavily criticized in recent years.

more jail time? How long it had been since they had eaten a meal. Appear before the judge soon after she's had a meal, and there was a roughly 65 percent chance of parole; appear a few hours after a meal, and there was close to a 0 percent chance.[*,28]

What's that about? It's not like judges would get light-headed by late afternoon, slurring their words, getting all confused, and jailing the court stenographer. Nobel laureate psychologist Daniel Kahneman, in discussing this study, suggests that as the hours since a meal creep by, and the PFC becomes less adept at focusing on the details of each case, the judge becomes more likely to default into the easiest, most reflexive thing, which is sending the person back to jail. Important support for this idea comes from a study in which subjects had to make judgments of increasing complexity; as this progressed, the more sluggish the dlPFC became during deliberating, the more likely subjects were to fall back on a habitual decision.[29]

Why is denying parole the easy, habitual response to fall back on? Because it's less demanding of the PFC. Someone is facing you who has done bad things but has been behaving himself in jail. It takes a mighty energetic PFC to try to understand, to *feel*, what the prisoner's life—filled with horrible luck—has been like, to view the world from his perspective, to search his face and see those hints of change and potential beneath the toughness. It takes a lot of frontal effort for a judge to walk in a prisoner's shoes before deciding on his parole. And reflecting that, across all those judicial decisions, judges averaged a longer length of time before deciding to parole the person rather than before sending them back to jail.[†,‡,30]

Thus, events in the world around you will be modulating the ability of

*The finding was challenged by some critics who suggested that it was a statistical artifact of the way parole hearings were carried out; the authors reanalyzed their data to control for these possibilities, convincingly showing that the effect was still there. An additional study showed the identical pattern: subjects read job applicant profiles from out-group minority members; the longer it had been since a meal, the less time was spent on each application.

†"My god, this guy is such a bleeding-heart liberal." Nah. *Way* beyond that—you'll see.

‡In the same vein, credit loan officers become more likely to turn down loan applications as the day progresses. Similarly, savvy actors know not to pick the time slot just before lunch or at the end of the day for auditioning.

your PFC to resist those M&M's, or a quick, easy judicial decision. Another relevant factor is the brain chemistry of just how tempting the temptation is. This has a lot to do with the neurotransmitter dopamine being released into the PFC from neurons originating back in the nucleus accumbens in the limbic system. What is the dopamine doing in the PFC? Signaling the salience of a temptation, how much your neurons are imagining how great M&M's taste. The more of a dopamine dump in the PFC, the stronger the salience signal of the temptation, the more of a challenge it is for the PFC to resist. Boost dopamine levels in your PFC, and you'll suddenly have trouble keeping a lid on your impulses.* And exactly as you'd expect, there's a whole world of factors out of your control influencing the amount of dopamine that is going to be soaking your PFC (i.e., understanding the dopamine system also requires a one-second-before, one-century-before . . . analysis).[31]

In those seconds to hours before, sensory information modulates PFC function without your awareness. Have a subject smell a vial of sweat from someone frightened, and her amygdala activates, making it harder for the PFC to rein it in.† How's this for rapidly altering frontal function—take an average heterosexual male and expose him to a particular stimulus, and his PFC becomes more likely to decide that jaywalking is a good idea. What's the stimulus? The proximity of an attractive woman. I know, pathetic.[‡,32]

*How was this learned? The hard way. Parkinson's disease, a movement disorder where initiating voluntary movements becomes difficult, is caused by a dearth of dopamine in an unrelated part of the brain. Well, let's treat that by raising the person's dopamine levels (done using a drug called L-DOPA; long story). You're not going to drill a hole in the person's head and infuse L-DOPA directly into that part of the brain. Instead, the person swallows an L-DOPA pill, resulting in more dopamine in that diseased part of the brain . . . as well as in the rest of the brain, including the PFC. Result? A side effect of high-dose L-DOPA regimes can be behaviors like compulsive gambling.

†Uh, what's this experiment about? The scared sweat came from swabbing the armpits of people after their first skydive. What's the control group? Sweat from happy people who have just had an enjoyable jog in the park. Science is the best; I love this stuff.

‡By the way, heterosexual women don't start acting in equivalently stupid ways because of the proximity of some hunk. Another study showed that male skateboarders did riskier tricks, with more crashes, when in proximity of an attractive woman. (Just to show that all the science was rigorous, attractiveness was assessed by teams of independent raters. And in the words of the authors, "attractiveness ratings were corroborated by many informal comments and phone number requests from the skateboarders").

Thus, all sorts of things often out of your control—stress, pain, hunger, fatigue, whose sweat you're smelling, who's in your peripheral vision—can modulate how effectively your PFC does its job. Usually without your knowing it's happening. No judge, if asked why she just made her judicial decision, cites her blood glucose levels. Instead, we're going to hear a philosophical discourse about some bearded dead guy in a toga.

To ask a question derived from the last chapter, do findings like these prove that there's no such thing as freely chosen grit? Even if the sizes of these effects were enormous (which they rarely are, although 65 percent versus nearly 0 percent parole rates in the judge/hunger study sure isn't minor), not on their own. We now zoom out more.

THE LEGACY OF THE PRECEDING HOURS TO DAYS

This lands us in the realm of what hormones have been doing to the PFC when you need to show what would be interpreted as some agentic grit.

As a reminder from the last chapters, elevations of testosterone during this time frame make people more impulsive, more self-confident and risk-taking, more self-centered, less generous or empathic, and more likely to react aggressively to a provocation. Glucocorticoids and stress make people poorer at executive function and impulse control and more likely to perseverate on a habitual response to a challenge that isn't working, instead of changing strategies. Then there's oxytocin, which enhances trust, sociality, and social recognition. Estrogen enhances executive function, working memory, and impulse control and makes people better at rapidly switching tasks when needed.[33]

Lots of these hormonal effects play out in the PFC. Have a horribly stressed morning, and by noon, glucocorticoids will have changed gene expression in the dlPFC, making it less excitable and less able to couple to the amygdala and calm it down. Meanwhile, stress and glucocorticoids make that emotional vmPFC more excitable and more impervious to negative feedback about social behavior. Stress also causes release in the PFC of a neurotransmitter called norepinephrine (sort of the brain's equivalent of adrenaline), which also disrupts the dlPFC.[34]

In that time span, testosterone will have changed the expression of genes in neurons in another part of the PFC (called the orbitofrontal cortex), making them more sensitive to an inhibitory neurotransmitter, quieting the neurons, and decreasing their ability to talk sense to the limbic system. Testosterone also reduces the coupling between one part of the PFC and a region implicated in empathy; this helps explain why the hormone makes people less accurate at assessing someone's emotions by looking at their eyes. Meanwhile, oxytocin has its prosocial effects by strengthening the orbitofrontal cortex and by changing the rates at which the vmPFC utilizes the neurotransmitters serotonin and dopamine. Then there's estrogen, which not only increases the number of receptors for the neurotransmitter acetylcholine but even changes the structure of neurons in the vmPFC.[*,35]

Please tell me that you haven't been writing down and starting to memorize these factoids. The point is the mechanistic nature of all this. Depending on where you are in your ovulatory cycle, if it's the middle of the night or day, if someone gave you a wonderful hug that's left you still tingling, or someone gave you a threatening ultimatum that's left you still trembling—gears and widgets in your PFC will be working differently. And, as before, rarely with large enough effects to spell doom for the myth of grit all on their own. Just another piece.

THE LEGACY OF THE PRECEDING DAYS TO YEARS

Chapter 3 covered how over this time span, the structure and function of the brain can change dramatically. Recall how years of depression can

*Minutia: not just in the ventromedial PFC but in the entire "medial PFC."

cause the hippocampus to atrophy, how the sort of trauma that produces PTSD can enlarge the amygdala. Naturally, neuroplasticity in response to experience occurs in the PFC as well. Suffer from major depression or, to a lesser extent, a major anxiety disorder for years, and the PFC atrophies; the longer the mood disorder persists, the greater the atrophy. Prolonged stress or exposure to stress levels of glucocorticoids accomplishes the same; the hormone suppresses the level or efficacy of a key neuronal growth factor called BDNF* in the PFC, causing dendritic spines and dendritic branches to retract so much that the layers of the PFC thin out. This impairs PFC function, including a really unhelpful twist: As noted, when activated, the amygdala helps initiate the body's stress response (including the secretion of glucocorticoids). The PFC works to end this stress response by calming down the amygdala. Elevated glucocorticoid levels impair PFC function; the PFC isn't as good at calming the amygdala, resulting in the person secreting ever higher levels of glucocorticoids, which then impair . . . A vicious cycle.[36]

The list of other regulators stretches out. Estrogen causes PFC neurons to form thicker, more complex branches connecting to other neurons; remove estrogen entirely and some PFC neurons die. Alcohol abuse destroys neurons in that orbitofrontal cortex, causing it to shrink; the more shrinkage, the more likely an abstinent alcoholic is to relapse. Chronic cannabis use decreases blood flow and activity in both the dlPFC and the vmPFC. Exercise aerobically on a regular basis, and genes related to neurotransmitter signaling are turned on in the PFC, more BDNF growth factor is made, and coupling of activity among various PFC subregions becomes tighter and more efficient; roughly the opposite happens with eating disorders. The list goes on and on.[37]

Some of these effects are subtle. If you want to see something unsubtle, watch what happens days to years after the PFC is damaged by a traumatic brain injury (TBI—à la Phineas Gage), or frontotemporal dementia redux. Extensive damage to the PFC increases the likelihood long after of

*Brain-Derived Neurotrophic Factor.

disinhibited behavior, antisocial tendencies, and violence, a phenomenon that has been called "acquired sociopathy"*—remarkably, such individuals can tell you that, say, murder is wrong; they know, but they just can't regulate their impulses. Roughly half the people incarcerated for violent antisocial criminality have a history of TBI, versus about 8 percent of the general population; having had a TBI increases the likelihood of recidivism in prison populations. Moreover, neuroimaging studies reveal elevated rates of structural and functional abnormalities in the PFC among prisoners with a history of violent, antisocial criminality.[†,38]

Then there's the effect of decades of experiencing racial discrimination, which is a predictor of poor health in every corner of the body. African Americans with more severe histories of suffering discrimination (based on the score from a questionnaire, after controlling for PTSD and trauma history) have greater *resting* levels of activity in the amygdala and greater coupling between the amygdala and the downstream brain regions that it activates. If the subjects in that miserable social-exclusion paradigm (where the other two players stop throwing the virtual ball to you) are African American, the more the ostracizing is attributed to racism, the more vmPFC activation there is. In another neuroimaging study, performance on a frontal task declined in subjects primed with pictures of spiders (versus birds); among African American subjects, the more of a history of discrimination, the more spiders activated the vmPFC and the more performance declined. What are the effects of a history of prolonged discrimination? A brain that is in a resting state of don't-let-your-guard-down vigilance, that is more reactive to perceived threat, and a PFC burdened by a torrent of reporting from the vmPFC about this constant state of dis-ease.[39]

*By the way, *psychopathy* and *sociopathy* are not the same, and I have the same challenge keeping them straight as I do with using *that* or *which*. There are crucial differences between the two. Nevertheless, barbarians that we are, we will focus on the similarities and use the terms interchangeably.

†Elevated rates compared with whom? Nobel Peace Prize winners? The comparison groups in this literature are demographically matched nonincarcerated subjects and/or matched controls in prison for nonviolent crimes.

To summarize this section, when you try to do the harder thing that's better, the PFC you're working with is going to be displaying the consequences of whatever the previous years have handed you.

THE LEGACY OF THE TIME OF PIMPLES

Take the previous paragraph, replace *the previous years* with *adolescence*, underline the entire section, and you're all set. Chapter 3 provided the basic facts: (a) when you're an adolescent, your PFC still has a ton of construction ahead of it; (b) in contrast, the dopamine system, crucial to reward, anticipation, and motivation, is already going full blast, so the PFC hasn't a prayer of effectively reining in thrill seeking, impulsivity, craving of novelty, meaning that adolescents behave in adolescent ways; (c) if the adolescent PFC is still a construction site, this time of your life is the last period that environment and experience will have a major role in influencing your adult PFC;* (d) delayed frontocortical maturation has to have evolved precisely so that adolescence has this influence—how else are we going to master discrepancies between the letter and the spirit of laws of sociality?

Thus, adolescent social experience, for example, will alter how the PFC regulates social behavior in adults. How? Round up all the usual suspects. Lots of glucocorticoids, lots of stress (physical, psychological, social) during adolescence, and your PFC won't be its best self in adulthood. There will be fewer synapses and less complex dendritic branching in the mPFC and orbitofrontal cortex, along with permanent changes in how PFC neurons respond to the excitatory neurotransmitter glutamate (due to persistent changes in the structure of one of the main glutamate re-

*Just to recall something from chapter 3, frontocortical maturation during adolescence doesn't consist of the last lap of building new synapses, neuronal projections, and circuits. Instead, the early-adolescent frontal cortex has *more* of those things, is proportionately bigger, than the adult frontal cortex. In other words, frontocortical maturation during this period consists of pruning away the superfluous, less efficient circuits and synapses, whittling down to your adult frontal cortex.

ceptors). The adult PFC will be less effective in inhibiting the amygdala, making it harder to unlearn conditioned fear and less effective at inhibiting the autonomic nervous system from overreacting to being startled. Impaired impulse control, impaired PFC-dependent cognitive tasks. The usual.[40]

Conversely, an enriched, stimulating environment during adolescence has great effects on the resulting adult PFC and can reverse some of the effects of childhood adversity. For example, an enriched environment during adolescence causes permanent changes in gene regulation in the PFC, producing higher adult levels of neuronal growth factors like BDNF. Furthermore, while prenatal stress causes reductions in BDNF levels in the adult PFC (stay tuned), adolescent enrichment can reverse this effect. All changes that impair the PFC's ability for impulse control and gratification postponement. So if you want to be better at doing the harder thing as an adult, make sure you pick the right adolescence.[41]

FURTHER BACK

Now go back to the paragraph you underlined, discussing "whatever adolescence has handed you," replace *adolescence* with *childhood*, and underline the paragraph eighteen more times. Whaddaya know, the sort of childhood you had shapes the construction of the PFC at the time and the sort of PFC you'll have in adulthood.*

For example, no surprise, childhood abuse produces kids with a smaller PFC, with less gray matter and with changes in circuitry: less communication among different subregions of the PFC, less coupling between the vmPFC and the amygdala (and the bigger the effect, the more prone the child is to anxiety). Synapses in the brain are less excitable; there are changes in the numbers of receptors for various neurotransmitters and changes in gene expression and patterns of epigenetic marking

*Even though PFC development is not completed until the midtwenties, construction on it begins in fetal life.

of genes—along with impaired executive function and impulse control in the child. Many of these effects occur in the first half decade or so of life. One might raise a cart-and-horse issue—the assumption in this section is that abuse causes these changes in the brain. What about the possibility that kids who already have these differences behave in ways that make them more likely to be abused? This is highly unlikely—the abuse typically precedes the behavioral changes.[42]

Unsurprising as well is that these changes in the PFC in childhood can persist into adulthood. Childhood abuse produces an adult PFC that is smaller, thinner, and with less gray matter, altered PFC activity in response to emotional stimuli, altered levels of receptors for various neurotransmitters, weakened coupling between both the PFC and dopaminergic "reward" regions (predicting increased depression risk), and weakened coupling with the amygdala as well, predicting more of a tendency to respond to frustration with anger ("trait anger"). And once again, all of these changes are associated with an adult PFC that isn't at its best.[43]

Thus, childhood abuse produces a different adult PFC. And grimly, having been abused as a child produces an adult with an increased likelihood of abusing their own child; at *one* month of age, PFC circuitry is already different in children whose mothers were abused in childhood.[44]

These findings concern two groups of people—abused in childhood or not. What about looking at the full spectrum of luck? How about the effects of childhood socioeconomic status on our realm of supposed grit?

No surprise, the socioeconomic status of a child's family predicts the size, volume, and gray matter content of the PFC in kindergarteners. Same thing in toddlers. In six-month-olds. In *four-week-olds*. You want to scream at how unfair life can be.[45]

All the individual pieces of these findings flow from that. Socioeconomic status predicts how much a young child's dlPFC activates and recruits other brain regions during an executive task. It predicts more responsiveness of the amygdala to physical or social threat, a stronger activation signal carrying this emotional response to the PFC via the vmPFC. And such status predicts every possible measure of frontal executive function in kids; naturally, lower socioeconomic status predicts worse PFC development.[46]

There are hints as to the mediators. By age six, low status is already predicting elevated glucocorticoid levels; the higher the levels, the less activity in the PFC on average.* Moreover, glucocorticoid levels in kids are influenced not only by the socioeconomic status of the family but by that of the neighborhood as well.† Increased amounts of stress mediate the relationship between low status and less PFC activation in kids. As a related theme, lower socioeconomic status predicts a less stimulating environment for a child—all those enriching extracurricular activities that can't be afforded, the world of single mothers working multiple jobs who are too exhausted to read to their child. As one shocking manifestation of this, by age three, your average high-socioeconomic status kid has heard about thirty million more words at home than a poor kid, and in one study, the relationship between socioeconomic status and the activity of a child's PFC was partially mediated by the complexity of language use at home.[47]

Awful. Given the start of constructing the frontal cortex during this period, it wouldn't be crazy to predict that childhood socioeconomic status predicts things in adults. Childhood status (independent of the status achieved in adulthood) is a significant predictor of glucocorticoid levels, the size of the orbitofrontal cortex, and performance of PFC-dependent tasks in adulthood. Not to mention incarceration rates.[48]

Miseries like childhood poverty and childhood abuse are incorporated in someone's Adverse Childhood Experiences (ACE) score. As we saw in

---

*Which means that the vicious cycle noted earlier about adults applies to kids as well—elevated glucocorticoid levels make for a weaker developing PFC; insofar as part of what the PFC does is turn off glucocorticoid stress responses, this weakened PFC adds to glucocorticoid levels rising even higher.

†Influences from the world outside a child's family are shown in a related literature: everything else being equal, growing up in an urban setting (versus suburban or rural) predicts less gray matter volume in the different parts of the PFC in adults, a more reactive amygdala, and more glucocorticoid secretion in response to social stress (where the bigger the size of the city in childhood, the more reactive the amygdala). Moreover, cortical brain development in newborns is predicted not only by familial social disadvantage but by neighborhood crime rates as well.

the last chapter, it queries whether someone experienced or witnessed physical, emotional, or sexual childhood abuse, physical or emotional neglect, or household dysfunction, including divorce, spousal abuse, or a family member mentally ill, incarcerated, or struggling with substance abuse. With each increase in someone's ACE score, there's an increased likelihood of a hyperreactive amygdala that has expanded in size and a sluggish PFC that never fully developed.[49]

Let's push the bad news one step further, into chapter 3's realm of prenatal environmental effects. Low socioeconomic status for a pregnant woman or her living in a high-crime neighborhood both predict less cortical development at the time of the baby's birth. Even back when the child was still in utero.* And naturally, high levels of maternal stress during pregnancy (e.g., loss of a spouse, natural disasters, or maternal medical problems that necessitate treatment with lots of synthetic glucocorticoids) predict cognitive impairment across a wide range of measures, poorer executive function, decreased gray matter volume in the dlPFC, a hyperreactive amygdala, and a hyperreactive glucocorticoid stress response when those fetuses become adults.[†,50]

An ACE score, a fetal adversity score, last chapter's Ridiculously Lucky Childhood Experience score—they all tell the same thing. It takes a certain kind of audacity and indifference to look at findings like these and still insist that how readily someone does the harder things in life justifies blame, punishment, praise, or reward. Just ask those fetuses in the womb of a low-socioeconomic-status woman, already paying a neurobiological price.

*The finding involved structural MRI imaging of the fetal brain. Note that these findings about fetuses and newborns consider only development of the cortex, rather than specifically in the frontal cortex. This is because it's just too hard to discern the subregions in brain imaging at that age.

†As a calming reminder, these are major maternal stressors, not everyday ones. Moreover, the magnitude of these effects are generally mild (with an exception being if the adversity that the fetus experiences includes maternal alcohol or drug abuse).

THE LEGACY OF THE GENES YOU WERE HANDED, AND THEIR EVOLUTION

Genes have something to do with the sort of PFC you have. Big shocker—as described in the last chapter, the growth factors, enzymes that generate or break down neurotransmitters, receptors for neurotransmitters and hormones, etc., etc., are all made of protein, meaning that they are coded for by genes.

The notion that genes have something to do with all this can be totally superficial and uninteresting. Differences between the type of genes possessed by particular species help explain why a frontal cortex occurs in humans but not in barnacles in the sea or heather on the hill. The types of genes possessed by humans help explain why the frontal cortex (like the rest of the cortex) consists of six layers of neurons and isn't bigger than your skull. However, the sort of genetics that interests us when "genes" come into the picture concerns the fact that that particular gene can come in different flavors, with these variants differing from one person to the next. Thus, in this section, we're not interested in genes that help form a frontal cortex in humans but don't exist in fungi. We're interested in the variation in versions of genes that helps explain variation in the volume of the frontal cortex, its level of activity (as detected with EEG), and performance on PFC-dependent tasks.* In other words, we're interested in the variants of those genes that help explain why two people differ in their likelihood of stealing a cookie.[51]

Nicely, the field has progressed to the point of understanding how variants of specific genes relate to frontal function. A bunch of them relate to the neurotransmitter serotonin; for example, there's a gene that codes for a protein that removes serotonin from the synapse, and which version of that gene you have influences the tightness of coupling between the PFC

*Note that the variability in a trait in a population is determined by the degree of variability in genes (i.e., a "heritability score"). This is a hugely controversial subject, often producing glass-half-empty/glass-half-full differences as to whether a result is indicating how important or how unimportant a gene is. For a detailed but nontechnical overview of the behavior genetics controversies, see chapter 8 in my book *Behave: The Biology of Humans at Our Best and Worst.*

and amygdala. Variation in a gene related to the breakdown of serotonin in the synapse helps predict people's performance on PFC-dependent reversal tasks. Variation in the gene for one of the serotonin receptors (there are a lot) helps predict how good people are at impulse control.* Those are just about the genetics of serotonin signaling. In a study of the genomes of thirteen thousand people, a complex cluster of gene variants predicted an increased likelihood of impulsive, risky behavior; the more of those variants someone had, the smaller their dlPFC.[52]

A crucial point about genes related to brain function (well, pretty much all genes) is that the same gene variant will work differently, sometimes even dramatically differently, in different environments. This interaction between gene variant and variation in environment means that, ultimately, you can't say what a gene "does," only what it does in each particular environment in which it has been studied. And as a great example of this, in variants in the gene for one type of serotonin receptor helps explain impulsivity in women . . . but only if they have an eating disorder.[53]

The section on adolescence considered why dramatic delayed maturation of the PFC evolved in humans and how that makes that region's construction so subject to environmental influences. How do genes code for freedom from genes? In at least two ways. The first, straightforward, way involves the genes that influence how rapidly PFC maturation occurs.† The second way is subtler and elegant—genes relevant to how *sensitive* the PFC will be to different environments. Consider an (imaginary) gene, coming in two variants, that influences how prone someone is to stealing. A person, on their own, has the same low likelihood, regardless of variant. However, if there's a peer group egging the person on, one variant results in a 5 percent increase in likelihood of succumbing, the other 50 percent. In other words, the two variants produce dramatic differences in sensitivity to peer pressure.

*For detail enthusiasts, the protein that removes serotonin is called the serotonin transporter; the protein that degrades serotonin is called MAO-alpha; the receptor is the 5HT2A receptor.

†Stress and adversity are bad for PFC development and, interestingly, this takes the form of *accelerated* maturation. Faster maturation equals the door being shut sooner on how much environment can foster optimal PFC growth.

Let's frame this sort of difference more mechanically. Suppose you have an electrical cord that plugs into a socket; when it's plugged in, you don't steal. The socket is made of an imaginary protein that comes in two variants, which determine how wide the slots are that the plug plugs into. In a silent, hermetically sealed room, a plug remains in the socket, regardless of variant. But if a group of taunting, peer-pressuring elephants thunders past, the plug is ten times more likely to vibrate out of the loose-slot socket than the tight one.

And that turns out to be something like a genetic basis for being freer from genes. Work by Benjamin de Bivort at Harvard concerns a gene coding for a protein called teneurin-A, which is involved in synapse formation between neurons. The gene comes in two variants that influence how tightly a cable from one neuron plugs into a teneurin-A socket on the other (to simplify enormously). Have the loose-socket variant, and the result will be more variability in synaptic connectiveness. Or stated our way, the loose-socket variant codes for neurons that are more sensitive to environmental influences during synapse formation. It's not known yet if teneurins work this way in our brains (these were studies of flies—yes, environmental influences even affect synapse formation in flies), but things conceptually similar to this have to be occurring in umpteen dimensions in our brains.[54]

THE CULTURAL LEGACY BEQUEATHED TO YOUR PFC BY YOUR ANCESTORS

As we saw in the previous chapter's overview, different sorts of ecosystems generate different sorts of cultures, which affects a child's upbringing from virtually the moment of birth, tilting the brain construction toward ways that make it easier for them to fit into the culture. And thus pass its values on to the next generation . . .

Of course, cultural differences majorly influence the PFC. Essentially all the studies done concern comparisons between Southeast Asian collectivist cultures valuing harmony, interdependence, and conformity, and North American individualist ones emphasizing autonomy, individual rights, and personal achievement. And their findings make sense.*

Here's one you couldn't make up—in Westerners, the vmPFC activates in response to seeing a picture of your own face but not your mother's; in East Asians, the vmPFC activates equally for both; these differences become even more extreme if you prime subjects beforehand to think about their cultural values. Study bicultural individuals (i.e., with one collectivist culture parent, one individualist); prime them to think about one culture or the other, and they then show that culture's typical profile of vmPFC activation.[55]

Other studies show differences in PFC and emotion regulation. A meta-analysis of thirty-five studies neuroimaging subjects during social-processing tasks showed that East Asians average higher activity in the dlPFC than Westerners (along with activation of a brain region called the temporoparietal junction, which is central to theory of mind); this is basically a brain more actively working on emotion regulation and understanding other people's perspectives. In contrast, Westerners present a picture of more emotional intensity, self-reference, capacity for strongly emotional disgust or empathy—higher levels of activity in the vmPFC, insula, and anterior cingulate. And these neuroimaging differences are greatest in subjects who most strongly espouse their cultural values.[56]

There are also PFC differences in cognitive style. In general, collectivist-culture individuals prefer and excel at context-dependent cognitive tasks, while it's context-independent tasks for individualistic-culture folks. And in both populations, the PFC must work harder when subjects struggle with the type of task less favored by their culture.

---

*A few of the studies focused on Western Europeans rather than North Americans, with the same general differences from East Asian cultures.

Where do these differences come from on a big-picture level?* As discussed in the last chapter, East Asian collectivism is generally thought to arise from the communal work demands of floodplain rice farming. Recent Chinese immigrants to the United States already show the Western distinction between activating your vmPFC when thinking about yourself and activating it when thinking about your mother. This suggests that people back home who were more individualistic were the ones more likely to choose to emigrate, a mechanism of self-selection for these traits.[57]

Where do these differences come from on a smaller-picture level? As covered in the last chapter, children are raised differently in collectivist versus individualist cultures, with implications for how the brain is constructed.

But in addition, there are probably genetic influences. People who are spectacularly successful at expressing their culture's values tend to leave copies of their genes. In contrast, fail to show up with the rest of the village during rice-harvesting day because you decided to go snowboarding, or disrupt the Super Bowl by trying to persuade the teams to cooperate rather than compete—well, such cultural malcontents, contrarians, and weirdos are less likely to pass on their genes. And if these traits are influenced at all by genes (which they are, as seen in the previous section), this can produce cultural differences in gene frequencies. Collectivist and individualist cultures differ in the incidence of gene variants related to dopamine and norepinephrine processing, variants of the gene coding for the pump that removes serotonin from the synapse, and variants of the gene coding for the receptor in the brain for oxytocin.[58]

In other words, there's coevolution of gene frequencies, cultural values, child development practices, reinforcing each other over the generations, shaping what your PFC is going to be like.

---

*Reminding once again that these are differences in *average* degrees of traits, populational differences with lots of individual exceptions.

THE DEATH OF THE MYTH OF FREELY CHOSEN GRIT

We're pretty good at recognizing that we have no control over the attributes that life has gifted or cursed us with. But what we do with those attributes at right/wrong crossroads powerfully, toxically invites us to conclude, with the strongest of intuitions, that we are seeing free will in action. But the reality is that whether you display admirable gumption, squander opportunity in a murk of self-indulgence, majestically stare down temptation or belly flop into it, these are all the outcome of the functioning of the PFC and the brain regions it connects to. And that PFC functioning is the outcome of the second before, minutes before, millennia before. The same punch line as in the previous chapter concerning the entire brain. And invoking the same critical word—*seamless*. As we've seen, talk about the evolution of the PFC, and you're also talking about the genes that evolved, the proteins they code for in the brain, and how childhood altered the regulation of those genes and proteins. A seamless arc of influences bringing your PFC to this moment, without a crevice for free will to lodge in.

Here's my favorite finding pertinent to this chapter. There's a task that can be done in two different ways: in version one, do some amount of work and you get some amount of reward, but if you do twice as much work you get three times as much of a reward. Version two: do some amount of work and you get some amount of reward, but if you do three times as much work, you get a hundred zillion times as much reward. Which version should you do? If you think you can freely choose to exercise self-discipline, choose version two—you're going to choose to do a little bit more work and get a huge boost in reward as a result. People usually prefer version two, independent of the sizes of the rewards. A recent study shows that activity in the vmPFC* tracks the degree of preference for version two. What does that mean? In this setting, the vmPFC is coding for how

*Plus one other region, the rostrolateral PFC.

much we prefer circumstances that reward self-discipline. Thus, this is the part of the brain that codes for how wisely we think we'll be exercising free will. In other words, this is the nuts-and-bolts biological machinery coding for a belief that there are no nuts or bolts.[59]

Sam Harris argues convincingly that it's impossible to successfully think of what you're going to think next. The takeaway from chapters 2 and 3 is that it's impossible to successfully wish what you're going to wish for. This chapter's punchline is that it's impossible to successfully will yourself to have more willpower. And that it isn't a great idea to run the world on the belief that people can and should.

5

A Primer on Chaos

Suppose that just before you started reading this sentence, you reached to scratch an itch on your shoulder, noted that it's becoming harder to reach that spot, thought of your joints calcifying with age, which made you vow to exercise more, and then you got a snack. Well, science has officially weighed in—each of those actions or thoughts, conscious or otherwise, and every bit of neurobiology underpinning it, was determined. Nothing just got it into its head to be a causeless cause.

No matter how thinly you slice it, each unique biological state was caused by a unique state that preceded it. And if you want to truly understand things, you need to break these two states down to their component parts, and figure out how each component comprising Just-Before-Now gave rise to each piece of Now. This is how the universe works.

But what if that isn't? What if some moments aren't caused by anything preceding them? What if some unique Nows can be caused by multiple, unique Just-Before-Nows? What if the strategy of learning how something works by breaking it down to its component parts is often useless? As it turns out, all of these are the case. Throughout the past century, the previous paragraph's picture of the universe was overturned, giving birth to the sciences of chaos theory, emergent complexity, and quantum indeterminacy.

To label these as revolutions is not hyperbolic. When I was a kid, I read a novel called *The Twenty-One Balloons*,* about a utopian society on the island of Krakatoa built on balloon technology, destined to be destroyed by the famed 1883 eruption of the volcano there. It was fantastic, and the second I got to the end, I immediately flipped to the front to reread it. And it was then almost a quarter century before I immediately flipped to the front to reread a different book,† an introduction to one of these scientific revolutions.

Staggeringly interesting stuff. This chapter, and the five after it, reviews these three revolutions, and how numerous thinkers believe that you can find free will in their crevices. I will admit that the previous three chapters have an emotional intensity for me. I am put into a detached, professorial, eggheady sort of rage by the idea that you can assess someone's behavior outside the context of what brought them to that moment of intent, that their history doesn't matter. Or that even if a behavior seems determined, free will lurks wherever you're not looking. And by the conclusion that righteous judgment of others is okay because while life is tough and we're unfairly gifted or cursed with our attributes, what we freely choose to do with them is the measure of our worth. These stances have fueled profound amounts of undeserved pain and unearned entitlement.

The revolutions in the next five chapters don't have that same visceral edge. As we'll see, there aren't a whole lot of thinkers out there citing, say, subatomic quantum indeterminacy when smugly proclaiming that free will exists and they earned their life in the top 1 percent. These topics don't make me want to set up barricades in Paris, singing revolutionary anthems from *Les Mis*. Instead, these topics excite me immensely because they reveal completely unexpected structure and pattern; this enhances rather than quenches the sense that life is more interesting than can be imagined. These are subjects that fundamentally upend how we think about how complex things work. But nonetheless, they are not where free will dwells.

This and the next chapter focus on chaos theory, the field that can

*By William Pène du Bois, Viking Books for Young Readers, 1947.

†James Gleick's *Chaos: Making a New Science* (first ed., Viking Press, 1987).

make studying the component parts of complex things useless. After a primer about the topic in this chapter, the next will cover two ways people mistakenly believe they've found free will in chaotic systems. First is the idea that if you start with something simple in biology and, unpredictably, out of that comes hugely complex behavior, free will just happened. Second is the belief that if you have a complex behavior that could have arisen from either of two different preceding biological states and there's no way to ever tell which one caused it, then you can get away with claiming that it wasn't caused by anything, that the event was free of determinism.

BACK WHEN THINGS MADE SENSE

Suppose that

X = Y + 1

If that is the case, then

X + 1 = ?

—and you were readily able to calculate that the answer is

(Y + 1) + 1.

Do **X + 3** and you've instantly got **(Y + 1) + 3**. And here's the crucial point—after solving **X + 1**, you were able to then solve **X + 3** *without first having to figure out X + 2*. You were able to extrapolate into the future without examining each intervening step. Same thing for **X + a gazillion**, or **X + sorta a gazillion**, or **X + a star-nosed mole.**

A world like this has a number of properties:

- As we just saw, knowing the starting state of a system (for example, **X = Y + 1**) lets you accurately predict what **X + whatever** will equal, without the intervening steps. This property runs in both directions. If you're given **(Y + 1) + whatever**, you know then that your starting point was **X − whatever.**

- Implicit in that, there is a unique pathway connecting the starting and ending states; it is also inevitable that **X + 1** cannot equal **(Y + 1) + 1** only some of the time.

- As shown dealing with something like "sorta a gazillion," the magnitude of uncertainty and approximation in the starting state is directly proportional to the magnitude at the other end. You can know what you don't know, can predict the degree of unpredictability.[1]

This relationship between starting states and mature states helped give rise to what has been the central concept of science for centuries. This is reductionism, the idea that to understand something complicated, break it down into its component parts, study them, add your insights about each component part together, and you will understand the complicated whole. And if one of those component parts is itself too complicated to understand, study its eensy subcomponent parts and understand them.

Reductionism like this is vital. If your watch, running on the ancient technology of gears, stops working, you apply a reductive approach to solving the problem. You take the watch apart, identify the one tiny gear that has a broken tooth, replace it, and put the pieces back together, and the watch runs. This approach is also how you do detective work—you arrive at a crime scene and interview the witnesses. The first witness observed only parts 1, 2, and 3 of the event. The second saw only 2, 3, and 4. The third, only 3, 4, and 5. Bummer, no one saw everything that happened. But thanks to a reductive mindset, you can solve the problem by taking the fragmentary component parts—each of the three witnesses' overlapping observations, and *combine them* to understand the complete sequence.* Or as another example, in the first season of the pandemic, the world waited for answers to reductive questions like what receptor on the surface of a lung cell binds the spike protein of SARS-CoV-2, allowing it to enter and sicken that cell.

---

*The same strategy was used to first sequence the human genome. Suppose a particular stretch of DNA is nine units of length too long to systematically figure out its sequence—the lab techniques just aren't up to it. Instead, cut that stretch into a series of fragments that are short enough to sequence, say, fragment 1/2/3, fragment 4/5/6, and fragment 7/8/9. Now take a second copy of the same stretch of DNA, and cut it into a different pattern: fragment 1, then fragment 2/3/4, then 5/6/7, then 8/9. Cut a third into 1/2, 3/4/5, and 6/7/8/9. Match up the overlapping fragments, and you now know the entire sequence.

Mind you, a reductive approach doesn't apply to everything. If there's a drought, the sky dotted with puffy clouds that haven't rained in a year, you don't first isolate a cloud, study its left half and then its right half and then half of each half, and so on, until you find the itty-bitty gear in the center that has a broken tooth. Nonetheless, a reductive approach has long been the gold standard for scientifically exploring a complex topic.

And then, starting in the early 1960s, a scientific revolution emerged that came to be called chaoticism, or chaos theory. And its central idea is that really interesting, complicated things are often not best understood, *cannot* be understood, on a reductive level. To understand, say, a human whose behavior is abnormal, approach the problem as if this were a cloud that does not rain, rather than as a watch that does not tick. And naturally, humans-as-clouds generate all sorts of nearly irresistible urges for concluding that you are observing free will in action.

CHAOTIC UNPREDICTABILITY

Chaos theory has its creation story. When I was a kid in the 1960s, inaccurate weather prediction was mocked with trenchant witticisms like "The weatherman on the radio [invariably, indeed, a man] said it's going to be sunny today, so better bring an umbrella." MIT meteorologist Edward Lorenz began using some antediluvian computer to model weather patterns in an attempt to increase prediction accuracy. Stick variables like temperature and humidity into the model and see how accurate the predictions became. See if additional variables, other variables, different weightings of variables,* improved predictability.

So Lorenz was studying a model on his computer using twelve variables. Time for lunch; halt the program in the middle of its cranking out a

*Weighting variables is the outcome of transitioning from "Add variables A and B together and you get decent prediction about whatever" to "Add variables A and B together . . . , and remember that variable A is more important than variable B" to "Add variable A and B together . . . and have variable A carry, say, 3.2 times as much weight in the equation as does variable B."

time course of predictions. Come back postlunch and, to save time, restart the program at a point before you stopped it, rather than starting all over. Punch in the values of those twelve variables at that time point, and let the model resume its predicting. That's what Lorenz did, which is when our understanding of the universe changed.

One variable at that time point had a value of 0.506127. Except that on the printout, the computer had rounded it down to 0.506; maybe the computer hadn't wanted to overwhelm this Human 1.0. In any case, 0.506127 became 0.506, and Lorenz, not knowing about this slight inaccuracy, ran the program with the variable at 0.506, thinking that it was actually 0.506127.

Thus, he was now dealing with a value that was a smidgen different from the real one. And we know just what should have happened now, in our supposedly purely linear, reductive world: the degree to which the starting state was off from what he thought it was (i.e., 0.506 rather than 0.506127) predicted how inaccurate his ending state would be—the program would generate a point that was only a smidgen different from that same point before lunch—if you superimposed the before- and after-lunch tracings, you'd barely see a difference.

Lorenz let the program, still depending on 0.506 instead of 0.506127, continue to run, and out came a result that was even more discrepant than he had expected from the prelunch run. Weird. And with each successive point, things got weirder—sometimes things seemed to have returned to the prelunch pattern but would then diverge again, with the divergences increasingly different, unpredictably, crazily so. And eventually rather than the program generating something even remotely close to what he saw the first time, the discrepancy in the two tracings was about as different as was possible.

This is what Lorenz saw—the pre- and postlunch tracings superimposed, a printout now with the status of a holy relic in the field (see figure on the next page).

Lorenz finally spotted that slight rounding error introduced after lunch and realized that this made the system unpredictable, nonlinear, and nonadditive.

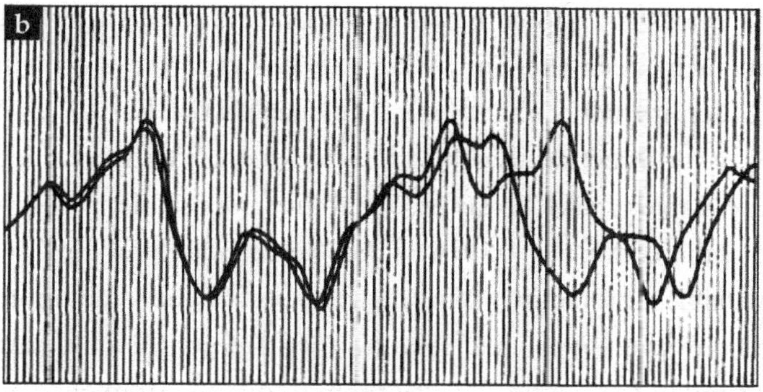

By 1963, Lorenz announced this discovery in a dense technical paper, "Deterministic Non-periodic Flow," in the highly specialized *Journal of Atmospheric Sciences* (and in the paper, Lorenz, while beginning to appreciate how these insights were overturning centuries of reductive thinking, still didn't forget where he came from. Will it ever be possible to perfectly predict all of future weather? readers of the journal plaintively asked. Nope, Lorenz concluded; the chance of this is "non-existent"). And the paper has since been cited in other papers a staggering 26,000+ times.[2]

If Lorenz's original program had contained only two weather variables, instead of the twelve he was using, the familiar reductiveness would have held—after a slightly wrong number was fed into the computer, the output would have been precisely as wrong at every step for the rest of time. Predictably so. Imagine a universe that consists of just two variables, the Earth and the Moon, exerting their gravitational forces on each other. In this linear, additive world, it is possible to infer precisely where they were at any point in the past and predict precisely where each will be at any point in the future;* if an approximation was accidentally introduced, the same magnitude of approximation would continue forever. But now add the Sun into the mix, and the nonlinearity happens. This is because the

*Which means that past and future are identical, that there is no direction of time, that events one second in the future are already the past of two seconds in the future. Which makes me feel queasy, reminding me that I've already died somewhere in the future.

Earth influences the Moon, which means that the Earth influences how the Moon influences the Sun, which means that the Earth influences how the Moon influences the Sun's influence on the Earth. . . . And don't forget the other direction, Earth to Sun to Moon. The interactions among the three variables make linear predictability impossible. Once you've entered the realm of what is known as the "three-body problem," with three or more variables interacting, things have inevitably become unpredictable.

When you have a nonlinear system, tiny differences in a starting state from one time to the next can cause them to diverge from each other enormously, even exponentially,* something since termed "sensitive dependence on initial conditions." Lorenz noted that the unpredictability, rather than hurtling off forever into the exponential stratosphere, is sometimes bounded, constrained, and "dissipative." In other words, the degree of unpredictability oscillates erratically around the predicted value, repeatedly a little more, a little less than predicted in the series of numbers you are generating, the degree of discrepancy always different, forever after. It's like each data point you are getting is sort of attracted to what the data point is predicted to be, but not enough to actually reach the predicted value. Strange. And thus, Lorenz named these strange attractors.[†,3]

*People in the field spend a lot of time debating whether exponential increases are occasional, probable, or inevitable, where the outcome depends on the finite-time Lyapunov exponent. I have no idea what that means, and this footnote is totally gratuitous. The differing opinions about exponentiality are reviewed by Wheaton College philosopher/mathematician Robert Bishop, who characterizes the view that chaotic systems always have exponential increases in unpredictability as laughable "folklore."

†The oscillations of unpredictability around the predicted answer in a strange attractor show some dizzyingly interesting properties:

A. The first is an extension of Lorenz's experience with his six decimal places. So the values in the chaotic oscillations never actually reach the attractor—they just keep dancing around it. You're dubious of this chaos stuff, know that at some point, this weirdo set of results you're getting will settle down to matching what is predicted. And that seems to happen—your nice linear predictions say that the observed value at some point should be, say, 27 units of something. And that's exactly what you measure. Aha, so much for this system being unpredictable. But then a chaoticist gives you a magnifying glass, and you look closely and see that the observed value wasn't 27. It was 27.1, in contrast to the predicted 27.0. "Okay, okay," you say. "I still don't believe this chaos theory stuff. All we've just learned is that we have to be precise out to one decimal place." And then at some point in the future, when you've predicted that the measure should be, say, 47.1, that's exactly what you actually observe;

So a tiny difference in a starting state can magnify unpredictably over time. Lorenz took to summarizing this idea with a metaphor about seagulls. A friend suggested something more picturesque, and by 1972 this was formalized into the title of a talk given by Lorenz. Here's another holy relic of the field (see figure on the next page).

Thus was born the symbol of the chaos theory revolution, the butterfly effect.*[4]

---

goodbye, chaos theory. But the chaoticist gives you an even bigger magnifying glass, and the observed value turns out to be 47.09 instead of the predicted 47.10. Okay, that doesn't prove that the mathematical world has chaotic elements; we just have to be precise out to two decimal places. And then you find a discrepancy three decimal places out And wait long enough, and you'll find one that's four decimal places out. And this goes on and on until you're dealing with an infinite number of decimal places, and the results are still not predictable (but if you could get past infinity, things would become perfectly predictable; in other words, chaos only superficially shows that Laplace was wrong—what it is mostly showing is how long infinity is). Thus, the relative magnitude of chaotic oscillations around a strange attractor stays the same, regardless of the magnification at which you're looking (something similar to the scale-free nature of fractals).

B. The oscillations around predicted values are the manifestation of their strange attraction to what is predicted. But the fact that the oscillations never actually precisely reach the predicted value (at a sufficient scale of magnification) shows that a strange attractor repels as well as attracts.

C. As a logical extension of these ideas, the pattern of oscillations around the predicted value never repeats either. Even if it looks like it oscillated to the same unpredicted point where it was at last week, look closer, and it will be slightly different. Same scale-free feature. When a dynamic pattern repeats over and over, it is referred to as being "periodic," and the pattern's infinity can be compressed into something far shorter, such as the statement "It goes like this forever" or "It alternates between these two patterns forever" (which is saying that the predictable shifting between multiple patterns *is* the pattern). In contrast, when the pattern of unpredictable oscillations around a strange attractor never repeats until the end of time, it is referred to as nonperiodic, as in the title of Lorenz's paper. And with nonperiodicity, the only possible description of an infinitely long pattern has to be as long itself. (Jorge Luis Borges wrote a very short story [i.e., one paragraph long], "On Exactitude in Science," in which a cartographer makes a perfect map of an empire, leaving out no detail; the map, of course, is as large as the empire.)

*Ray Bradbury anticipated all of this with his 1952 short story "A Sound of Thunder." A man travels sixty million years back in time, being careful not to alter anything while there. Inevitably he does alter something, and returns to the present to find the world a different place—as Bradbury framed it, the man had knocked over a small domino that led to big dominoes falling and, eventually, gigantic ones. What was the infinitesimally small impact that he had in the past? He stepped on a butterfly. Mere coincidence that this was the metaphor suggested by Lorenz's friend? I think not.

AMERICAN ASSOCIATION FOR THE ADVANCEMENT OF SCIENCE, 139th MEETING

Subject............................Predictability; Does the Flap of a Butterfly's wings in Brazil Set Off a Tornado in Texas?

Author.............................Edward N. Lorenz, Sc.D.
Professor of Meteorology

Address............................Massachusetts Institute of Technology
Cambridge, Mass. 02139

Time...............................10:00 a.m., December 29, 1972

Place..............................Sheraton Park Hotel, Wilmington Room

Program............................AAAS Section on Environmental Sciences
New Approaches to Global Weather: GARP
(The Global Atmospheric Research Program)

Convention Address.................Sheraton Park Hotel

RELEASE TIME
10:00 a.m., December 29

CHAOTICISM YOU CAN DO AT HOME

Time to see what chaoticism and sensitive dependence on initial conditions look like in practice. This makes use of a model system that is so cool and fun that I've even fleetingly wished that I could do computer coding, as it would make it easier to play with it.

Start off with a grid, like the one on a piece of graph paper, where the first row is your starting condition. Specifically, each of the boxes in the row can be in one of two states, either open or filled (or, in binary coding, either zero or one). There are 16,384 possible patterns for that row;* here's our randomly chosen one:

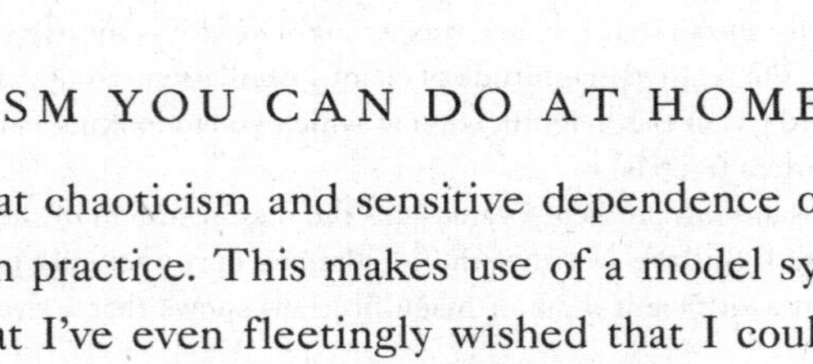

*The grid is 14 boxes wide; each box can be in one of 2 states; therefore, the total number of patterns possible is 2^{14}, or 16,384.

Time now to generate the second row of boxes that are open or filled, that new pattern determined* by the pattern in row 1. We need a rule for how to do this. Here's the most boring possible example: in row 2, a box that is underneath a filled box gets filled; a box underneath an open box remains open. Applying that rule over and over, using row 2 as the basis for row 3, 3 for 4, and so on, is just going to produce some boring columns. Or impose the opposite rule, such that if a box is filled, the one below it in the next row becomes open, while an open box spawns a filled one, and the outcome isn't all that exciting, producing sort of a lopsided checkered pattern:

As the main point, starting with either of these rules, if you know the starting state (i.e., the pattern in row 1), you can accurately predict what a row anywhere in the future will look like. Our linear universe again.

Let's go back to our row 1:

Now whether a particular row 2 box will be open or filled is determined by the state of three boxes—the row 1 box immediately above and the row 1 box's neighbor on each side.

Here's a random rule for how the state of a trio of adjacent row 1 boxes determines what happens in the row 2 box below: *A row 2 box is filled if*

*A word pregnant with significance.

and only if one of the trio of boxes above it is filled in. Otherwise, the row 2 box will remain open.

Let's start with the second box from the left in row 2. Here is the row 1 trio immediately above it (i.e., the first three boxes of row 1):

One of three boxes is filled, meaning that the row 2 box we're considering will get filled:

Look at the next trio in row 1 (i.e., boxes 2, 3, and 4). Only one box is filled, so box 3 in row 2 will also be filled:

In the row 1 trio of boxes 3, 4, and 5, two boxes (4 and 5) are filled, so the next row 2 box is left open. And so on. The rule we are working with—if and only if one box of the trio is filled, fill in the row 2 box in question—can be summarized like this:

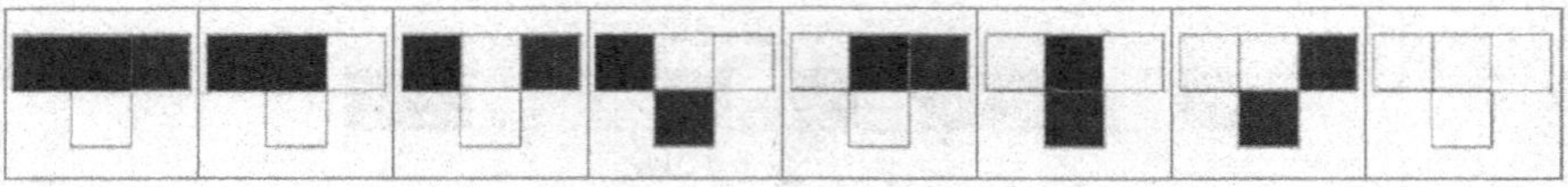

There are eight possible trios (two possible states for the first box of a trio times two possible for the second box times two for the third), and only trios 4, 6, and 7 result in the row 2 box in question being filled.

Back to our starting state, and using this rule, the first two rows will look like this:

But wait—what about the first and last boxes of row 2, where the box above has only one neighbor? We wouldn't have that problem if row 1 were infinitely long in both directions, but we don't have that luxury. What do we do with each of them? Just look at the box above it and the single neighbor, and use the same rule—if one of those two is filled, fill in the row 2 box; if both or neither of the two is filled, row 2 box is open. Thus, with that addendum in place, the first 2 rows look like this:

Now use the same rule to generate row 3:

Keep going, if you have nothing else to do.

Now let's use this starting state with the same rule:

The first 2 rows will look like this:

Complete the first 250 or so rows and you get this:

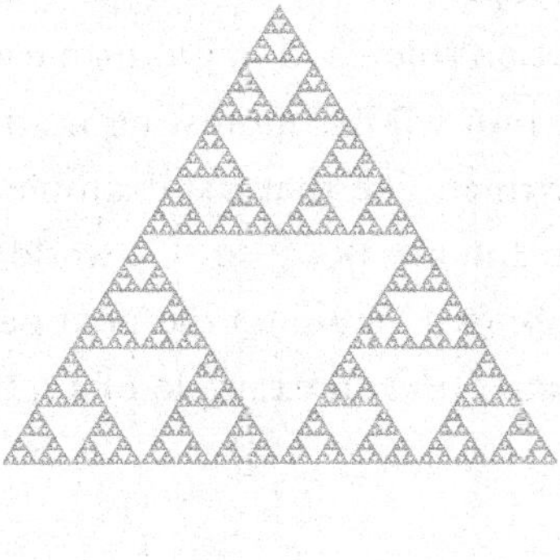

Take a different, wider random starting state, apply the same rule over and over, and you get this:

Whoa.

Now try this starting state:

By row 2, you get this:

Nothing. With this particular starting state, row 2 is all open boxes, as will be the case in every subsequent row. Row 1's pattern is snuffed out.

Let's describe what we've learned so far in a metaphorical way, rather than using terms like *input*, *output*, and *algorithm*. With some starting states and the reproduction rule used to produce each subsequent generation, things can evolve into wildly interesting mature states, but you can also get some that go extinct, like that last example.

Why the biology metaphors? Because this world of generating patterns like this applies to nature (see figure on the next page).

We have just been exploring an example of a *cellular automaton*, where

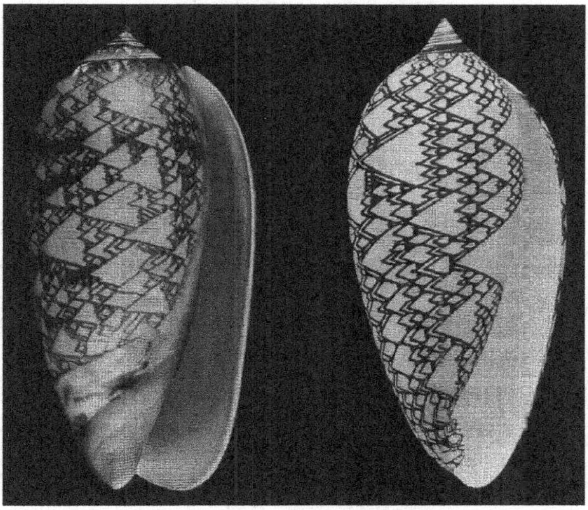

An actual shell on the left, a computer-generated pattern on the right

you start with a row of cells that are either open or filled, supply a reproduction rule, and let the process iterate.*[,5]

The rule we've been following (if and only if one box of the trio above is filled . . .) is called rule 22 in the cellular automata universe, which consists of 256 rules.† Not all of these rules generate something interesting—depending on the starting state, some produce a pattern that just repeats for infinity in an inert, lifeless sort of way, or that goes extinct by the second row. Very few generate complex, dynamic patterns. And of the few

*Cellular automata were first studied and named by the Hungarian American mathematician / physicist / computer scientist John von Neumann in the 1950s. It's virtually required by law to call him a genius. He was wildly precocious—at age six, he could divide eight-digit numbers in his head and was fluent in ancient Greek. One day when von Neumann was six, he found his mother daydreaming and he asked her, "What are you calculating?" (This contrasts with the daughter of a friend of mine, who, finding her father lost in thought, asked, "Daddy, which candy are you thinking about?")

†Back to our set of instructions for rule 22: Just look at the first row. As we saw, there are eight possible trios. Each trio can result in two possible states in the next generation, namely open or filled. For example, our first trio, where all three boxes in the trio are filled, could lead to either an open row 2 box (as we would get when applying rule 22) or one that is filled (as with other rules). Thus, two possible states for each of the eight trios means 2^8, which equals 256, the total number of possible rules in this system.

that do, rule 22 is one of the favorites. People have spent their careers studying its chaoticism.

What is chaotic about rule 22? We've now seen that, depending on the starting state, by applying rule 22 you can get one of three mature patterns: (a) nothing, because it went extinct; (b) a crystallized, boring, inorganic periodic pattern; (c) a pattern that grows and writhes and changes, with pockets of structure giving way to anything but, a dynamic, organic profile. And as the crucial point, *there is no way to take any irregular starting state and predict what row 100, or row 1,000, or row any-big-number will look like.* You have to march through every intervening row, simulating it, to find out. It is impossible to predict if the mature form of a particular starting state will be extinct, crystalline, or dynamic or, if either of the latter two, what the pattern will be; people with spectacular mathematical powers have tried and failed. And this limit, paradoxically, extends to showing that you can't prove that somewhere a few baby steps before reaching infinity, that the chaotic unpredictability will suddenly calm down into a sensible, repeating pattern. We have a version of the three-body problem, with interactions that are neither linear nor additive. You cannot take a reductive approach, breaking things down to its component parts (the eight different possible trios of boxes and their outcomes), and predict what you're going to get. This is not a system for generating clocks. It's for generating clouds.[6]

So we've just seen that knowing the irregular starting state gives you no predictive power about the mature state—you'll just have to simulate each intervening step to find out.

Now consider rule 22 applied to each of these four starting states (see top figure on the next page).

Two of these four, once taken out ten generations, produce an identical pattern for the rest of time. I dare you to stare at these four and correctly predict which two it is going to be. It cannot be done.

Get some graph paper and crank through this, and you'll see that two of these four *converge.* In other words, knowing the mature state of a system like this gives you no predictive power as to what the starting state was, or

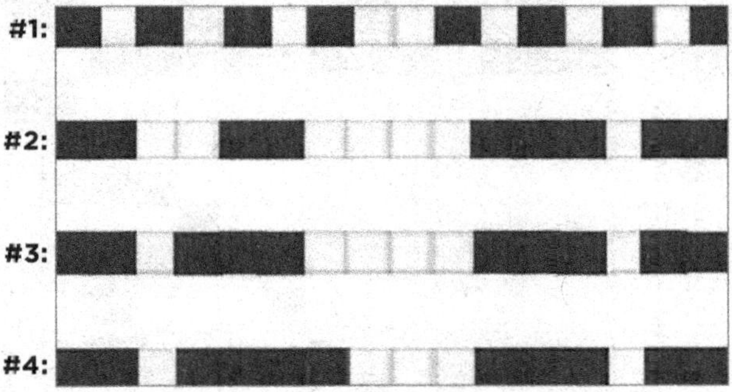

if it could have arisen from multiple different starting states, another defining feature of the chaoticism of this system.

Finally, consider the following starting state:

Which goes extinct by row 3:

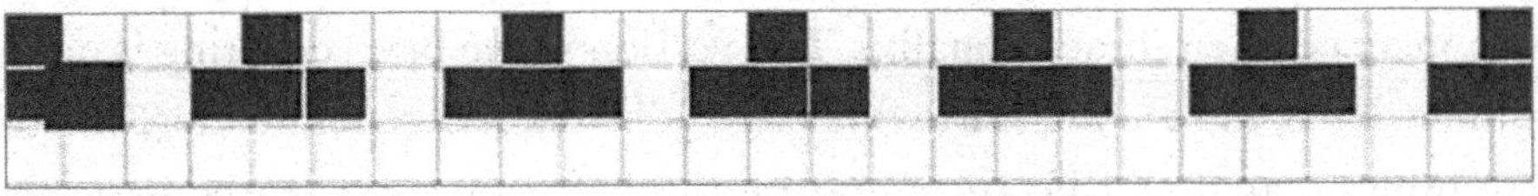

Introduce a smidgen of a difference in this nonviable starting state, namely that the open/filled status of just one of the twenty-five boxes differs—box 20 is filled instead of open:

And suddenly, life erupts into an asymmetrical pattern (see figure on the next page).

Let's state this biologically: a single *mutation*, in box 20, can have major consequences.

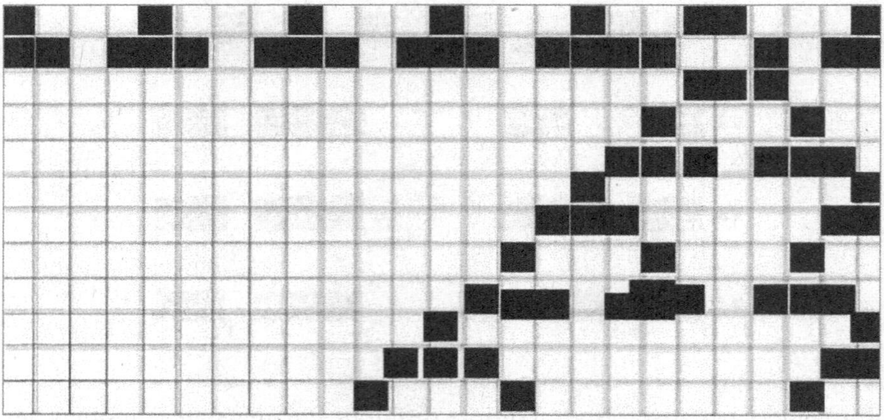

Let's state this with the formalism of chaos theory: this system shows sensitive dependence on the initial condition of box 20.

Let's state it in a way that is ultimately most meaningful: a butterfly in box 20 either did or didn't flap its wings.

I love this stuff. One reason is because of the ways in which you can model biological systems with this, an idea explored at length by Stephen Wolfram.* Cellular automata are also inordinately cool because you can increase their dimensionality. The version we've been covering is one-dimensional, in that you start with a line of boxes and generate more lines. Conway's Game of Life (invented by the late Princeton mathematician

*As with von Neumann, it is impossible to mention Wolfram without noting that he is a major-league genius. Wolfram had written three books on particle physics by the time he was fourteen years old, was a professor at Caltech by age twenty-one, produced a computer language and a computing system called Mathematica that is widely used, helped create the language that the aliens communicated with in the movie *Arrival*, generated Wolfram's atlas of cellular automata, which allows you to play with the 256 rules, etc., etc. In 2002, he published a book called *A New Kind of Science*, which explores how computational systems like cellular automata are foundational to everything from philosophy to evolution, from biological development to postmodernism. This generated a great deal of controversy, built around the question of whether these computational systems are good ways to generate *models* of things in the real world, or to actually generate the complicated things themselves (as one piece of the critique, things in nature don't progress in discrete, synchronized "time steps" as in these models). Lots of people also weren't thrilled about the grandiosity of the claims in the book (starting with the title) or about a perceived tendency of Wolfram's to claim every idea in the book as his own. Everyone bought a copy of it and discussed it endlessly (and hardly ever actually read the entire thing, as it was 1,192 pages long—yeah, me included).

John Conway) is a two-dimensional version where you start with a grid of boxes and generate each subsequent generation's grid. And produce absolutely astonishingly dynamic, chaotic patterns that are typically described as involving individual boxes that are "living" or "dying." All with the usual properties—you can't predict the mature state from the starting state—you have to simulate every intervening step; you can't predict the starting state from the mature state because of the possibility that multiple starting states converged into the same mature one (we're going to return to this convergence feature in a big way); the system shows sensitive dependence on initial conditions.[7]

(There's an additional realm classically discussed when introducing chaoticism. I've sidestepped covering it here, however, because I've learned the hard way from my classrooms that it is very difficult and/or I'm very bad at explaining it. If interested, read up about Lorenz's waterwheel, period doubling, and the significance of period 3 for the onset of chaos.)

With this introduction to chaoticism in hand, we can now appreciate the next chapter of the field—unexpectedly, the concepts of chaos theory became *really* popular, sowing the seeds for a certain style of free-will belief.

6

Is Your Free Will Chaotic?

THE AGE OF CHAOS

The upheaval in the early 1960s caused by chaos theory, strange attractors, and sensitive dependence on initial conditions was rapidly felt throughout the world, fundamentally altering everything from the most highfalutin philosophical musings to the concerns of everyday life.

Actually, not at all. Lorenz's revolutionary 1963 paper was mostly met with silence. It took years for him to begin to collect acolytes, mostly a group of physics grad students at UC Santa Cruz who supposedly spent a lot of time stoned and studied things like the chaoticism of how faucets drip.* Mainstream theorists mostly ignored the implications.

Part of the neglect reflected the fact that *chaos theory* is a horrible name, insofar as it is about the opposite of nihilistic chaos and is instead about the patterns of structure hidden in seeming chaos. The more fundamental reason for chaoticism getting off to a slow start was that if you have a reductive mindset, unsolvable, nonlinear interactions among a large number of variables is a total pain to study. Thus, most researchers tried to study

*This study produced the now legendary 1984 paper by Robert Shaw, *The Dripping Faucet as a Model Chaotic System*, Science Frontier Express Series (Aerial Press, 1984).

complicated things by limiting the number of variables considered so that things remained tame and tractable. And this guaranteed the incorrect conclusion that the world is mostly about linear, additive predictability and nonlinear chaoticism was a weird anomaly that could mostly be ignored. Until it couldn't be anymore, as it became clear that chaoticism lurked behind the most interesting complicated things. A cell, a brain, a person, a society, was more like the chaoticism of a cloud than the reductionism of a watch.[1]

By the eighties, chaos theory had exploded as an academic subject (this was around the time that the pioneering generation of renegade stoner physicists began to be things like a professor at Oxford or the founder of a company using chaos theory to plunder the stock market). Suddenly, there were specialized journals, conferences, departments, and interdisciplinary institutes. Scholarly papers and books appeared about the implications of chaoticism for education, corporate management, economics, the stock market, art and architecture (with the interesting idea that we find nature to be more beautiful than, say, modernist office buildings, because the former has just the right amount of chaos), literary criticism, cultural studies of television (with the observation that, like chaotic systems, television "dramas are both complex and simple at the same time"), neurology and cardiology (in both of which, interestingly, too *little* chaoticism was appearing to be a bad thing*). There were even scholarly articles about the relevance of chaos theory to theology (including one with the wonderful title "Chaos at the Marriage of Heaven and Hell," in which the author wrote, "Those of us who seek to engage modern culture in our theological reflection cannot afford to overlook chaos theory").[2]

Meanwhile, interest in chaos theory, accurate or otherwise, burst into

*In cardiology, healthier cardiovascular systems show more chaotic variability in the time intervals between heartbeats; in neurology, insufficient chaoticism is a marker of neurons that wind up firing at abnormally high rates in abnormally synchronized waves—a seizure. At the same time, other neuroscientists have explored how chaoticism can be exploited by the brain for enhancing some types of information transmission.

the general public's consciousness as well—who could have predicted that? There were the ubiquitous wall calendars of fractals. Novels, books of poetry, multiple movies, TV episodes, numerous bands, albums, and songs commandeered *strange attractor* or *the butterfly effect* in their titles.* According to a *Simpsons* fandom site, in one episode during her baseball-coaching period, Lisa is seen reading a book called *Chaos Theory in Baseball Analysis*. And as my favorite, in the novel *Chaos Theory*, part of the Nerds of Paradise Harlequin romance series, our protagonist has her eyes on handsome engineer Will Darling. Despite his unbuttoned shirt, six-pack, and insouciant bedroom eyes, it is understood that Will must still be a nerd, since he wears glasses.[3]

The growing interest in chaos theory generated the sound of a zillion butterfly wings flapping. Given that, it was inevitable that various thinkers began to proclaim that the unpredictable, chaotic cloud-ness of human behavior is where free will runs free. Hopefully, the material already covered, showing what chaoticism is and isn't, will help show how this cannot be.

The giddy conclusion that chaoticism proves free will takes at least two forms.

*The popularization of the latter has also led to a proliferation that I've noted in the locations of the butterfly effect, with the different citations placing the butterfly in the likes of the Congo, Sri Lanka, the Gobi Desert, Antarctica, and Alpha Centauri. In contrast, the tornado almost always seems to be in Texas, Oklahoma, or, evoking Dorothy and Toto too, Kansas.

WRONG CONCLUSION #1: THE FREELY CHOOSING CLOUD

For free-will believers, the crux of the issue is lack of predictability—at innumerable junctures in our lives, including highly consequential ones, we choose between X and not-X. And even a vastly knowledgeable observer could not have predicted every such choice.

In this vein, physicist Gert Eilenberger writes, "It is simply improbable that reality is completely and exhaustively mappable by mathematical constructs." This is because "the mathematical abilities of the species *Homo sapiens* are in principle limited because of their biological basis. . . . Because of [chaoticism], the determinism of Laplace* cannot be absolute and the question of the possibility of chance and freedom is open again!" The exclamation mark at the end is Eilenberger's: a physicist means business if he's putting exclamation marks in his writing.[4]

Biophysicist Kelly Clancy makes a similar point concerning chaoticism in the brain: "Over time, chaotic trajectories will gravitate toward [strange attractors]. Because chaos can be controlled, it strikes a fine balance between reliability and exploration. Yet because it's unpredictable, it's a strong candidate for the dynamical substrate of free will."[5]

Doyne Farmer weighs in as well in a way I found disappointing, given that he was one of the faucet-drip apostles of chaos theory and should know better. "On a philosophical level, it struck me [that chaoticism was] an operational way to define free will, in a way that allowed you to reconcile free will with determinism. The system is deterministic, but you can't say what it's going to do next."[6]

As a final example, philosopher David Steenburg explicitly links the supposed free will of chaos with morality: "Chaos theory provides for the

*As a reminder from earlier in the book, Laplace was the eighteenth-century philosopher who stated the rallying cry of scientific determinism, namely that if you understand the physical laws shaping the universe and know the exact position of every particle in it, you could accurately predict what had happened during every moment since the start of time, and what would happen in every subsequent moment until the end of time. Which means that whatever happens in the universe was destined to happen (in a mathematical rather than theological sense).

reintegration of fact and value by opening each to the other in new ways." And to underline this linkage, Steenburg's paper wasn't published in some science or philosophy journal. It was in the *Harvard Theological Review*.[7]

So a bunch of thinkers find free will in the structure of chaoticism. Compatibilists and incompatibilists debate whether free will is possible in a deterministic world, but now you can skip the whole brouhaha because, according to them, chaoticism shows that the world isn't deterministic. As Eilenberger summarizes, "But since we now know that the slightest, immeasurably small differences in the initial state can lead to completely different final states (that is, decisions), physics cannot empirically prove the impossibility of free will."[8] In this view, the indeterminism of chaos means that, although it doesn't help you prove that there is free will, it lets you prove that you can't prove that there isn't.

But now to the critical mistake running through all of this: determinism and predictability are very different things. Even if chaoticism is unpredictable, *it is still deterministic.* The difference can be framed a lot of ways. One is that determinism allows you to explain why something happened, whereas predictability allows you to say what happens next. Another way is the woolly-haired contrast between ontology and epistemology; the former is about what is going on, an issue of determinism, while the latter is about what is knowable, an issue of predictability. Another is the difference between "determined" and "determinable" (giving rise to the heavy-duty title of one heavy-duty paper, "Determinism Is Ontic, Determinability Is Epistemic," by philosopher Harald Atmanspacher).[9]

Experts tear their hair out over how fans of "chaoticism = free will" fail to make these distinctions. "There is a persistent confusion about determinism and predictability," write physicists Sergio Caprara and Angelo Vulpiani. The first name–less philosopher G. M. K. Hunt of the University of Warwick writes, "In a world where perfectly accurate measurement is impossible, classical physical determinism does not entail epistemic determinism." The same thought comes from philosopher Mark Stone: "Chaotic systems, even though they are deterministic, are not predictable [they are not *epistemically* deterministic]. . . . To say that chaotic systems are unpredictable is not to say that science cannot explain them." Philosophers Vadim Batitsky and Zoltan Domotor, in their wonderfully titled paper, "When Good Theories Make Bad Predictions," describe chaotic systems as "deterministically unpredictable."[10]

Here's a way to think about this extremely important point. I just went back to that fantastic pattern in the last chapter, on page 138, and estimated that it is around 250 rows long and 400 columns wide. This means that the figure consists of about 100,000 boxes, each now either open or filled. Get a hefty piece of graph paper, copy the row 1 starting state from the figure, and then spend the next year sleeplessly applying rule 22 to each successive row, filling in the 100,000 boxes with your #2 pencil. And you will have generated the same exact pattern as in the figure. Take a deep breath and do it a second time, same outcome. Have a trained dolphin with an extraordinary capacity for repetition go at it, same result. Row eleventy-three would not be what it is because at row eleventy-two, you or the dolphin just happened to choose to let the open-or-filled split in the road depend on the spirit moving you or on what you think Greta Thunberg would do. That pattern was the outcome of a completely deterministic system consisting of the eight instructions comprising rule 22. At none of the 100,000 junctures could a different outcome have resulted (unless a random mistake occurred; as we'll see in chapter 10, constructing an edifice of free will on random hiccups is quite iffy). Just as the search for an uncaused neuron will prove fruitless, likewise for an uncaused box.

Let's frame this in the context of human behavior. It's 1922, and you're presented with a hundred young adults destined to live conventional lives. You're told that in about forty years, one of the hundred is going to diverge from that picture, becoming impulsive and socially inappropriate to a criminal extent. Here are blood samples from each of those people, check them out. And there's no way to predict which person is above chance levels.

It's 2022. Same cohort with, again, one person destined to go off the rails forty years hence. Again, here are their blood samples. This time, this century, you use them to sequence everyone's genome. You discover that one individual has a mutation in a gene called MAPT, which codes for something in the brain called the tau protein. And as a result, you can accurately predict that it will be that person, because by age sixty, he will be showing the symptoms of behavioral variant frontotemporal dementia.[11]

Back to the 1922 cohort. The person in question has started shoplifting, threatening strangers, urinating in public. Why did he behave that way? Because he chose to do so.

Year 2022's cohort, same unacceptable acts. Why will he have behaved that way? Because of a deterministic mutation in one gene.*

According to the logic of the thinkers just quoted, the 1922 person's behavior resulted from free will. Not "resulted from behavior we would *erroneously attribute* to free will." It *was* free will. And in 2022, it is *not* free will. In this view, "free will" is *what we call* the biology that we don't understand on a predictive level yet, and when we do understand it, it stops being free will. Not that it stops being mistaken for free will. It literally stops being. There is something wrong if an instance of free will exists only until there is a decrease in our ignorance. As the crucial point, our intuitions about free will certainly work that way, but free will itself can't.

We do something, carry out a behavior, and we *feel* like we've chosen, that there is a Me inside separate from all those neurons, that agency and volition dwell there. Our intuitions scream this, because we don't know about, can't imagine, the subterranean forces of our biological history that brought it about. It is a huge challenge to overcome those intuitions when you still have to wait for science to be able to predict that behavior precisely. But the temptation to equate chaoticism with free will shows just how much harder it is to overcome those intuitions when science will *never* be able to predict precisely the outcomes of a deterministic system.

---

*With a reminder from chapter 3 that it is very rare for a single gene to be deterministic in this way. To reiterate, almost all genes are about potential and vulnerability, rather than inevitability, interacting in nonlinear ways with environment and other genes.

WRONG CONCLUSION #2: A CAUSELESS FIRE

Most of the fascination with chaoticism comes from the fact that you can start with some simple deterministic rules for a system and produce something ornate and wildly unpredictable. We've now seen how mistaking this for indeterminism leads to a tragic downward spiral into a cauldron of free-will belief. Time now for the other problem.

Go back to the figure at the top of page 141 with its demonstration with rule 22 that two different starting states can turn into the identical pattern and thus, it is not possible to know *which* of those two was the actual source.

This is the phenomenon of convergence. It's a term frequently used in evolutionary biology. In this instance, it's not so much that you can't tell which of two different possible ancestors a particular species arose from (e.g., "Was the ancestor of elephants three-legged or five-legged? Who can tell?"). It's more when two very different sorts of species have converged on the same solution to the same sort of selective challenge.* Among analytical philosophers, the phenomenon is termed *overdetermination*—when two different pathways could each separately determine the progression to the same outcome. Implicit in this convergence is a loss of information. Plop down in some row in the middle of a cellular automaton, and not only can't you predict what is *going to* happen, but you can't know what *did* happen, which possible pathway led to the present state.

This issue of convergence has a surprising parallel in legal history.

*I've observed a great example of this. Near the equator in Kenya is Mount Kenya, the second-highest mountain in Africa, at more than seventeen thousand feet. Among the cool things about it, the climate is equatorial African at the base and glacial on top (at least it's glacial for a little while longer—melting fast), with completely different ecosystems every few thousand feet higher. There are some odd-looking plant species in the montane zone at about fifteen thousand feet. I was once chatting with a plant evolutionary biologist in his office, and there were some pictures of one of those plants. "Hey, nice, I see you've been up Mount Kenya," I said. "No, I took those in the Andes." The Andean plant was completely unrelated to the Kenyan one yet looked virtually the same. Apparently, there are only a few ways to be a high-altitude plant on the equator, and these very different plant species, on opposite sides of the globe, had converged on these solutions. Implicit in this is a great quote from Richard Dawkins: "However many ways there may be of being alive, it is certain that there are vastly more ways of being dead"—there's a very finite number of ways to be alive, with each living species having converged on one of them.

Thanks to negligence, a fire starts in building A. Nearby, completely unrelated, separate negligence gives rise to a fire in building B. The two fires spread toward each other and converge, burning down building C in the center. The owner of building C sues the other two owners. But which negligent person was responsible for the fire? Not me, each would argue in court—if my fire hadn't happened, building C would still have burned down. And it worked, in that neither owner would be held responsible. This was the state of things until 1927, when the courts ruled in *Kingston v. Chicago and NW Railroad* that it is possible to be partially responsible for what happened, for there to be fractions of guilt.[12]

Similarly, consider a group of soldiers lining up in a firing squad to kill someone. No matter how much one is pulling a trigger in glorious obedience to God and country, there's often some ambivalence, perhaps some guilt about mowing down someone or worry that fortunes will shift and you'll wind up in front of a firing squad. And for centuries, this gave rise to a cognitive manipulation—one soldier at random was given a blank rather than a real bullet. No one knew who had it, and thus every shooter knew that they might have gotten the blank and thus weren't actually a killer. When lethal injection machines were invented, some states stipulated that there'd be two separate delivery routes, each with a syringe full of poison. Two people would press each of two buttons, and a randomizer in the machine would infuse the poison from one syringe into the person and dump the contents of the other into a bucket. And not keep a record of which did which. Each person thus knew that they might not have been the executor. Those are nice psychological tricks for defusing a sense of responsibility.[13]

Chaoticism pulls for a related type of psychological trick. The feature of chaoticism where knowing a starting state doesn't allow you to predict what will happen is a crushing blow to classic reductionism. But the inability to ever know what happened in the past demolishes what's called *radical eliminative reductionism*, the ability to rule out every conceivable cause of something until you've gotten down to *the* cause.

So you can't do radical eliminative reductionism and decide what single thing caused the fire, which button presser delivered the poison, or what prior state gave rise to a particular chaotic pattern. But *that doesn't mean that the fire wasn't actually caused by anything, that no one shot the bullet-riddled prisoner, or that the chaotic state just popped up out of nowhere.* Ruling out radical eliminative reductionism doesn't prove indeterminism.

Obviously. But this is subtly what some free-will supporters conclude—if we can't tell what caused X, then you can't rule out an indeterminism that makes room for free will. As one prominent compatibilist writes, it is unlikely that reductionism will rule out the possibilities of free will, "because the chain of cause and effect contains breaks of the type that undermine radical reductionism and determinism, at least in the form required to undermine freedom." God help me that I've gotten to the point of examining the split hair of *and*, but chaotic convergence does not undermine radical reductionism *and* determinism. Just the former. And in the view of that writer, this supposed undermining of determinism is relevant to "policies upon which we hinge responsibility." Just because you can't tell which of two towers of turtles propping you up goes all the way down doesn't mean that you're floating in the air.[14]

CONCLUSION

Where have we gotten at this point? The crushing of knee-jerk reductionism, the demonstration that chaoticism shows just the opposite of chaos, the fact that there's less randomness than often assumed and, instead, unexpected structure and determinism—all of this is wonderful. Ditto for butterfly wings, the generation of patterns on sea shells, and Will Darling. But to get from there to free will requires that you mistake a failure of reductionism that makes it impossible to precisely describe the past or predict the future as proof of indeterminism. In the face of complicated things, our intuitions beg us to fill up what we don't understand, even can never understand, with mistaken attributions.

On to our next, related topic.

7

A Primer on Emergent Complexity

The previous two chapters can basically be distilled to the following:

—"Break it down to its component parts" reductionism doesn't work for understanding some vastly interesting things about us. Instead, in such chaotic systems, minuscule differences in starting states amplify enormously in their consequences.

—This nonlinearity makes for fundamental unpredictability, suggesting to many that there is an essentialism that defies reductive determinism, meaning that the "there can't be free will because the world is deterministic" stance goes down the drain.

—Nope. Unpredictable is not the same thing as undetermined; reductive determinism is not the only kind of determinism; chaotic systems are purely deterministic, shutting down that particular angle of proclaiming the existence of free will.

This chapter focuses on a related domain of amazingness that seems to defy determinism. Let's start with some bricks. Granting ourselves some artistic license, they can crawl around on tiny invisible legs. Place one brick in a field; it crawls around aimlessly. Two bricks, ditto. A bunch, and some start bumping in to each other. When that happens, they interact in boringly simple ways—they can settle down next to each other and stay that way, or one can crawl up on top of another. That's all. Now scatter a hundred zillion of these identical bricks in this field, and they slowly crawl around, zillions sitting next to each other, zillions crawling on top of others . . . and they slowly construct the Palace of Versailles. The amazingness is not that, wow, something as complicated as Versailles can be built out of simple bricks.* It's that once you made a big enough pile of bricks, all those witless little building blocks, operating with a few simple rules, without a human in sight, *assembled themselves* into Versailles.

This is not chaos's sensitive dependence on initial conditions, where these identical building blocks actually all differed when viewed at a high magnification, and you then butterflew to Versailles. Instead, put enough of the same simple elements together, and they spontaneously self-assemble into something flabbergastingly complex, ornate, adaptive, functional, and cool. With enough quantity, extraordinary quality just . . . emerges, often even unpredictably.†,[1]

As it turns out, such *emergent complexity* occurs in realms very pertinent to our interests. The vast difference between the pile of gormless, identical building blocks and the Versailles they turned themselves into seems to defy conventional cause and effect. Our sensible sides think (incorrectly . . .) of words like *indeterministic.* Our less rational sides think of words like *magic.* In either case, the "self" part of self-assembly seems so agentive, so rife with "be the palace of bricks that you wish to be," that dreams of free will beckon. An idea that this and the next chapter will try to dispel.

---

*Note to self: check to see if Versailles is made of bricks.

†This concept was invoked by chess grand master Garry Kasparov in 1996 when he famously lost a match to IBM's chess-playing computer, Deep Blue. Referring to the sheer power of the computer, arising from its ability to evaluate two hundred million positions on the board per second, he explained, "What I discovered yesterday was that we are now seeing for the first time what happens when quantity becomes quality" (B. Weber, "In Kasparov vs. Computer, the Chess Scorecard Is 1–1," *New York Times*, February 12, 1996). This principle was first stated by Hegel and greatly influenced Marx.

WHY WE'RE NOT TALKING ABOUT MICHAEL JACKSON MOONWALKING

Let's start with what *wouldn't* count as emergent complexity.

Put a beefy guy in a faux military uniform carrying a sousaphone in the middle of a field. His behavior is simple—he can walk forward, to the left, or to the right, and does so randomly. Scatter a bunch of other instrumentalists there, and the same thing happens, all randomly moving, collectively making no sense. But toss three hundred of them onto the field and out of that emerges a giant Michael Jackson moonwalking past the fifty-yard line during the halftime performance.*

There are all these interchangeable, fungible marching band marchers with the same minuscule repertoire of movements. Why doesn't this count as emergence? Because there's a master plan. Not inside the sousaphonist but in the visionary who fasted in the desert, hallucinating pillars of salt moonwalking, then returned to the marching band with the Good News. This is not emergence.

Here's real emergent complexity: Start with one *ant*. It wanders aimlessly on the field. As do ten of them. A hundred interact with vague hints of patterns. But put thousands of them together and they form a society with job specialization, construct bridges or rafts out of their bodies that float for weeks, build flood-proof underground nests with passageways paved with leaves, leading to specialized chambers with their own microclimates, some suited for farming fungi and others for brood rearing. A society that even alters its functions in response to changing environmental demands. No blueprint, no blueprint maker.[2]

What makes for emergent complexity?

> —There is a huge number of ant-like elements, all identical or coming in just a few different types.
>
> —The "ant" has a very small repertoire of things it can do.

---

*Check out Ohio State's marching band doing the Michael Jackson shtick at www.youtube.com/watch?v=RhVAga3GhNM.

—There are a few simple rules based on chance interactions with immediate neighbors (e.g., "walk with this pebble in your little ant mandibles until you bump into another ant holding a pebble, in which case, drop yours"). No ant knows more than these few rules, and each acts as an autonomous agent.

—Out of the hugely complicated phenomena this can produce emerge irreducible properties that exist only on the collective level (e.g., a single molecule of water cannot be wet; "wetness" emerges only from the collectivity of water molecules, and studying single water molecules can't predict much about wetness) and that are *self-contained* at their level of complexity (i.e., you can make accurate predictions about the behavior of the collective level without knowing much about the component parts). As summarized by Nobel laureate physicist Philip Anderson, "More is different."[*,3]

—These emergent properties are robust and resilient—a waterfall, for example, maintains consistent emergent features over time despite the fact that no water molecule participates in waterfall-ness more than once.[4]

—A detailed picture of the maturely emergent system can be (but is not necessarily) unpredictable, which should have echoes of the previous two chapters. Knowing the starting state and reproduction rules (à la cellular automata) gives you the means to *develop* the complexity but not the means to *describe* it. Or, to use a word offered by a leading developmental neurobiologist of the past century, Paul Weiss, the starting state can never contain an "itinerary."[†,5]

—Part of this unpredictability is due to the fact that in emergent systems, the road you are traveling on is being constructed at the same

---

*Anderson gives a wonderful example of this idea, quoting an exchange between F. Scott Fitzgerald and Ernest Hemingway: "Fitzgerald: The rich are different from us. Hemingway: Yes, they have more money. Everything else about rich-ness just emerges from that."

†Neurobiologist Robin Hiesinger, whose work will be covered later in the chapter, gives a wonderful example of this idea. You're learning a piece on the piano, and you make a mistake and grind to a halt. Rather than being able to resume two measures earlier, akin to resuming on the highway, most of us need to let the complexity unfold again—we go back to the beginning of the section.

time and, in fact, your being on it is influencing the construction process by constituting feedback on the road-making process.* Moreover, the goal you are traveling toward may not even exist yet—you are destined to interact with a target spot that may not exist yet but, with any luck, will be constructed in time. In addition, unlike last chapter's cellular automata, emergent systems are also subject to randomness (jargon: "stochastic events"), where the sequence of random events makes a difference.†

—Often the emergent properties can be breathtakingly adaptive and, despite that, there's no blueprint or blueprint maker.[6]

Here's a simple version of the adaptiveness: Two bees leave their hive, each flying randomly until finding a food source. They both do, with one source being better. Each returns to the hive, neither bee knowing anything about *both* food sources. Nonetheless, all the bees fly straight to the better site.

Here's a more complex example: An ant forages for food, checking eight different places. Little ant legs get tired, and ideally the ant visits each site only once, and in the shortest possible path of the 5,040 possible ones (i.e., seven factorial). This is a version of the famed "traveling salesman problem," which has kept mathematicians busy for centuries, fruitlessly searching for a general solution. One strategy for solving the problem is with brute force—examine *every* possible route, compare them all, and pick the best one. This takes a ton of work and computational power—by the time you're up to ten places to visit, there are more than 360,000 possible ways to do it, more than 80 billion with fifteen places to visit. Impossible. But take the roughly ten thousand ants in a typical colony, set them loose on the eight-feeding-site version, and they'll come up with something close

*The early-twentieth-century essayist Lu Xun captured the essence of this, writing, "The world has no road at the beginning; once enough people walked on it, the road appeared" (Liqun Luo, personal communication).

†For example, suppose you share a sequence of ten items, nine of which are roughly similar. There is one glaring exception, and your overall assessment of the properties of this sequence can change depending on whether randomness resulted in the exception being the second example you see or the tenth.

to the optimal solution out of the 5,040 possibilities in a fraction of the time it would take you to brute-force it, with no ant knowing anything more than the path that it took plus two rules (which we'll get to). This works so well that computer scientists can solve problems like this with "virtual ants," making use of what is now known as swarm intelligence.*[,7]

There's the same adaptiveness in the nervous system. Take a microscopic worm that neurobiologists love;† the wiring of its neurons shows close to traveling-salesman optimization, in terms of the cost of wiring them all up; same in the nervous system of flies. And in primate brains as well; examine the primate cortex, identify eleven different regions that wire up with each other. And of several million possible ways of doing it, the developing brain finds the optimal solution. As we'll see, in all these cases, this is accomplished with rules that are conceptually similar to what the traveling-salesmen ants do.[8]

Other types of adaptiveness also abound. A neuron "wants" to spread its array of thousands of dendritic branches as efficiently as possible for receiving inputs from other neurons, even competing with neighboring cells. Your circulatory system "wants" to spread its thousands of branching arteries as efficiently as possible in delivering blood to every cell in the body. A tree "wants" to branch skyward most efficiently to maximize the sunlight its leaves are exposed to. And as we'll see, all three solve the challenge with similar emergent rules.[9]

How can this be? Time to look at examples of how emergence actually emerges, using simple rules that work in similar ways in solving optimization challenges for, among other things, ants, slime molds, neurons, humans, and societies. This process will easily dispose of the first temptation: to decide that emergence demonstrates indeterminacy. Same answer as in

*Crossing my t's and dotting my i's: As I noted, the traveling salesman problem is formally unsolvable, in that it is not possible mathematically to prove or disprove that a particular solution is the most optimal. This is closely related to what are called "minimal spanning tree problems," where mathematical proofs are possible. The latter are relevant to things like telecommunication companies figuring out how to connect a bunch of transmission towers in a way that minimizes the total distance of cable needed.

†The worm, called *Caenorhabditis elegans*, is beloved because every worm has exactly 302 neurons, wired up in the same way in every worm. It's a dream for studying how neuronal circuits form.

the last chapter—unpredictable is not the same thing as undetermined. Disposing of the second temptation is going to be more challenging.

INFORMATIVE SCOUTS FOLLOWED BY RANDOM ENCOUNTERS

Many examples of emergence involve a motif that requires two simple phases. In the first, "scouts" in a population explore an environment; when they find some resource, they broadcast the news.* The broadcast must include information about the quality of the resource, such as better resources producing louder or longer signals. In the second phase, other individuals wander randomly in their environment with a simple rule regarding their response to the broadcast.

Back to honey bees as an example. Two bee scouts check out the neighborhood for possible food sources. They each find one, come back to the hive to report; they broadcast their news by way of the famed bee waggle dance, where the features of the dance communicate the direction and distance of the food. Crucially, the better the food source a scout found, the longer it carries out one part of the dance—this is how quality is being broadcast.† As the second phase, other bees wander about randomly in the hive, and if they bump into a dancing scout, they fly away to check out the food source the scout is broadcasting about . . . and then return to dance the news as well. And because a better potential site = longer dancing, it's more likely that one of those random bees bumps into the great-news bee than the good-news one. Which increases the odds that soon there will be two great-news dancers, then four, then eight . . . until the entire colony

*This is a very abstract, dimensionless sort of "environment," so that the likes of an ant leaving its nest to forage, a neuron extending a cable toward another one to form a connection, and someone doing an online search can be reduced to their similarities.

†The information contained in the waggle dance was first fully decoded by Karl von Frisch early in the twentieth century; the work was seminal to the founding of the field of ethology and won von Frisch a Nobel Prize in Physiology or Medicine, to the utter bafflement of most scientists—what do dancing bees have to do with physiology or medicine? A lot, as one point of this chapter.

converges on going to the optimal site. And the original good-news scout will have long since stopped dancing, bumped into a great-news dancer, and been recruited to the optimal solution. Note—*there is no decision-making bee that gets information about both sites, compares the two options, picks the better one, and leads everyone to it.* Instead, longer dancing recruits bees that will dance longer, and the comparison and optimal choice emerge implicitly; this is the essence of swarm intelligence.[10]

Similarly, suppose the two scout bees discover two potential sites that are equally good, but one is half as far from the hive as the other one. It will take the local-news bee half the time to get to and back from its food source that it takes the distant-news bee—meaning that the two, four, eight doubling starts sooner, exponentially swamping the signal of distant-news bee. Everyone soon heads to the closer source. Ants find the optimal site for a new colony this way. Scouts go out, and each finds a possible site; the better the site, the longer they stay there. Then the random wanderers spread out with the rule that if you bump into an ant standing at a possible site, maybe check the site out. Once again, better quality translates into a stronger recruitment signal, which becomes self-reinforcing. Work by my pioneering colleague Deborah Gordon shows an additional layer of adaptiveness. A system like this has various parameters—how far do ants wander, how much longer do you stay at a good site versus a mediocre one, and so on. She shows that these parameters vary in different ecosystems as a function of how abundant food sources are, how patchily they are distributed, and how costly foraging is (for example, foraging is more expensive, in terms of water loss, for desert ants than for forest ants); the better a colony has evolved to get these parameters just right for its particular environment, the more likely it is to survive and leave descendants.[*,†,11]

*Thus, colonies differ as to how evolutionarily "fit" they are at getting self-organized swarm intelligence just right. One paper exploring this has an all-time best title in a science journal: "Honeybee Colonies Achieve Fitness through Dancing." Presumably, this paper pops up regularly in Google searches for Zumba classes.

†This approach isn't perfect and can produce the wrong consensus decision. Ants living on a plain want a really good lookout from the top of a hill. There are two nearby hills, one twice the height of the other. Two scouts go out, each heads up their hill, and the one on the shorter hill gets there and starts broadcasting in half the time that it takes the loftier hill ant to start. Meaning it starts

The two steps of scout broadcasters followed by recruitment of random wanderers explains virtual ant traveling-salesman optimization. Place a bunch of ants at each of the virtual foraging sites; each ant then picks a route at random that involves visiting each site once, and leaves a pheromone trail in the process.* How does better quality translate into a stronger broadcast? The shorter the route, the thicker the pheromone trail that is laid down by a scout; pheromones evaporate, and thus shorter, thicker pheromone trails last longer. A second generation of ants shows up; they wander randomly, with the rule that if they encounter a pheromone trail, they join it, adding their own pheromones. As a result, the thicker and therefore longer-lasting the trail, the more likely another ant is to join it and amplify its recruiting message. And soon the less efficient routes for connecting the sites evaporate away, leaving the optimized solution. No need to gather data about the length of every possible route and have a centralized authority compare them and then direct everyone to the best

---

the recruitment doubling earlier than the other ant, and soon the colony has chosen . . . the shorter hill. In this case, the problem arises because the strength of the recruitment signal is inversely correlated with the quality of the resource. Sometimes a process can be completely out of whack. There are all sorts of cases of machine-learning algorithms that come up with a bizarre solution to a problem because the programmer underspecified the instructions, not informing it of all the things it was not allowed to do, what information it was not supposed to pay attention to, and so on. For example, one AI seemingly learned to diagnose melanomas but learned instead that lesions photographed with a ruler next to them are likely to be malignant. In another case, an algorithm was designed to evolve a simulated organism that was very fast; the AI simply grew an organism that was incredibly tall and thus reached high velocities when it would fling itself over. In another, the AI was supposed to design a Roomba that could move around without bumping into things—as assessed by its bumper being hit—and learned instead to simply stagger around leading with its back, where there was no bumper. For more examples, see: "Specification Gaming Examples in AI—Master List: Sheet1," docs.google.com/spreadsheets/d/e/2PACX-1vRPiprOaC3HsCf5Tuum8bRfzYUiKLRqJmbOoC-32JorNdfyTiRRsR7Ea5ezWtvsWzuxo8bjOxCG84dAg/pubhtml.

*Pheromones are chemical signals released into the air—odorants—that carry information; in the case of ants, they have glands for this particular pheromone in their rears, which they dip down to the ground, leaving a trail of droplets of the stuff. So these virtual ants are leaving virtual pheromones. If there is a constant amount of pheromone in the gland at the beginning, the shorter the total walk, the thicker the amount of pheromone that gets laid down per unit of distance.

solution. Instead, something that comes close to the optimal solution emerges on its own.*

(Something worth pointing out: As we'll see, these rich-get-richer recruitment algorithms explain optimized behavior in us as well, along with other species. But "optimal" is not meant in the value-laden sense of "good." Just consider rich-get-richer scenarios where, thanks to the recruitment signaling of economic inequality, it's literally the rich who get richer.)

Next we turn to how emergence helps slime molds solve problems.

Slime molds are these slimy, moldy, fungal, amoeboid, single-cell protists, just to make a bunch of taxonomic errors, that grow and spread like a carpet over surfaces, looking for microorganisms to eat.

In a slime mold, zillions of single-cell amoebas have joined forces by merging into a giant, cooperative single cell that oozes over surfaces in search of food, apparently an efficient food-hunting strategy† (and as a hint of the emergence pending, a single, independent slime mold cell can no more ooze than a molecule of water can be wet). What used to be the individual cells are interconnected by tubules that can stretch or contract, depending on the direction of oozing (see figure on the next page).

Out of these collectivities emerge problem-solving capabilities. Spritz a dollop of slime mold into a little plastic well that leads to two corridors, one with an oat flake at the end, the other with two oat flakes (beloved by

---

*This search algorithm was first proposed by the AI researcher Marco Dorigo in 1992, giving rise to "ant colony optimization" strategies, with virtual ants, in computer science. This is such a beautiful example of quantity producing quality; when I first grasped it, I felt dizzy with its elegance. And as a result, the quality of this approach is reflected in the loudness of my broadcasting about it—I drone on about this more frequently in lectures than about less cool subjects, making it more likely that my students will grasp it and tell their parents about it at Thanksgiving, increasing the odds that parents will tell neighbors, clergy, and elected representatives about it, leading to the optimized emergent behavior of everyone naming their next child Dorigo.

Note that, as stated, this is an ideal way to get close to the optimal solution. If you require the optimal solution, you're going to need to brute-force it with a slow and expensive centralized comparator. Moreover, ants and bees obviously don't follow these algorithms precisely, as individual differences and chance creep in.

†As a dichotomy, in cellular slime mold species, the collective forms only temporarily; in plasmodial slime molds, it's permanent.

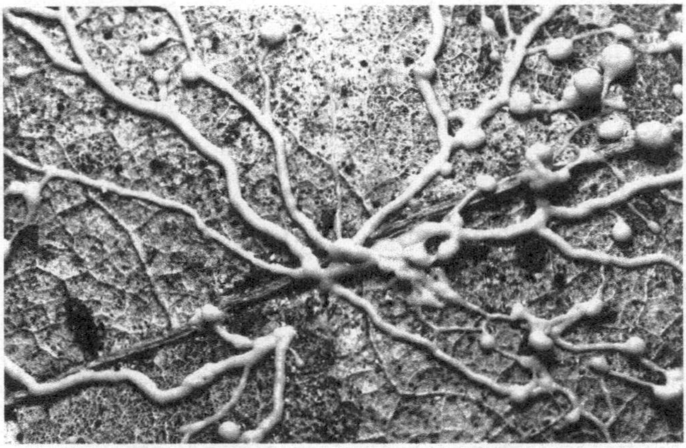

slime molds). Rather than sending out scouts, the entire slime mold expands to fill both corridors, reaching both food sources. And within a few hours, the slime mold retracts from the one–oat flake corridor and accumulates around the two oats. Have two pathways of differing lengths leading to the same food source; the slime mold initially fills both paths but eventually takes only the shortest route. Same with a maze with multiple routes and dead ends.[*,12]

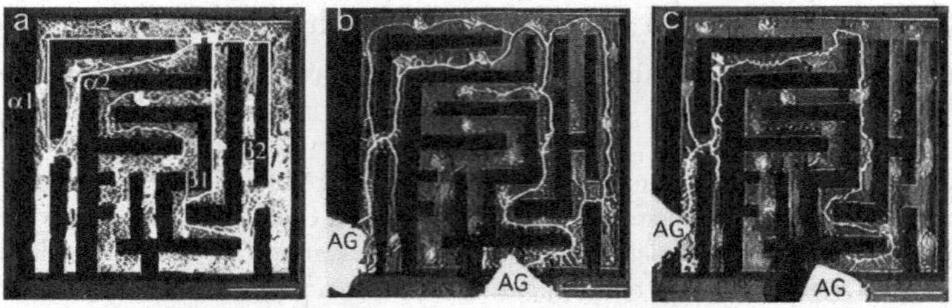

Initially, the slime mold fills every path (panel a); it then begins retracting from superfluous paths (panel b), until eventually reaching the optimal solution (panel c). (Ignore the various markings.)

As the tour de force of slime mold intelligence, Atsushi Tero at Hokkaido University plopped a slime mold down into a strangely shaped

*Raising the question of when the optimized behavior of those ex–individual cells constitute "intelligence," in the same way that the optimized function of vast numbers of neurons can constitute an intelligent person.

walled-off area with oat flakes at very specific locations. Initially, the mold expanded, forming tubules connecting all the food sources to each other in multiple ways. Eventually, most tubules retracted, leaving something close to the shortest total path length of tubules connecting food sources. The Traveling Slime Mold. Here's the thing that makes the audience shout for more—the wall outlines the coastline around Tokyo; the slime was plopped onto where Tokyo would be, and the oat flakes corresponded to the suburban train stations situated around Tokyo. And out of the slime mold emerged a pattern of tubule linkages that was statistically similar to the actual train lines linking those stations. A slime mold without a neuron to its name, versus teams of urban planners.[13]

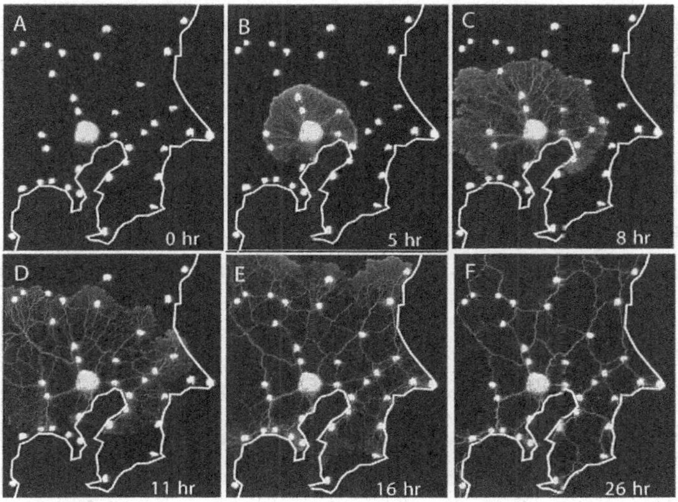

How do slime molds pull this off? A lot like ants and bees. Take the two corridors leading to either one or two oat flakes. The slime mold initially oozes into both corridors, and when food is found, tubules contract in the direction of the food, pulling the rest of the slime mold toward it. Crucially, the better the food source, the greater the contractile force generated on the tubules. Then the tubules a bit farther away dissipate the force by contracting in the same orientation, increasing the force of contraction, spreading outward until the whole slime mold has been pulled into the optimal pathway. No part of the slime mold compares the two options and makes a decision. Instead, the slime mold extensions into the two corridors act as scouts, with the better route broadcast in a way that causes rich-get-richer recruiting via mechanical forces.[14]

Now let's consider a growing neuron. It extends a projection that has branched into two scout arms ("growth cones") heading toward two neurons. Simplifying brain development to a single mechanism, each target neuron is attracting the growth cone by secreting a gradient of "attractant" molecules. One target is "better," thus secreting more of the attractant, resulting in a growth cone reaching it first—which causes a tubule inside that growing neuron's projection to bend in that direction, to be attracted to that direction. Which makes the parallel tubule adjacent to it more likely to do the same. Which increases the mechanical forces recruiting more and more of these tubules. The other scout arm is retracted, and our growing neuron has connected up with the better target.[*,15]

Let's look at our ant / bee / slime mold motif as applied to the developing brain forming the cortex, the fanciest, most recently evolved part of the brain.

The cortex is a six-layer-thick blanket over the surface of the brain, and cut into cross section, each layer consists of different types of neurons (see figure on the next page).

The multilayered architecture has lots to do with cortical function. In the picture, think of that slab of cortex as being divided into six vertical

---

*To *vastly* simplify things, the two growth cones have receptors on their surfaces for the attractant molecule. As those receptors fill up with the attractant, a different type of attractant molecule is released within the growth-cone branch, forming a gradient down to the trunk that pulls the tubules toward that branch. More extracellular attractant broadcasting, by way of more receptors filled, and more of an intracellular broadcast signal recruiting tubules. As one complexity in real nervous systems, different target neurons might be secreting *different* attractant molecules, making it possible to be broadcasting qualitative as well as quantitative information. As another complexity, sometimes a growth cone has a specific address in mind for the neuron it wants to connect with. In contrast, sometimes there's relative positional coding, where neuron A wants to connect with the target neuron that is adjacent to the target neuron that has connected up with the neuron next to neuron A. Implicit in all this is that the growth cones are secreting signals that repulse each other, so that the scouts scout different areas. I thank my departmental colleagues Liqun Luo and Robin Hiesinger, two pioneer scouts in this field, for generous and helpful discussions about this topic.

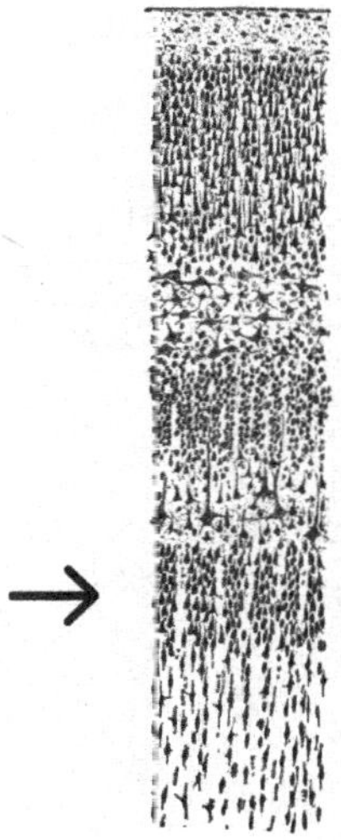

columns (best seen as the six dense clusters of neurons at the level of the arrow). The neurons within any of these mini columns send lots of vertical projections (i.e., axons) to each other, collectively working as a unit; for example, in the visual cortex, one mini column might decode the meaning of light falling on one spot of the retina, with the mini column next to it decoding light on an adjacent spot.*

It's ants redux in building a cortex. The first step in cortical development is when a layer of cells at the bottom of each cross section of cortex sends long, straight projections to the surface, serving as vertical scaffolding. These are our ant scouts, called radial glia (ignore the letters in the diagram on the next page). There is initially an excess of them, and the ones that have blazed the less optimal, less direct paths are eliminated (through a controlled type of cell death). As such, we have our first genera-

*As an aside, there are also horizontal connections within the same layer between different mini columns. This produces a thoroughly cool piece of circuitry. Consider a cortical mini column responding to light stimulating a small patch of retina. As just noted, the mini columns surrounding it respond to light stimulating patches on either side of that first patch. As a great circuitry trick, when a mini column is being stimulated, it uses its horizontal projections to silence the surrounding mini columns. Result? An image that is sharper around its edges, a phenomenon called lateral inhibition. This stuff is the best.

tion of explorers, with the ones with the more optimal solution to cortex building persisting longer.[16]

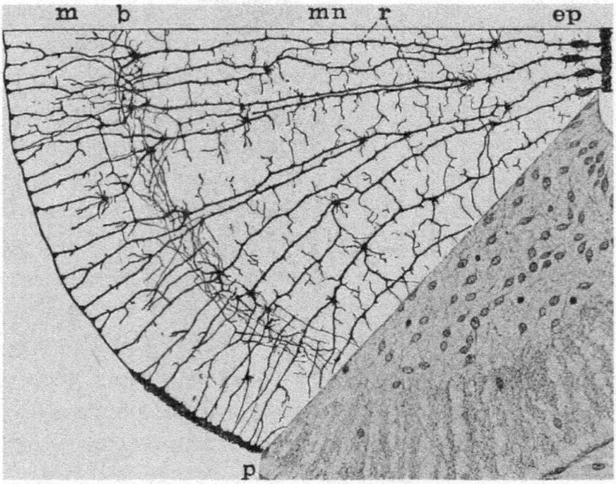

Radial glia radiating outward from the center of a cross section

You know what's coming next. Newly born neurons wander randomly at the base of the cortex until they bump into a radial glia. They then migrate upward along the glial guide rail, leaving behind chemoattractant signals that recruit more newbies to join the soon-to-be mini column.[*,17]

Scouts, quality-dependent broadcasting, and rich-get-richer recruiting, from insects and slime molds to your brain. All without a master plan, or constituent parts knowing anything beyond their immediate neighborhood, or any component comparing options and choosing the best one. With remarkable prescience about these ideas in 1874, the biologist Thomas Huxley wrote about the mechanistic nature of organisms,

---

*As each new neuron arrives at the scene, it forms its synapses in sequence, one at a time, which is a way for a neuron to keep track of whether it has made the desired number of synapses. Inevitably, among the various growth cones spreading outward, looking for dendritic targets to start forming a synapse, one growth cone will have more of a "seeding" growth factor than the others, just by chance. Lots of the seeding factor causes the growth cone to recruit even more of the seeding factor and to suppress the process in neighboring growth cones. This rich-get-richer scenario results in one synapse forming at a time.

such that they "only simulate intelligence as a bee simulates a mathematician."[18]

Time for another motif in emergent systems.

FITTING INFINITELY LARGE THINGS INTO INFINITELY SMALL SPACES

Consider the figure below. The top row consists of a single straight line. Remove its middle third, producing the two lines that constitute the second row; the length of those two together is two thirds the length of the original line. Remove the middle third from each of those, producing four lines that, collectively, are four ninths the total length of the original line. Do this forever, and you generate something that seems impossible—an infinitely large number of specks that have an infinitely short cumulative length.

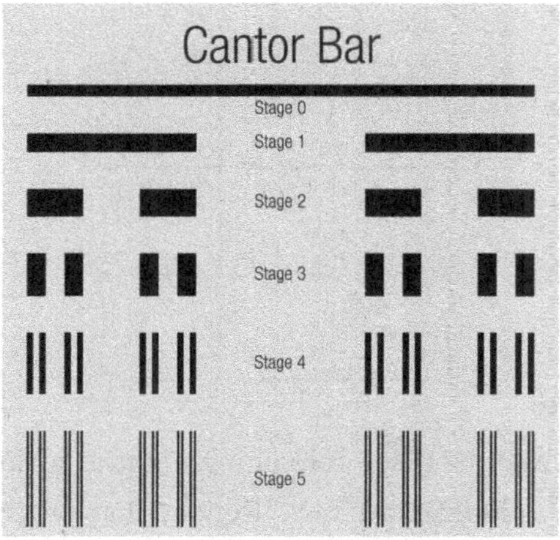

Let's do the same thing in two dimensions (below). Take an equilateral triangle (#1). Generate another equilateral triangle on each face, using the middle third as the base for the new triangle, resulting in a six-pointed star (#2). Do the same to each of those points, producing an eighteen-pointed star (#3), then a fifty-four-pointed star (#4), over and over. Do this forever and you'll generate a two-dimensional version of the same impossibility, namely a shape whose increase in area from one iteration to the next is infinitely small, while its perimeter is infinitely long:

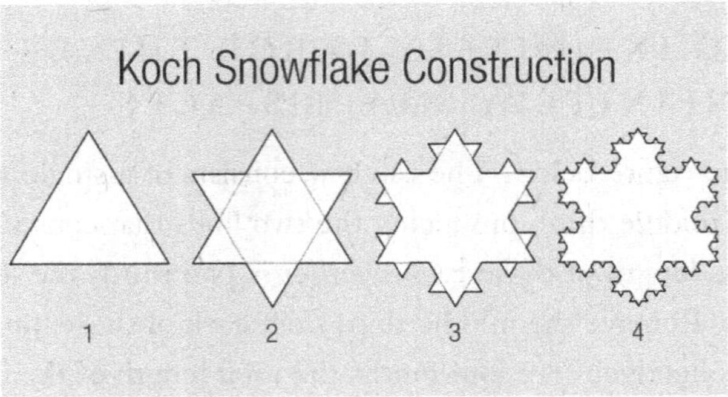

Now three dimensions. Take a cube. Each of its faces can be thought of as being a three-by-three grid of nine boxes. Take out the middle-most of those nine boxes, leaving eight:

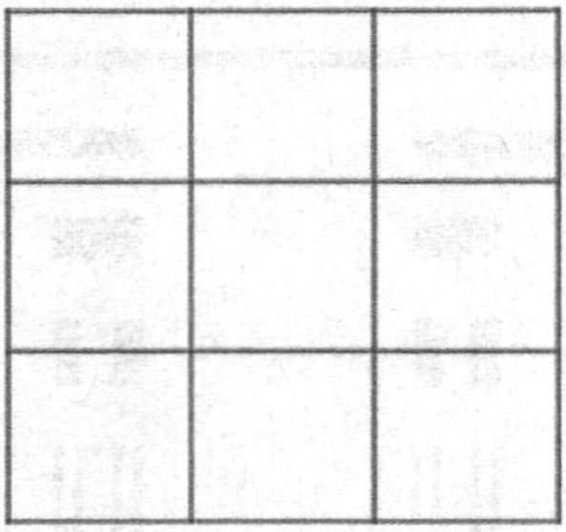

Now think of each of those remaining eight as a three-by-three grid, and take out the middle-most box. Repeat that process forever, on all six faces of the cube. And the impossibility achieved when you reach infinity is a cube with infinitely small volume but infinitely large surface area (see figure on the next page).

These are, respectively, called a Cantor set, a Koch snowflake, and a Menger sponge. These are mainstays of fractal geometry, where you

iterate the same operation over and over, eventually producing something impossible in traditional geometry.[19]

Which helps explain something about your circulatory system. Each cell in your body is at most only a few cells away from a capillary, and the circulatory system accomplishes this by growing around forty-eight thousand miles of capillaries in an adult. Yet that ridiculously large number of miles takes up only about 3 percent of the volume of your body. From the perspective of real bodies in the real world, this begins to approach the circulatory system being everywhere, infinitely present, while taking up an infinitely small amount of space.[20]

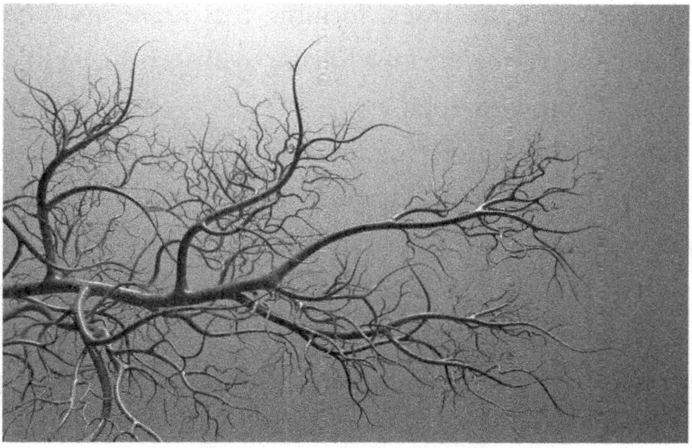

Branching patterns in capillary beds

A neuron has a similar challenge, in that it wants to send out a tangle of dendritic branches that can accommodate inputs at ten thousand to fifty thousand synapses, all with the dendritic "tree" taking up as little space as possible and costing as little as possible to construct:

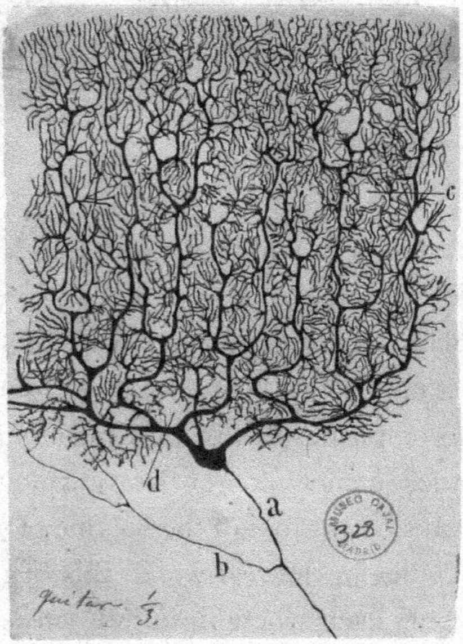

A classic textbook drawing of an actual neuron

And of course, there are trees, forming real branches to generate the maximal amount of surface area for foliage to absorb sunlight, while minimizing the costs of growing it all.

The similarities and underlying mechanisms would be obvious to Cantor, Koch, or Menger,* namely iterative bifurcation—something grows a distance and splits in two; those two branches grow some distance and each splits in two; those four branches . . . over and over, going from the aorta down to forty-eight thousand miles of capillaries, from the first dendritic branch in a neuron to two hundred thousand dendritic spines, from a tree trunk to something like fifty thousand leafy branch tips.

How are bifurcating structures like these generated in biological sys-

*Lest we get overly familiar, that's Georg Cantor, nineteenth-century German mathematician; Helge von Koch, turn-of-the-century Swedish mathematician; and Karl Menger, twentieth-century Austrian American mathematician.

tems, on scales ranging from a single cell to a massive tree? Well, I'll tell you one way it *doesn't* happen, which is to have specific instructions for each bifurcation. In order to generate a bifurcating tree with 16 branch tips, you have to generate 15 separate branching events. For 64 tips, 63 branchings. For 10,000 dendritic spines in a neuron, 9,999 branchings. You can't have one gene dedicated to overseeing each of those branching events, because you'll run out of genes (we only have about twenty thousand). Moreover, as pointed out by Hiesinger, building a structure this way requires a blueprint as complicated as the structure itself, raising the turtles question: How is the blueprint generated, and how is the blueprint that generated that blueprint generated . . . ? And it's these sorts of problems writ large and larger for the circulatory system and for actual trees.

Instead, you need instructions that work the same way at every scale of magnification. Scale-free instructions like this:

> **Step #1.** Start with a tube of diameter Z (a tube because geometrically, a blood vessel branch, a dendritic branch, and a tree branch can all be thought of that way).
>
> **Step #2.** Extend that tube until it is, to pull a number out of a hat, four times longer than its diameter (i.e., 4Z).
>
> **Step #3.** At that point, the tube bifurcates, splits in two. Repeat.

This produces two tubes, each with a diameter of 1/2Z. And when those two tubes are four times longer than that diameter (i.e., 2Z), they split in two, producing four branches, each 1/4Z diameter, which will split in two when each is 1Z (see figure on the following page).

While a mature tree sure seems immensely complex, the idealized coding for it can be compressed into three instructions requiring only a handful of genes to pull this off, rather than half your genome.* You can even have the effects of those genes interact with the environment. Say you're a

*With it being likely that dendrites, blood vessels, and trees would differ as to how many multiples of the diameter branches grow before splitting.

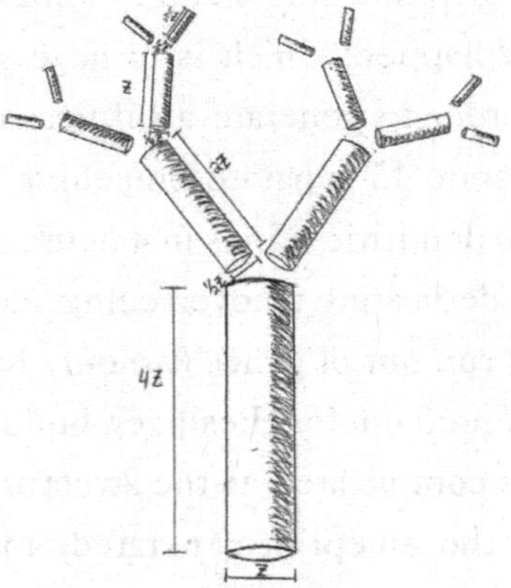

fetus inside someone living at high altitude, with low levels of oxygen in the air and thus in your fetal circulation. This triggers an epigenetic change (back to chapter 3) so that tubes in your circulation grow only 3.9 times the width, instead of 4.0, before splitting. This will produce a bushier spread of capillaries (I'm not sure if that would solve the high-altitude problem—I'm making this up).*

So you can do this with just a handful of genes that can even interact with the environment. But let's turn this into the reality of real biological tubes and what genes actually do. How can your genes code for something abstract like "grow four times the diameter and then split, regardless of scale"?

Various models have been proposed; here's a totally beautiful one. Let's consider a fetal neuron that is about to generate a bifurcating tree of dendrites (although this could be any of the other bifurcating systems we've been covering). We start with a stretch of the neuron's surface membrane that is destined to be where the tree starts growing (see figure below, left). Note that in this very artificial version, the membrane is made of two layers, and in between the layers is some Growth Stuff (hatched), coded

*Lurking in here is the need for a fourth rule, namely to know when to stop the bifurcating. With neurons, or the circulatory or pulmonary systems, it's when cells reach their targets. With growing, branching trees . . . I don't know.

for by a gene. The Growth Stuff triggers the area of the neuron just below to start constructing a trunk that will rise from there (right):[21]

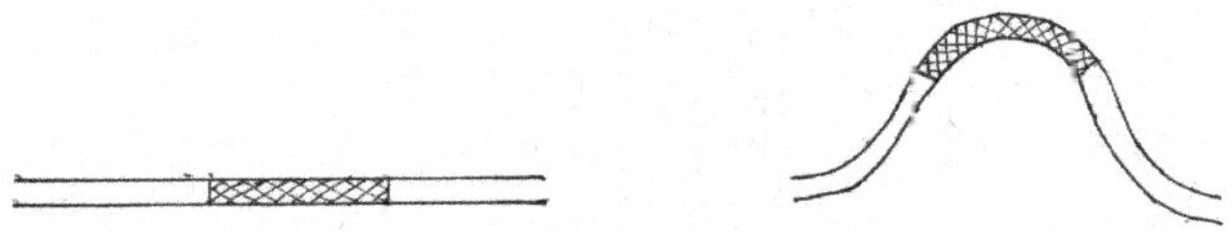

How much Growth Stuff was there at the beginning? 4Zs' worth, which will make the trunk grow 4Z in length before stopping. Why does it stop? Critically, the inner layer of the growing front of the neuron grows a little faster than the outer layer, such that right around a length of 4Z, the inner layer touches the outer layer, splitting the pool of Growth Stuff in half. No more Growth Stuff in the tip; things stop at 4Z. But crucially, there's now 2Zs' worth of Growth Stuff pooled on each side of the tip of the trunk (left). Which triggers the area underneath to start growing (right):

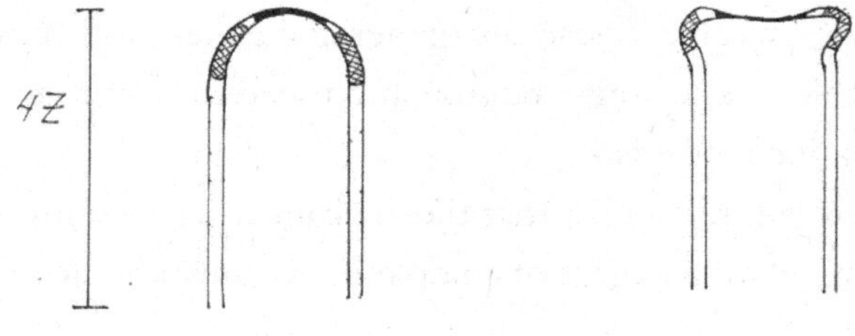

Because these two branches are narrower, the inner layers touch the outer layers after a length of only 2Z (below left), which splits the Growth Stuff into four pools, each with 1Z's worth. And so on (below right).*[,22]

*Chapter 10 will cover where randomness comes into biology, in this case in the form of the Growth Stuff not splitting *exactly* in half (i.e., 50 percent of the molecules going each way) every time. Those small differences mean that there can be some variability tolerated in a bifurcating system; in other words, the real world is messier than these beautiful, clear models. As emphasized by Hungarian biologist Aristid Lindenmayer, this is why everyone's brains (or neurons, or

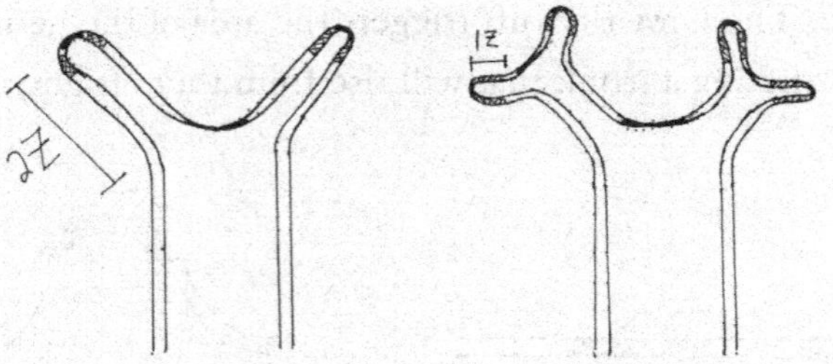

The key to this "diffusion-based geometry" model is the speed of growth of the two layers differing. Conceptually, the outer layer is about growing, the inner about stopping growing. Numerous other models produce bifurcations just as emergently, with similar themes.* Wonderfully, two genes, coding for molecules with growth and stopping-growth properties, respectively, have been identified that are central to bifurcation in the developing lung.[†,23]

And the intensely cool thing is that these very different physiological systems—neurons, blood vessels, the pulmonary system, and lymph nodes—use some of the *same* genes, coding for the same proteins in the construction process (a menagerie of proteins such as VEGF, ephrins, netrins, and semaphorins). These are not genes used for, say, generating the circulatory system. These are genes for generating bifurcating systems, applicable to one single neuron and to vascular and pulmonary systems using billions of cells.[24]

Aficionados will recognize that these bifurcating systems all form fractals, where the relative degree of complexity is constant, no matter at what

---

circulatory system . . .) look similar but are never identical (even in identical twins). This is symbolically represented by the asymmetry in the final drawing of the 1Z level (which wasn't what I planned, but which I messed up while drawing it).

*One model is called a Turing mechanism, named after Alan Turing, one of the founders of computer science and the source of the Turing test and Turing machines. When he wasn't busy accomplishing all that, Turing generated the math showing how patterns (e.g., bifurcations in neurons, spots in leopards, stripes in zebras, fingerprints in us) can be generated emergently with a small number of simple rules. He first theorized about this in 1952; it then took a mere sixty years for biologists to prove that his model was correct.

†A recent study has shown that two genes pretty much account for the branching pattern in Romanesco cauliflower. If you don't know what one of those looks like, stop reading right now and go Google a picture of it.

scale of magnification you are considering the system (with the recognition that unlike the fractals of mathematics, fractals in the body don't bifurcate forever—physical reality asserts itself at some point). We're now in very strange terrain, having to consider the molecules of the sort mentioned in the previous paragraph being coded for by "fractal genes." Which means that there must be fractal mutations, disrupting normal branching in everything from single neurons to entire organ systems; there are some hints of these out there.[25]

These principles apply to nonbiological complexity as well—for example, why rivers emptying into the sea bifurcate into river deltas. And it even applies to cultures. Let's consider one last emergent bifurcating tree, one that shows either the deeply abstract ubiquity of the phenomenon or how I'm running too far with a metaphor.

Look at the intensely bifurcated diagram below; don't worry about what the branch tips are—just note the branchings all over the place.

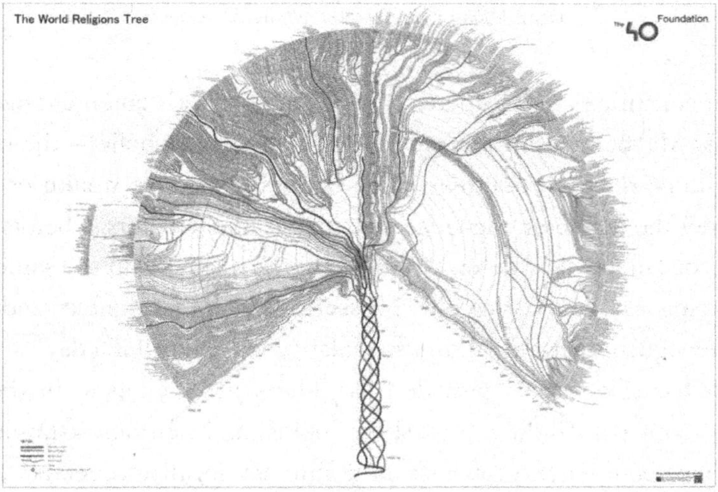

What is this tree? The perimeter represents the present. Each ring represents one hundred years back into the past, reaching the year 0 AD at the center, with a trunk going back millennia from there. And the branching pattern? The history of the emergence of earth's religions—a mass of bifurcations, trifurcations, dead-end side branches, and so on. A partial magnification:[26]

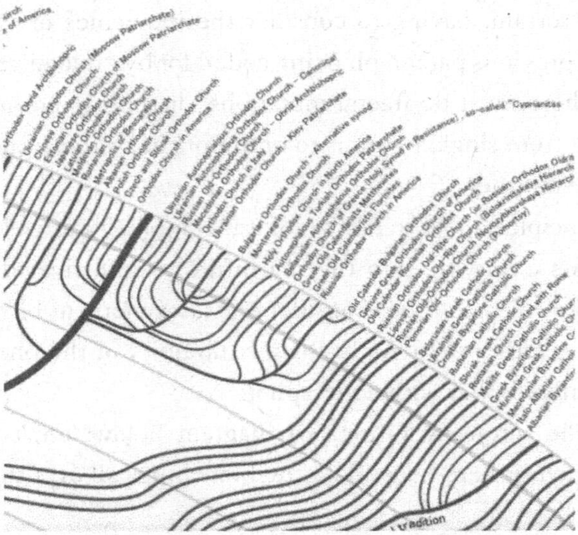

One tiny piece of the history of religious branching

What constitutes the diameter of each "tube" in this emergent history of religions? Maybe measures of the intensity of religious belief—the number of adherents, their cultural homogeneity, their collective wealth or power. The wider the diameter, the longer the tube is likely to persist before destabilizing, but in a scale-free way.* Would this be adaptive, in the same sense as analyzing, say, bifurcating blood vessels? I think that right around now, I should recognize that I'm on thin speculative ice and call it a day.

What has this section provided us? The same themes as in the prior section about pathfinding ants, slime molds, and neurons—simple rules about how components of a system interact locally, repeated a huge

*With historical events providing some of that instability. Think of Martin Luther getting fed up with the corruption of Rome, leading to the Catholic/Protestant schism; a disagreement as to whether Abu Bakr or Ali should be Muhammad's successor, resulting in Sunnis and Shi'ites going their separate Islamic ways; Central European Jews being allowed to assimilate into Christian society, in contrast to Eastern European Jews, giving rise to the former's more secular Reform Judaism.

number of times with huge numbers of those components, and out emerges optimized complexity. All without centralized authorities comparing the options and making freely chosen decisions.*

LET'S DESIGN A TOWN

You're on the planning board for a new town, and after endless meetings, you've collectively decided where it will be built, how big it will be. You've laid out a grid of the streets, decided on locations for the schools, hospitals, and bowling alleys. Time now to figure out where the stores will go.

The Stores Committee first proposes that stores be randomly scattered throughout town. Uh, that's not ideal; people want stores conveniently clustered. Right, says the committee, and then proposes that all the stores be in a single cluster in the middle of town.

Uh, not quite right either. With this single cluster, there won't be convenient parking, and the stores in the center of this megamall will be so inaccessible that they'll go out of business—they'll die from some commercial equivalent of insufficient oxygen.

Next plan—have six malls of the same size, set equal distances from each other. That's good, but someone notices that all dozen coffee shops are in the same mall; these shops will drive each other out of business, while five malls will have no coffee shops.

Back to planning, paying attention now not just to "store-ness" but to the type of store. In each mall, one pharmacy, one market, two coffee shops. Consider interactions between different types of stores. Separate the candy shop and the dentist. The optometrist goes next to the bookstore. Get the correct ratio of places for sinning—a gelato shop, a bar—to those for repenting—a fitness center, a church. And whatever you do,

*A reminder, once again, that the real world of cells and bodies isn't as clean as these highly idealized models.

don't put the store selling "God Bless America" sweatshirts next to the store selling "God-Less America" ones.

Once that is implemented, there's one last step, which is building major thoroughfares that connect the malls to each other.

At last, the commercial districts in your town are planned, after all these urban *planning* meetings filled with individuals with differing expertise, careerism, personal agendas, cooperation taking a hit because one person resents another for taking the last doughnut.

Take a beaker full of neurons. They're newly born, so no axons or dendrites yet, just rounded-up little cells destined for glory. Pour the contents into a petri dish filled with a soup of nutrients that keep neurons happy. The cells are now randomly scattered everywhere. Go away for a few days, come back, look at those neurons under a microscope, and this is what you see:

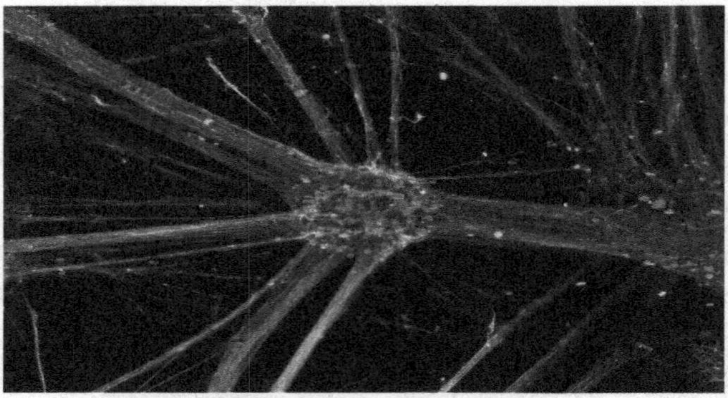

A bunch of neurons in a mall, er, I mean clumped together; to the far right is the start of another cluster of cell bodies, with major thoroughfares of projections linking the two, as well as to distant clusters outside the picture.

No committee, no planning, no experts, no choices freely taken. Just the same pattern as for the planned town, emerging from some simple rules:

—Each neuron that has been thrown randomly into the soup secretes a chemoattractant signal; they're all trying to get the others to migrate to them. Two neurons happen to be closer than average to each other by chance, and they wind up being the first pair to be clumped together in their neighborhood. This doubles the power of the attractant signal emanating from there, making it more likely that they'll attract a third neuron, then a fourth . . . Thus, through a rich-get-richer scenario, this forms a nidus, the starting point of a local cluster growing outward. Growing aggregates like these are scattered throughout the neighborhood.

—Each clump of neurons reaches a certain size, at which point the chemoattractant stops working. How would that work? Here's one mechanism—as a ball of clumping neurons gets bigger, the ones in the center are getting less oxygen, triggering them to start secreting a molecule that inactivates chemoattractant molecules.

—All along, neurons have been secreting a second type of attractant signal in minuscule amounts. It's only when enough neurons have migrated into an optimally sized cluster that there is collectively enough of the stuff to prompt the neurons in the cluster to start forming dendrites, axons, and synapses with each other.

—Once this local network is wired up (detectable by, say, a certain density of synapses), a chemo*repellent* is secreted, which now causes neurons to stop making connections to their neighbors, and to instead start sending long projections to other clusters, following a chemoattractant gradient to get there, forming the thoroughfares between clusters.*

This is a motif of how complex, adaptive systems, like neuronal shopping malls, can emerge thanks to control over space and time of attractant

*And there's an additional level of rules like these with different attractant and repellent signals that sculpt what *types* of neurons wind up in each cluster, rules like "Only two coffee shops per mall."

and repellent signals. This is the fundamental yin/yang polarity of chemistry and biology—magnets attracting or repelling each other, positively charged or negatively charged ions, amino acids attracted to or repelled by water.* Long strings of amino acids form proteins, each with a distinctive shape (and therefore function) that represents the most stable formation for balancing the various attraction and repulsion forces.†

As just shown, constructing neuronal shopping malls in the developing brain entailed two different types of attractant signals and one repellent one. And things get fancier: Have a variety of attractant and repellent signals that work individually or in combinations. Have emergent rules for which part of a neuron a growing neuron forms a connection with. Have growth cones with receptors that respond to only a subset of attractant or repellent signals. Have an attractant signal pulling a growth cone toward it; however, when it gets close, the attractant starts working as a repellent; as a result, the growth cone swoops past—it's how neurons make long-distance projections, doing flybys of one signpost after another.[27]

Most neurobiologists spend their time figuring out minutiae like, say, the structure of a particular receptor for a particular attractant signal. And then there are those marching superbly to their own drummer, like Robin Hiesinger, quoted earlier, who studies how brains develop with simple, emergent informational rules like we've been looking at. Hiesinger, whose review papers have puckish section titles like "The Simple Rules That Can," has shown things like the three simple rules needed for neurons in the eye of a fly to wire up correctly. Simple rules about the duality of attraction and repulsion, and no blueprints.‡ Time now for one last style of emergent patterning.[28]

*What are termed hydrophobic or hydrophilic amino acids—whether the amino acid is attracted to or repelled by water. I once heard a scientist mention in passing how she didn't like to swim, referring to herself as being hydrophobic.

†Think biochemistry's equivalent of domes being most stable for the smallest cost when geodesic.

‡How do those various neurons know, say, *which* attractant or repellent signals to secrete and when to do it? Thanks to other emergent rules that came earlier, and earlier before that, and . . . turtles.

TALK LOCALLY, BUT DON'T FORGET TO ALSO TALK GLOBALLY NOW AND THEN

Suppose you live in a thoroughly odd community. There is a total of 101 people in it, each in their own house. The houses are arranged in a straight line, say, along a river. You live in the first house of this 101-house-long line; how often do you interact with each of your 100 neighbors?

There are all sorts of potential ways. Maybe you talk only to your next-door neighbor (figure A). Maybe, as a contrarian, you interact only with the neighbor the farthest from you (figure B). Maybe the same amount with each person (figure C), maybe randomly (figure D). Maybe you interact the most with your immediate neighbor, X percent less with the neighbor after that, and X percent of that less with the neighbor after that, decreasing at a constant rate (figure E).

Then there's a particularly interesting distribution where around 80 percent of your interactions occur with the twenty closest neighbors and

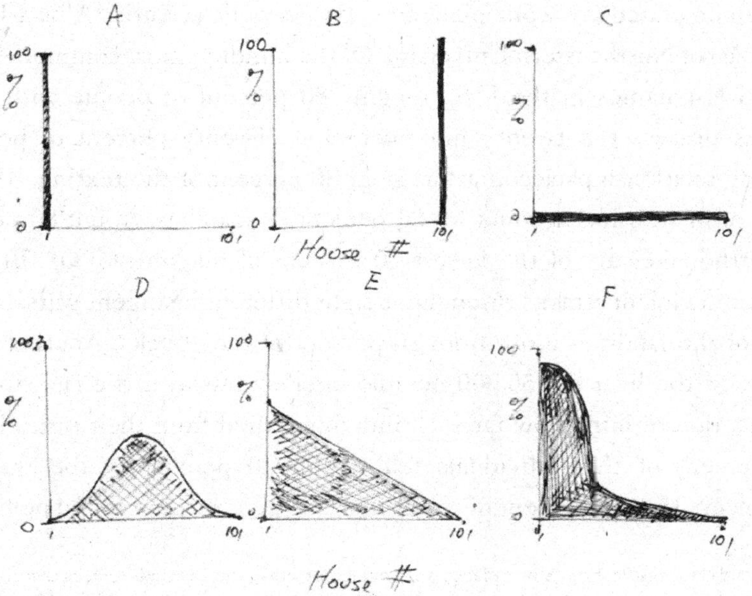

the remainder spread out across everyone else, with interactions a little less likely with each step farther out (figure F).

This is the 80:20 rule—approximately 80 percent of interactions occur among approximately 20 percent of the population. In the commercial world, it's sardonically stated as 80 percent of complaints come from 20 percent of the customers. Eighty percent of crime is caused by 20 percent of the criminals. Eighty percent of the company's work is due to the efforts of 20 percent of the employees. In the early days of the pandemic, a large majority of COVID-19 infections were caused by the small subset of infected super-spreaders.[29]

The 80:20 descriptor captures the spirit of what is known as a Pareto distribution, of a type mathematicians call a "power law." While it is formally defined by features of the curve, it's easiest to understand in plain English: a power-law distribution is when the substantial majority of interactions are very local, with a steep drop-off after that, and as you go out further, interactions become rarer.

All sorts of weird things turn out to have power-law distributions, as demonstrated by work pioneered by network scientist Albert-László Barabási of Northeastern University. Of the hundred most common Anglo-Saxon last names in the U.S., roughly 80 percent of people with those names possess the twenty most common. Twenty percent of people's texting relationships account for about 80 percent of the texting. Twenty percent of websites account for 80 percent of searches. About 80 percent of earthquakes are of the lowest 20 percent of magnitude. Of fifty-four thousand violent attacks throughout eight different insurgent wars, 80 percent of the fatalities arose from 20 percent of the attacks. Another study analyzed the lives of 150,000 notable intellectuals over the last two millennia, determining how far each individual died from their birthplace—80 percent of the individuals fell within 20 percent of the maximal distance.* Twenty percent of words in a language account for 80 percent of

*The study was fascinating. Some places were net exporters of intellectuals, places that they were more likely to move away from than move to—Liverpool, Glasgow, Odessa, Ireland, the Russian Empire, and my simple village of Brooklyn. This is the "please get me out of here" scenario. And

the usage. Eighty percent of craters on the Moon are in the smallest twentieth percentile of size. Actors get a Bacon number, where if you were in a movie with the prolific Kevin Bacon (1,600 people), your Bacon number is 1; if you were in a movie with someone who was in a movie with him, yours is 2; in a movie with someone who was in a movie with someone who was in a movie with Bacon, 3 (the most common Bacon number, held by ~350,000 actors), and so on. And starting with that modal number and increasing the Bacon number from there, there is a power-law distribution to the smaller and smaller number of actors.[*,30]

I'd be hard-pressed to see something adaptive about power-law distributions in Bacon numbers or the size of lunar craters. However, power-law distributions in the biological world display can be highly adaptive.[†,31]

For example, when there's lots of food in an ecosystem, various species forage randomly, but when food is spare, roughly 80 percent of foraging forays (i.e., moving in one direction looking for food, before trying a

---

then there are the net importers, magnets like Manhattan, Paris, Los Angeles, London, Rome. One of those magnets where intellectuals clustered, living out the rest of their (short) lives, was Auschwitz.

*Bacon numbers show what the long tail of unlikelihood looks like in a power-law distribution. There are approximately one hundred thousand actors with a Bacon number of 4 (84,615), about ten thousand with 5 (6,718), about one thousand with 6 (788), about one hundred with 7 (107), and eleven with a Bacon number of 8—with each step further out in the distribution, the event becomes roughly ten times rarer.

Mathematicians have "Erdös numbers," named for the brilliant, eccentric mathematician Paul Erdös, who published 1,500+ papers with 504 collaborators; a low Erdös number is a point of pride among mathematicians. There is, of course, only one person with an Erdös number of 0 (i.e., Erdös); the most common Erdös number is 5 (with 87,760 mathematicians), with the frequency declining with a power-law distribution after that.

Get this—there are people with both a low Bacon number *and* a low Erdös number. The record, 3, is shared by two people. There's Daniel Kleitman (who published with Erdös and appeared in the movie *Good Will Hunting* as an MIT mathematician, which is, well, what he is; Minnie Driver, with a Bacon number of 1, costarred). And there's mathematician Bruce Reznick (also a 1-Erdös-er who, oddly, was an extra in what was apparently an appallingly bad movie, with a Rotten Tomatoes score of 8 percent, called *Pretty Maids All in A Row*, which included 1-Baconist Roddy McDowall). As long as we're at it, MIT mathematician John Urschel has a combined Flacco/Erdös number of 5, due to an Erdös number of 4 and a Flacco number of 1; Urschel played in the NFL alongside quarterback Joe Flacco, who apparently is/was extremely important.

†Most, but not all, show this property. The exceptions are important, showing that cases with the distribution were selected for, evolutionarily, rather than being just inevitable features of networks.

different direction) are within 20 percent of the maximal distance ever searched—this turns out to optimize the energy spent searching relative to the likelihood of finding food; cells of the immune system show the same when searching for a rare pathogen. Dolphins show an 80:20 distribution of within-family and between-family social interactions; the 80-ness means that family groups remain stable even after an individual dies, while the 20-ness allows for the flow of foraging information between families. Most proteins in our bodies are specialists, interacting with only a handful of other types of proteins, forming small, functional units. Meanwhile, a small percentage are generalists, interacting with scores of other proteins (generalists are switch points between protein networks—for example, if one source of energy is rare, a generalist protein switches to using a different energy source).[*,32]

Then there are adaptive power-law relationships in the brain. What counts as adaptive or useful in how neuronal networks are wired? It depends on what kind of brain you want. Maybe one where every neuron synapses onto the maximal possible number of other neurons while minimizing the miles of axons needed. Maybe one that optimizes solving familiar, easy problems quickly or being creative in solving rare, difficult ones. Or maybe one that loses the minimal amount of function when the brain is damaged.

You can't optimize more than one of those attributes. For example, if your brain cares only about solving familiar problems quickly, thanks to neurons being wired up in small, highly interconnected modules of similar neurons, you're screwed the first time something unpredictable demands some creativity.

While you can't optimize more than one attribute, you can optimize how differing demands are *balanced*, what trade-offs are made, to come up with the network that is ideal for the balance between predictability and

*As an example of a generalist, the mutation in Huntington's disease produces an abnormal version of a particular protein. How does this explain the symptoms of the disease? Who knows. The protein interacts with more than *one hundred* other types of protein.

novelty in a particular environment.* And this often turns out to have a power-law distribution where, say, the vast majority of neurons in cortical mini columns interact only with immediate neighbors, with an increasingly rare subset wandering out increasingly longer distances.† Writ large, this explains "brain-ness," a place where the vast majority of neurons form a tight, local network—the "brain"—with a small percentage projecting all the way out to places like your toes.[33]

Thus, on scales ranging from single neurons to far-flung networks, brains have evolved patterns that balance local networks solving familiar problems with far-flung ones being creative, all the while keeping down the costs of construction and the space needed. And, as usual, without a central planning committee.‡,[34]

EMERGENCE DELUXE

We've now seen a number of motifs that come into play in emergent systems—rich-get-richer phenomena where higher-quality solutions give off stronger recruiting signals, iterative bifurcation that inserts near-infinity into finite places, spatiotemporal control of attraction and repulsion rules, mathematical optimizing of the balance between different wiring needs—and there are many more.§,[35]

---

*A contrast that has been framed as choosing between maximizing strength versus robustness, or maximizing evolvability versus flexibility, or maximizing stability versus maneuverability.

†The brain contains "small-world networks," a particular type of power-law distribution that emphasizes the balance between optimizing the interconnected nature of clusters of functionally related nodes, on one hand, and optimizing the fewest average number of steps linking any given node to another.

‡Due diligence footnote: Not everyone is thrilled with the notion of the brain being chock-full of power-law distributions. For one thing, as some techniques improve for detecting thin axonal projections, many of the scant long-distance projections turn out to be less scant than expected. Next, there is a difference between power-law distributions and "truncated" power-law distributions. And mathematically, other "heavy-tailed" distributions are incorrectly labeled as power-law ones in many cases. This is where I gave up on reading this stuff.

§"Many more" including an emergent phenomenon called stigmergy, which, among other things, explains how termites move more than a quarter ton of soil to build thirty-foot-high mounds that do gas exchange like your lungs do; back-propagating neural networks that computer scientists

Here are two last examples of emergence that incorporate a number of these motifs. One is startling in its implications; one is so charming that I can't omit it.

Charm first. Consider a toenail that is a perfect Platonic rectangle X units in height (after ignoring the curvature of a nail) (diagram A). Savage the perfection with some scissors, cutting off a triangle of toenail (diagram B). If the toenail universe did not involve emergent complexity, the toenail would now regrow as in diagram C. Instead, you get diagram D.

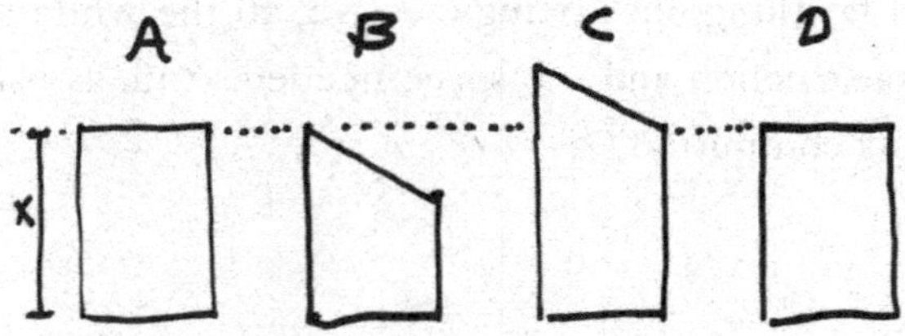

How? The top of a toenail thickens from bearing the brunt of contacting the outside world (e.g., the inside of your sock; a boulder; that damn coffee table, why don't we get rid of it, all we do is pile up junk on it), and once it thickens, it stops growing. After the cutting, only point a, at the original length (next diagram), retains the thickening. And as point b's regrowth brings it to the same height as point a, it now bears the brunt of the outside worlds and thickens (its further growth is probably also constrained by the thickness of point a adjacent to it). The same process occurs when point c arrives. . . . There's no comparative information involved; point c doesn't have to choose between emulating point b or emulating point d. Instead, the optimal solution emerges from the nature of toenail regrowth.

copy in order to make machines that learn; wisdom-of-the-crowd emergence where a group of individuals with average expertise about something outperforms a single *extreme* expert; and bottom-up curation systems that, when utilized by Wikipedia, generate accuracy on the scale of the *Encyclopedia Britannica* (Wikipedia has become the major source of medical information used by doctors).

What inspired me to include this example? A man named Bhupendra Madhiwalla, then age eighty-two, living in Mumbai, India, did that experiment with a toenail of his, repeatedly photographed the regrowth process and then emailed pictures to me from out of the blue. Which made me immensely happy.

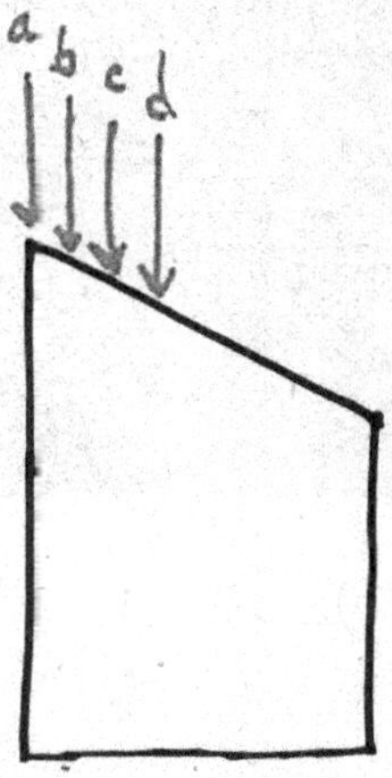

Now the awesome final example. As a tautology, studying the function of neurons in the brain tells you about the function of neurons in the brain. But sometimes more detailed information can be found by growing neurons in petri dishes. These are typically two-dimensional "monolayer" cultures, where a slurry of individual neurons is plated down randomly, then begin to connect with each other as a carpet. However, some fancy techniques make it possible to grow three-dimensional cultures, where the slurry of a few thousand neurons is suspended in a solution. And these neurons, each floating on its own, find and connect up with each other, forming clumps of brain "organoids." And after months, these organoids, barely large enough to be visible without a microscope, self-organize into brain structures. A slurry of human cortical neurons starts making radiating scaffolding,* constructing a primitive cortex with the beginnings of separate layers, even the beginnings of cerebrospinal fluid. And these organoids eventually produce *synchronized brain waves that mature similarly to the way they do in fetal and neonatal brains.* A random bunch of neurons, perfect strangers floating in a beaker, spontaneously build themselves into the starts of our brains.† Self-organized Versailles is child's play in comparison.[36]

*Which seems important, as the differences in patterns of genes expressed in these cells when comparing human brain organoids with those of other apes are really dramatic.

†A number of labs now are making human brain organoids with neurons containing Neanderthal genes. Other research allows cortical organoids to communicate with organoids of muscle cells, making them contract. And another group has been making organoid/robot interfaces, each communicating with the other.

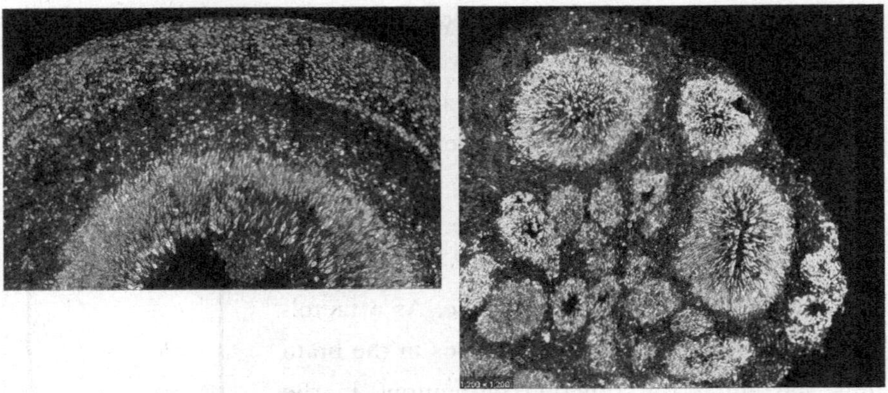

What has this tour shown us? (A) From molecules to populations of organisms, biological systems generate complexity and optimization that match what computer scientists, mathematicians, and urban planners achieve (and where roboticists explicitly borrow swarm intelligence strategies of insects[37]). (B) These adaptive systems emerge from simple constituent parts having simple local interactions, all without centralized authority, overt comparisons followed by decision-making, a blueprint, or a blueprint maker.* (C) These systems have characteristics that exist only at the emergent level—a single neuron cannot have traits related to circuitry—and whose behavior can be predicted without having to resort

Okay, is it time to freak out? Are these things on their way to consciousness, feeling pain, dreams, aspirations, and love/hate feelings about us, their creators? As framed in the title of one relevant paper, time for a "reality check." These are model systems of brains, rather than brains themselves (useful for understanding, say, why Zika virus causes massive structural abnormalities in human fetal brains); to give a sense of scale, organoids consist of a few thousand neurons, while insect brains range in the hundreds of thousands. Nonetheless, all this must give one pause ("Can Lab-Grown Brains Become Conscious?" asks another paper as its title), and legal scholars and bioethicists are starting to weigh in about what kinds of organoids might not be okay to make.

*There's a wonderful quote often used about emergence: "The locusts have no King, yet all of them march in rank." I like the irony of this, since it's found in a book that extols the putative individual who gains the most if the world runs on centralized top-down authority—it comes from the Old Testament (Proverbs 30:20). Oh, and by the way, why do locusts march? Each locust marches forward because the locust immediately behind is trying to eat it.

to reductive knowledge about the component parts. (D) Not only does this explain emergent complexity in our brains, but our nervous systems use some of *the same* tricks used by the likes of individual proteins, ant colonies, and slime molds. All without magic.

Well, that's nice. Where does free will come into this?

8

Does Your Free Will Just Emerge?

FIRST, WHAT ALL OF US CAN AGREE ON

So emergence is about reductive piles of bricks producing spectacular emergent states, ones that can be thoroughly unpredictable or that can be predicted based on properties that exist only at the emergent level. Reassuringly, no one thinks that free will lurks in the neuronal equivalent of individual bricks (well, almost no one; wait for the next chapter). This is nicely summarized by philosopher Christian List of Ludwig Maximilian University in Munich: "If we look at the world solely through the lens of fundamental physics or even that of neuroscience, we may not find agency, choice, and mental causation," and people rejecting free will "make the mistake of looking for free will at the wrong level, namely the physical or neurobiological one—a level at which it cannot be found." Robert Kane states the same: "We think we have to become originators at the micro-level [to explain free will] . . . and we realize, of course, that we cannot do that. But we do not have to. It is the wrong place to look. We do not have to micro-manage our individual neurons one by one."[1]

So these free-will believers accept that an individual neuron cannot defy the physical universe and have free will. But a bunch of them can; to quote List, "free will and its prerequisites are emergent, higher-level phenomena."[2]

Thus, a lot of people have linked emergence and free will; I will not consider most of them because, to be frank, I can't understand what they're suggesting, and to be franker, I don't think the lack of comprehension is entirely my fault. As for those who have more accessibly explored the idea that free will is emergent, I think there are broadly three different ways in which they go wrong.

PROBLEM #1: CHAOTIC MISSTEPS REDUX

We know the drill. Compatibilists and free-will-skeptic incompatibilists agree that the world is deterministic but disagree about whether free will can coexist with that. But if the world is indeterministic, you've cut the legs out from under free-will skeptics. The chaos chapter showed how you get there by confusing the unpredictability of chaotic systems with indeterminism. You can see how folks drive off a cliff with the same mistake about the unpredictability of many instances of emergent complexity.

A great example of this is found in the work of List, a philosophy heavyweight who made a big splash with his 2019 book, *Why Free Will Is Real.* As noted, List readily recognizes that individual neurons work in a deterministic way, while holding out for higher-level, emergent free will. In this view, "the world may be deterministic at some levels and indeterministic at others."[3]

List emphasizes unique evolution, a defining feature of deterministic systems, where any given starting state can produce only one given outcome. Same starting state, run it over and over, and not only should you get one mature outcome each time, but it better be the same one. List then ostensibly proves the existence of emergent indeterminism with a model that appears in various forms in a number of his publications:

The top panel represents a reductive, fine-grain scenario where (progressing from left to right) five similar starting states each produce five distinct outcomes. We then turn to the bottom panel, which is a state that List says displays emergent indeterminism. How does he get there? The

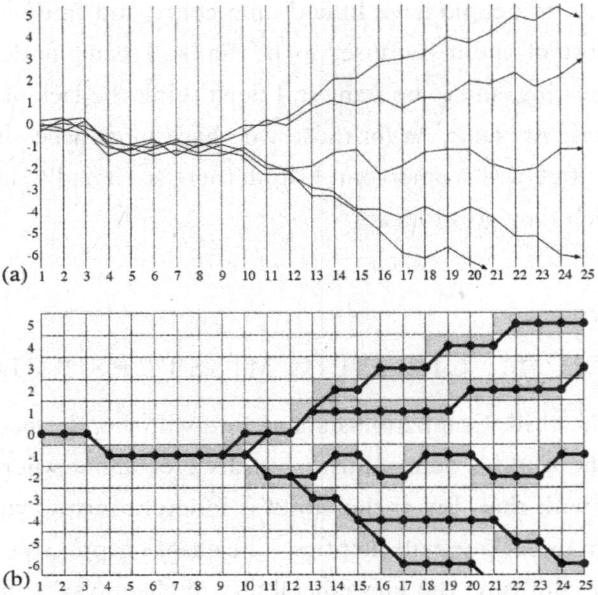

Emergent indeterminism

bottom panel "shows the same system at a higher level of description, obtained by *coarse-graining* the state space," making use of "the usual rounding convention." And when you do that, those five different starting states become the same, and that singular starting state can produce five completely different paths, proving that it is indeterministic and unpredictable.[4]

Er, maybe not. Sure, a system that is deterministic at the micro level can be indeterministic at the macro in this way, *but only if you're allowed to decide that five different (though similar) starting states are all actually the same*, merging them into a single higher-order simulation. This is the last chapter all over again—when you're Edward Lorenz, come back from lunch and coarse-grain your computer program, decide that the morning's parameters can be rounded off with *the usual rounding convention*, and you're bit in the rear by a butterfly. Two things that are similar are not identical, and you can't decide that they are simply because that represents the conventions of thinking.

Reflecting my biological roots, here's a demonstration of the same point:

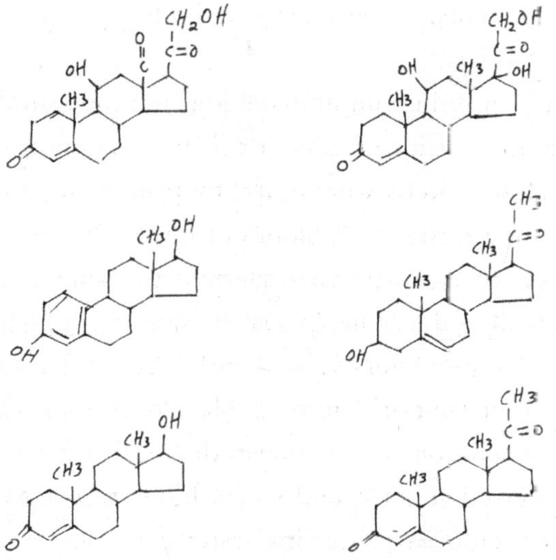

Here are six different molecules, all with similar structures.* Now let's coarse-grain 'em, decide that they are similar enough that we can consider them to be the same, by the usual scale of rounding convention, and therefore, they can be used interchangeably when we inject one of them into someone's body and see what happens. And if there isn't always the same exact effect, yeah, you've supposedly just demonstrated emergent indeterminism.

But they're not all the same. Consider the middle and bottom structures in the first column. Majorly similar—just try remembering their structural differences for a final exam. But if you coarse-grain them into being the same, rather than just very similar, things are going to get really messy—because the top molecule of the two is a type of estrogen, and the

*Jargon: they all have a "steroidal ring structure."

bottom is testosterone. Ignore sensitive dependence on initial conditions, decide the two molecules are the same by whatever you've deemed the usual conventional rounding, and sometimes you get someone with a vagina, sometimes a penis, sometimes sort of both. Supposedly proving emergent indeterminism.*

It's the last chapter redux; unpredictable is not the same thing as indeterministic. Disperse armies of ants at ten feeding spots, and you can't predict just how close (and by what route) they are going to get to *the* solution to the traveling-salesman problem out of the 360,000+ possibilities. Instead, you'll have to simulate what happens to their cellular automaton step by step. Do it all again, same ants at the same starting points but with one of those ten feeding spots in a slightly different location, and you might get a different (but still remarkably close) approximation of the traveling-salesman solution. Do it repeatedly, each time with one of the feeding stations moved slightly, and you're likely to get an array of great solutions. Small differences in starting states can generate very different outcomes. But an identical starting state can't do that and supposedly prove indeterminacy.

PROBLEM #2: ORPHANS RUNNING WILD

So much for the idea that in emergent systems the same starting state can give rise to multiple outcomes. The next mistake is a broader one—the idea that emergence means the reductive bricks that you start with can give rise to emergent states that can then do whatever the hell they want.

This has been stated in a variety of ways, where terms like *brain*, *cause and effect*, or *materialism* stand in for the reductive level, while terms like *mental states*, *a person*, or *I* imply the big, emergent end product. According to philosopher Walter Glannon, "although the brain generates and sustains our mental states, it does not determine them, and this leaves enough

*For completeness: The top hormone in the left column is aldosterone. Starting at the top of the right column, the hormones are cortisol, a neurosteroid called pregnenolone, and progesterone.

room for individuals to 'will themselves to be' through their choices and actions." "Persons," he concludes, "are constituted by but not identical to their brains." Neuroscientist Michael Shadlen writes of emergent states having a special status as a "consequence of their emergence as *entities orphaned from* the chain of cause and effect that led to their implementation in neural machinery" (italics mine). Adina Roskies relatedly writes, "Macrolevel explanations are independent of the truth of determinism. These same arguments suffice to explain why an agent still makes a choice in a deterministic world, and why he or she is responsible for it."[5]

This raises an important dichotomy. Philosophers with this interest discuss "weak emergence," which is where no matter how cool, ornate, unexpected, and adaptive an emergent state is, it is still constrained by what its reductive bricks can and can't do. This is contrasted with "strong emergence," where the emergent state that emerges from the micro can no longer be deduced from it, even in chaoticism's sense of a stepwise manner.

The well-respected philosopher Mark Bedau, of Reed College, considers the strong emergence that can do as it pleases with happy-go-lucky free will to be close to theoretically impossible.* Strong emergence claims "heighten the traditional worry that emergence entails illegitimately getting something from nothing," which is "uncomfortably like magic."† The influential philosopher David Chalmers of New York University weighs in as well, considering that the only thing that comes close to qualifying as a case of strong emergence is consciousness; likewise with another major contributor to this field, Johns Hopkins physicist Sean Carroll, who thinks that while consciousness is the only real reason to be interested in strong emergence, it's sure not a case of it.

With a limited role, if any, for strong emergence (and thus for its being the root of free will), we are left with weak emergence, which, in Bedau's

*Twentieth-century philosophy pretty much only considered the hypotheticals of strong emergence, and Bedau makes an eye-catching plea for why philosophers should become interested in weak emergence—because it's how the real world actually works.

†Brazilian philosopher Gilberto Gomes, defensively disavowing magic, writes that in his compatibilist viewpoint, "this I is not an abstract or supernatural entity outside the realm of natural causality. The I is a self-organizing and self-steering system."

words, "is no universal solvent." You can be out of your mind but not out of your brain; no matter how emergently cool, ant colonies are still made of ants that are constrained by whatever individual ants can or can't do, and brains are still made of brain cells that function like brain cells.[6]

Unless you resort to one last trick to pull free will from emergence.

PROBLEM #3: DEFYING GRAVITY

The place where a final mistake creeps in is the idea that an emergent state can reach down and change the fundamental nature of the bricks comprising it.

We all know that an alteration at the brick level can change the emergent end product. If you're injected with many copies of a molecule that activates six of the fourteen subtypes of serotonin receptors,* your macro level is likely to include perceiving vivid images that other people don't, plus maybe even some religious transcendence. Dramatically drop the number of glucose molecules in someone's bloodstream, and their resulting macro level will have trouble remembering whether Grover Cleveland was president before or after Benjamin Harrison.† Even if consciousness qualifies as the closest thing to true strong emergence, induce unconsciousness by infusing a molecule like phenobarbital, and you'll have shown that it isn't remotely free from its building blocks.

Good, we all agree that altering the little can change the emergent big. And the reverse certainly holds true. Sit here and press button A or B, and which motor neurons tell your arm muscles to shift this way or that will be manipulated by the emergent macrophenomenon called aesthetics, if you're asked which painting you prefer, the one of a Renaissance woman with a half smile or the one of Campbell's soup cans. Or press the button indicating which of two people you deem more likely to be destined for

---

*I.e., LSD.

†Trick question.

hell, or whether 1946's *Call Me Mister* or 1950's *Call Me Madam* is the more obscure musical.

A 2005 study concerning social conformity shows a particularly stark, fascinating version of the emergent level manipulating the reductive business of individual neurons. Sit a subject down and show them three parallel lines, one clearly shorter than the other two. Which is shorter? Obviously that one. But put them in a group where everyone else (secretly working on the experiment) says the longest line is actually the shortest—depending on the context, a shocking percentage of people will eventually say, yeah, that long line is the shortest one. This conformity comes in two types. In the first, go-along-to-get-along public conformity, you know which line is shortest but join in with everyone else to be agreeable. In this circumstance, there is activation of the amygdala, reflecting the anxiety driving you to go along with what you know is the wrong answer. The second type is "private conformity," where you drink the Kool-Aid and truly believe that somehow, weirdly, you got it all wrong with those lines and everyone else really was correct. And in this case, there is also activation of the hippocampus, with its central role in learning and memory—conformity trying to rewrite the history of what you saw. But even more interesting, there's activation of the visual cortex—"Hey, you neurons over there, the line you foolishly thought was longer at first is actually shorter. Can't you just see the truth now?"[*,7]

Think about this. When is a neuron in the visual cortex supposed to activate? Just to wallow in minutiae that can be ignored, when a photon of light is absorbed by rhodopsin in disc membranes within a retinal photoreceptive cell, causing the shape of the protein to change, changing transmembrane ion currents, thus decreasing the release of the neurotransmitter glutamate, which gets the next neuron in line involved, starting a sequence

*This experimental approach alludes to classic research by Solomon Asch in the 1950s showing that an unnervingly large percentage of people will conform in particular settings to something they know is wrong (with the full range of what *wrong* can mean, ranging from "Which line is shortest?" to "Should these people be exterminated?"). Little surprise that this and other classic conformity and obedience studies were prompted by World War II: Did all those Germans actually believe that stuff, or were they just being team players?

culminating in that visual cortical neuron having an action potential. One big micro-level blowout of reductionism.

And what's happening instead during private conformity? That same Mr. Machine little neuron in the visual cortex activates because of the macro-level emergent state that we'd call an urge toward fitting in, a state built out of the neurobiological manifestations of the likes of cultural values, a desire to seem likable, adolescent acne having left scars of low self-esteem, and so on.[*,8]

So some emergent states have *downward causality,* which is to say that they can alter reductive function and convince a neuron that long is short and war is peace.

The mistake is the belief that once an ant joins a thousand others in figuring out an optimal foraging path, downward causality causes it to suddenly gain the ability to speak French. Or that when an amoeba joins a slime mold colony that is solving a maze, it becomes a Zoroastrian. And that a single neuron, normally being subject to gravity, stops being so once it holds hands with all the other neurons producing some emergent phenomenon. That the building blocks work *differently* once they're part of something emergent. It's like believing that when you put lots of water molecules together, the resulting wetness causes each molecule to switch from being made of two hydrogens and one oxygen to two oxygens and one hydrogen. But the whole point of emergence, the basis of its amazingness, is that those idiotically simple little building blocks that only know a few rules about interacting with their immediate neighbors remain *precisely as idiotically simple* when their building-block collective is outperforming urban planners with business cards. Downward causation doesn't

---

*Another fascinating example or macro influencing micro concerns something covered in chapter 3—on the average, people from individualist cultures look at the person in the center of a picture, while those from collectivist ones scan the entire scene. Reflect on this: Culture is as emergent as things get, influencing what foods are sacred, what kinds of sex are taboo, what counts as heroism or villainy in stories. And all this determines the microfunction of neurons that control your unconscious eye movements. Hmm, why'd you look at that part of the picture first? Because of my neuronal circuitry. Because of what happened to my people five centuries ago in the Battle of Wherever. Because . . .

cause individual building blocks to acquire complicated skills; instead, it determines the contexts in which the blocks are doing their idiotically simple things. Individual neurons don't become causeless causes that defy gravity and help generate free will just because they're interacting with lots of other neurons.

And the core belief among this style of emergent free-willers is that emergent states can in fact change how neurons work, and that free will depends on it. It is the assumption that emergent systems "have base elements that behave in novel ways when they operate as part of the higher-order system." But no matter how unpredicted an emergent property in the brain might be, neurons are not freed of their histories once they join the complexity.[9]

This is another version of our earlier dichotomy. There's *weak* downward causality, where something emergent like conformity can make a neuron fire the same way as it would in response to photons of light—the workings of this component part have not changed. And there's *strong* downward causality, where it can. The consensus among most philosophers and neurobiologists thinking about this is that strong downward causality, should it exist, is irrelevant to this book's focus. In a critique of this approach to discovering free will, psychologists Michael Mascolo of Merrimack College and Eeva Kallio of the University of Jyväskylä write, "While [emergent systems] are irreducible, they are not autonomous in the sense of having causal powers that override those of their constituents," a point emphasized as well by Spanish philosopher Jesús Zamora Bonilla in his essay "Why Emergent Levels Will Not Save Free Will." Or stated in biological terms by Mascolo and Kallio, "while the capacities for experience and meaning are emergent properties of biophysical systems, the capacity for behavioral regulation is not. The capacity for self-regulation is an already existing capacity of living systems." There's still gravity.[10]

AT LAST, SOME CONCLUSIONS

Thus, in my view, emergent complexity, while being immeasurably cool, is nonetheless not where free will exists, for three reasons:

a. Because of the lessons of chaoticism—you can't just follow convention and say that two things are the same, when they are different, and in a way that matters, regardless of how seemingly minuscule that difference; unpredictable doesn't mean undetermined.

b. Even if a system is emergent, that doesn't mean it can choose to do whatever it wants; it is still made up of and constrained by its constituent parts, with all their mortal limits and foibles.

c. Emergent systems can't make the bricks that built them stop being brick-ish.[*,11]

These properties are all intrinsic to a deterministic world, whether chaotic, emergent, predictable, or unpredictable. But what if the world isn't really deterministic after all? On to the next two chapters.

* Despite the fact that, to quote the architect Louis Kahn, "even a brick wants to be something."

9

A Primer on Quantum Indeterminacy

I really do not want to write this chapter, or the next one. I've been dreading it, in fact. When friends ask me how the book writing is going, I grimace and say, "Well, okay, but I'm still postponing doing the chapters on indeterminacy." Why the dread? To start, (a) the chapters' subject rests on profoundly bizarre and counterintuitive science (b) that I barely understand and (c) that even the people who you'd think understand it admit that they don't, but with a profound noncomprehension, compared with my piddly cluelessness, and (d) the topic exerts a gravitational pull upon crackpot ideas as surely as does a statue upon defecating pigeons, a pull that constitutes a "What are they talking about?" strange attractor. Nonetheless, here goes.

This chapter examines some foundational domains of the universe in which extremely tiny stuff operates in ways that are not deterministic. Where unpredictability does not reflect the limitations of humans tackling math, or the wait for an even more powerful magnifying glass, but instead reflects ways in which the physical state of the universe does *not* determine it. And the next chapter is about reining in the free-willers in this playground of indeterminacy.

Were I to chicken out and end this pair of chapters right here, the conclusions would be that, yes, Laplacian determinism really does appear to fall apart down at the subatomic level; however, such eensy-weensy indeterminism is vastly unlikely to influence anything about behavior; even if it did, it's even more unlikely that it would produce something resembling free will; scholarly attempts to find free will in this realm frequently strain credulity.

UNDETERMINED RANDOMNESS

What exactly do we mean by "randomness"? Suppose we have a particle that moves "randomly." To qualify, it would show these properties:

—If at time 0 a particle is in spot X, the most likely place you'd expect to find that randomly moving particle for the rest of time is back at spot X. And if at some point after time 0, the particle happens to be in spot Z, now for the rest of time, spot Z is where it's most likely to be. The best predictor of where a randomly moving particle is likely to be is wherever it is right now.

—Take any unit of time—say, one second. The amount of variability in the particle's movement in the next second will be as much as during one second a million years from now.

—The pattern of movement at time 0 has zero correlation with time 1 or –1.

—If it looks as if the particle has moved in a straight line, get that magnifying glass and look closer and you'll see that it isn't really a straight line. Instead, the particle zigzags, regardless of the scale of magnification.

—Because of that zigzagging, when magnified infinitely, a particle will have moved an infinitely long distance between any two points.

These are stringent features for a particle to qualify as undetermined.* These requirements, especially that spacey Menger-sponge business about something infinitely long fitting into a finite space, show how capital-*R* Randomness differs from random channel surfing.

So what does a particle being random have to do with your being the agentive captain of your fate?

LOW-RENT RANDOMNESS: BROWNIAN MOTION

We start with the Jane and Joe Lunchbucket version of indeterminism, one that is rarely contemplated at meditation retreats.

Sit in an otherwise dark room that has a shaft of light coming in from a window, and look at what is being illuminated along the way by the shaft (i.e., not the spot on the wall being lit up but the air illuminated between the window and the lit wall). You'll see minuscule dust particles that are in constant motion, vibrating, jerking this way or that. Behaving randomly.

People (e.g., Robert Brown, in 1827) had long noted the phenomenon, but it wasn't until the last century that random (aka "stochastic") movement was identified to occur among particles suspended in a fluid or gas. Tiny particles oscillate and vibrate as a result of being hit randomly by photons of light, which transfer energy to the particle, producing the vibratory phenomenon of kinetic energy. Which causes particles to bump into each other randomly. Which causes them to bump into other particles. Everything moving randomly, the unpredictability of the three-body problem on steroids.

Mind you, this isn't the unpredictability of cellular automata, where every step is deterministic but not determinable. Instead, the state of a particle in any given instant is *not* dependent on its state an instant before.

*In this case, a "particle" is anything from subatomic particles to atoms, molecules, and macroscopic things like dust motes.

Laplace is vibrating disconsolately in his grave. The features of such stochasticity were formalized by Einstein in 1905, his annus mirabilis when he announced to the world that he was not going to be a patent clerk forever. Einstein explored the factors that influence the extent of Brownian motion of suspended particles (note the plural on particles—any given particle is random, and predictability is probabilistic only on the aggregate level of lots of particles). One thing that increases Brownian motion is heat, which increases kinetic energy in particles. In contrast, it's decreased when the surrounding fluid or gas environment is sticky or viscous or when the particle is bigger. Think of this last one this way: The bigger a particle, the bigger the bull's-eye, the more likely it is to be bumped into by lots of other particles, on all its sides. Which increases the odds of all those bumps canceling each other out and the big particle staying put. Thus, the smaller the particle, the more exciting the Brownian motion that it shows—while the Great Pyramid of Giza may be vibrating, it isn't doing it much.*

So that's Brownian motion, particles bumping into each other randomly. How does that relate to biology (a first step toward seeing its relevance to behavior)? Lots, as it turns out. One paper explores how a type of Brownian motion explains the distribution of populations of axon terminals. Another concerns how copies of the receptor for the neurotransmitter acetylcholine randomly aggregate into clusters, something important to their function. Another example concerns abnormality in the brain—some mostly mysterious factors increase the production of a weirdly folded fragment called the beta-amyloid peptide. If one copy of this fragment randomly bumps into another one, they stick together, and this clump of aggregated protein crud grows bigger. These soluble amyloid aggregates are the most likely killers of your neurons in Alzheimer's disease. And

*These factors influencing Brownian motion are formalized in the Stokes-Einstein equation (named for Sir George Stokes, a viscosity savant who died shortly before Einstein burst on the scene). The numerator in the equation concerns the main force that increases motion, namely temperature; the denominator concerns the forces countering the particles, namely high viscosity of the surrounding environment and large average size of particles.

Brownian motion helps explain probabilities of fragments bumping into each other.[1]

I like teaching one example of Brownian motion, because it undermines myths of how genes determine everything interesting in living systems. Take a fertilized egg. When it divides in two, there is random Brownian splitting of the stuff floating around inside, such as thousands of those powerhouses-of-the-cell mitochondria—it's never an exact 50:50 split, let alone the same split each time. Meaning those two cells already differ in their power-generating capacity. Same for vast numbers of copies of proteins called transcription factors, which turn genes on or off; the uneven split of transcription factors when the cell divides means the two cells will differ in their gene regulation. And with each subsequent cell division, randomness plays that role in the production of all those cells that eventually constitute you.[*,2]

Now, time to scale up and see where Brownian-esque randomness plays into behavior. Consider some organism—say, a fish—looking for food. How does it find food most efficiently? If food is plentiful, the fish forages in little forays anchored around this place of easy eating.[†] But if food is diffuse and sparse, the most efficient way to bump into some is to switch to a random, Brownian foraging pattern called a "Levy walk." So if you're the only thing worth eating in the middle of the ocean, the predator that grabs you will probably have gotten there by a Levy walk. And logically, many prey species move randomly and unpredictably in evading predators. The same math describes another type of predator hunting for prey—a white blood cell searching for pathogens to engulf. If the cell is in the middle of a cluster of pathogens, it does the same sort of home-based forays as a killer whale feasting in the middle of a bunch of seals. But when

---

[*]Which is why identical twins, with identical genes, don't have identical cells even when each twin consists of just two cells, with the differences magnifying up from there. This is part of why identical twins aren't identical people with brains supposedly sculpted identically by their identical genes.

[†]With the movement pattern showing a power-law distribution. Back to chapter 7—around 80 percent of foraging forays are within 20 percent of the maximal foraging distance.

the pathogens are sparse, white blood cells switch to a random Levy-walk hunting strategy, just like a killer whale. Biology is the best.[3]

To summarize, the world is filled with instances of indeterministic Brownian motion, with various biological phenomena having evolved to optimally exploit versions of this randomness. Are we talking free will here?* Before addressing this question, time to face the inevitable and tackle the mother of all theories.[4]

QUANTUM INDETERMINACY

Here goes. The classical physical picture of how the universe works, invariably attributed to Newton, tanked in the early twentieth century with the revolution of quantum indeterminacy, and nothing has been the same since. The subatomic world turns out to be deeply weird and still can't be fully explained. I'll summarize here the findings that are most pertinent to free-will believers.

WAVE/PARTICLE DUALITY

The start of the most foundational weirdness was the immeasurably cool, landmark double-slit experiment first carried out by Thomas Young in 1801 (another one of those polymaths who, when he wasn't busy with physics, or outlining the biology of how color vision works, helped translate the Rosetta stone). Shoot a beam of light at a barrier that has two vertical slits in it. Behind it is a wall that can detect where the light is hitting it. This shows that the light travels through the two slits as waves. How is this detected? If there was a wave emanating from each slit, the two waves would wind up overlapping. And there's a characteristic signature when a

*In the small-world category, one of the contributors to this topic, favoring a free-will stance for both humans and other animals, is neurobiologist Martin Heisenberg. Yes, son of Werner Heisenberg. Apparently, the tree freely wills an apple to drop locally.

pair of waves does this—when the peaks of two waves converge, you get an immensely strong signal; when the troughs of the two converge, the opposite; when a peak and a trough meet, they cancel each other out. Surfers understand this.

So light travels as a wave—classical knowledge. Shoot a stream of electrons at the double-slit barrier, and there's the same punch line—a wave function. Now, shoot *one* electron at a time, recording where it hits the detector wall, and the individual electron, the individual particle, passes through as a wave. Yup, the single electron passes through both slits simultaneously. It's in two places at once.

Turns out that it's more than just two places. The exact location of the electron is indeterministic, distributed probabilistically across a cloud of locations at once, something termed superposition.

Accounts of this now usually say something to the effect of "Now things get weird"—as if a single particle being in multiple places at once weren't weird. Now things get weirder. Build a recording device into the double-slit wall, to document the passage of each electron. You already know what will happen—each individual electron passes through both slits at once, as a wave. But no; each electron now passes through one slit or the other, randomly. The mere process of measuring, documenting what happens at the double-slit wall causes the electrons (and, as it turns out, streams of light, made up of photons) to stop acting as waves. The wave function "collapses," and each electron passes through the double-slit wall as a singular particle.

Thus, electrons and photons show particle/wave duality, with the process of measurement turning waves into particles. Now measure the properties of the electron *after* it passes through the slits but *before* it hits the detector wall, and as a result, each electron passes through one of the slits as a single particle. It "knows" that it is *going to be* measured in a bit, which collapses its wave function. Why the process of measuring collapses wave functions—the "measurement problem"—remains mysterious.[5]

(To jump ahead for a moment, you can guess that things are going to get very New Agey if you assume that the macroscopic world—big things like, say, you—also works this way. You can be in multiple places at once; you are nothing but potential. Merely observing something can change it;* your mind can alter the reality around it. Your mind can determine your future. Heck, your mind can change your past. More jabberwocky to come.)

Particle/wave duality generates a key implication. When an electron is moving past a spot as a wave, you can know its momentum, but you obviously can't know its exact location, since it's indeterministically everywhere. And once the wave function collapses, you can measure where that particle now is, but you can't know its momentum, since the process of measurement changes everything about it. Yup, it's Heisenberg's uncertainty principle.†

The inability to know both location and momentum, the fact of superposition and things being in multiple places at once, the impossibility of knowing which slit an electron will pass through once a wave has collapsed into a particle—all introduce a fundamental indeterminism into the universe. Einstein, despite upending the reductive, deterministic world of Newtonian physics, hated this type of indeterminism, famously declaring, "God does not play dice with the universe." This began a cottage industry of physicists trying to slip some form of determinism in the back door. Einstein's version is that the system actually is deterministic, thanks to some still-undiscovered factor(s), and things will go back to making sense once this "hidden variable" is identified. Another backdoor move is the very opaque "many-world" idea, which posits that waves don't really collapse into a singularity; instead their wave-ness continues in an infinite number of universes, making for a completely deterministic world(s), and it just looks singular if you're looking from only one universe at a time. I think. My sense is that the hidden-variable dodge is most doubters'

*And note here how the New Age interpretation has just jumped from considering the consequences of the formal process of "measurement" to the highly personal process of "observing."

†Which has not only caught the public's imagination but also generated endless Heisenberg uncertainty jokes (Heisenberg, speeding down the freeway, is stopped by a cop. "Do you know how fast you were going?" the cop asks. "No, but I know where I am," Heisenberg replies. "You were driving eighty miles per hour," says the cop. "Oh, great," says Heisenberg, "now I'm lost.").

favorite. However, the majority of physicists accept the indeterministic picture of quantum mechanics—known as the Copenhagen interpretation, reflecting its being championed by the Copenhagen-based Niels Bohr. In his words, "Those who are not shocked when they first come across quantum theory cannot possibly have understood it."[*,6]

ENTANGLEMENT AND NONLOCALITY

Next weirdness.[†] Two particles (say, two electrons in different shells of an atom) can become "entangled," where their properties (such as their direction of spin) are linked and perfectly correlated. The correlation is always negative—if one electron spins in one direction, its coupled partner spins the opposite way. Fred Astaire steps forward with his left leg; Ginger Rogers steps back with her right.

But it's stranger than that. For starters, the two electrons don't have to be in the same atom. They can be a few atoms apart. Okay, sure. Or, it turns out, they can be even farther apart. The current record is particles nearly *nine hundred miles* apart, at two ground stations linked by a quantum satellite.[‡] Moreover, if you alter the property of one particle, the other changes as well, implying a causality that isn't local. There is no theoretical limit for how far apart entangled particles can be. An electron in the Crab Nebula in the constellation Taurus can be entangled with an electron in the piece of broccoli stuck between your incisors. And as the strangest feature, when the state of one particle is altered, the complementary change in the other occurs *instantaneously*[§]—meaning that the

[*]Bohr also supplied one of my favorite quotes about the scientific venture: "The opposite of fact is falsehood, but the opposite of one profound truth may very well be another profound truth."

[†]I thank physicist Sean Carroll for guiding me through much of this. By the way, research on entanglement was the basis of the 2022 Nobel Prize in Physics, awarded to John Clauser, Alain Aspect, and Anton Zeilinger.

[‡]Implicit in this is that you can experimentally induce entanglement in two particles, which seems to involve pointing lasers at things.

[§]Or at least way faster than the experimental limits of time resolution, on the scale of quadrillionths of a second. Which is at least nine orders of magnitude faster than the speed of light. By broccoli and the Crab Nebula are influencing each other faster than the speed of light.[7]

Einstein was not amused (and labeled the phenomenon with a sarcastic German equivalent of *spooky*).* In 1935, he and two collaborators published a paper that challenged the possibility of this instantaneous entanglement, again positing hidden variables that explained things without invoking faster-than-the-speed-of-light mojo. In the 1960s, the Irish physicist John Stewart Bell showed that there was something off in the math in that paper of Einstein's. And in the decades since, extraordinarily difficult experiments (like the one with that satellite) have confirmed that Bell was right when he said that Einstein was wrong when *he* said that the interpretation of entanglement was wrong. In other words, the phenomenon is for real, although it still remains basically unexplained, nonetheless generating highly accurate predictions.[8]

Since then, scientists have explored the potential of using quantum entanglement in computing (with people at Apple apparently making significant progress), in communication systems, maybe even in automatically receiving a widget from Amazon the instant you think that you'll be happier owning one. And the weirdness just won't stop—entanglement over long enough distances can also show nonlocality over time. Suppose you have two entangled electrons a light-year apart; alter one of them and the other particle is altered at the same instant . . . a year ago. Scientists

---

the way, if I understand things correctly, superposition of a single particle can be thought of as involving entanglement—an electron is entangled with itself as it passes through two slits at the same time.

*In 1905, Einstein was the most glamorous, dashing revolutionary since Che (if time flows backward). As he aged, though, Einstein led some rearguard reactions against subsequent physics revolutions. This is a familiar pattern with many revolutionary thinkers. The psychologist Dean Simonton has shown that this closing to novel ideas is a function not so much of someone's chronological age as of their disciplinary age—it's being acclaimed in a particular field for a long time (after all, all anything new and revolutionary can do is knock you and your buddies out of the textbooks). Years ago, I did a quasiscientific study (published in that esteemed technical journal *The New Yorker*), showing the ways in which most people, acclaimed thinkers or otherwise, close to novelty in music, food, and fashion as they age. Learning of Einstein as an aged counterrevolutionary disappointed all of us who had the obligatory poster of him sticking out his tongue on our dorm room walls.

have also shown quantum entanglement in living systems, between a photon and the photosynthetic machinery of bacteria.* You better bet that we've got free-will speculations coming that invoke time travel, entanglement between neurons in the same brain, and, as long as we're at it, between brains.[9]

QUANTUM TUNNELING

This one is a piece of cake conceptually, after all the preceding strangeness. Shoot a stream of electrons at a wall. As we know, each travels as a wave, superposition dictating that until you measure its location, each electron is probabilistically in numerous places at once. Including the really, really unlikely but theoretically possible outcome of one of those numerous places being on the *other* side of the wall, because the electron has tunneled through it. And, as it turns out, this can happen.

That's it for this pitiful tour of quantum mechanics. For our purposes, the main points are that in the view of most of the savants, the subatomic universe works on a level that is fundamentally indeterministic on both an ontic and epistemic level. Particles can be in multiple places at once, can communicate with each other over vast distances faster than the speed of light, making both space and time fundamentally suspect, and can tunnel through solid objects. As we'll now see, that's plenty enough for people to run wild when proclaiming free will.

---

*The study is controversial, though, as some scientists suggest nonentanglement mechanisms as explanations. The study involved bacteria that were placed between two mirrors that were less than a hair's width apart. And the phenomenon was demonstrated in *six* individual bacteria. One is accustomed to things like "neuroimaging was carried out on six adults carrying the mutation" or "epidemiological surveys were carried out in six countries." A study using six bacteria seems charming and commensurate with all this weirdness. But given this tiny number of bacteria, one has to ask questions like what each one had eaten that morning; when they were fetuses, whether their moms had regular wellness checks; what kind of culture these bacteria's ancestors grew up in.

10

Is Your Free Will Random?

QUANTUM ORGASMIC-NESS: ATTENTION AND INTENTION ARE THE MECHANICS OF MANIFESTATION

The previous chapter revealed some truly weird things about the universe that introduce a fundamental indeterminism into the proceedings. And from virtually the first moment this news got around, some believers in free will have attributed all sorts of mystical gibberish to quantum mechanics.* There are now proponents of quantum metaphysics, quantum philosophy, quantum psychology. There's quantum theology and quantum Christian realism; in one tract in that vein, quantum mechanics is cited as proving that humans cannot be reduced to predictable machines, making for human uniqueness that aligns with the biblical claim that God

*Interestingly, I've seen none of the same done with the indeterminacy of Brownian motion—for example, no one is making a bundle running Brownian transcendence seminars. This isn't surprising—quantum indeterminacy is about being in multiple places at once, while Brownian motion is about dust particles being random. Thus, I suspect that New Agers view Brownian motion as dead-White-male-ish, like union guys who nonetheless vote Republican, while quantum indeterminacy is about love, peace, and multiple orgasms. (This pretty picture is complicated by the fact that quantum patriarch Werner Heisenberg labored to make an atomic bomb for the Nazis. Historians are divided as to whether Heisenberg's postwar claim that the bomb didn't happen because he was quietly sabotaging the effort is a redemptive truth or Heisenberg covering his ass.)

loves each person in a unique manner. For the "I don't believe in organized religion, but I'm a very spiritual person" crowd, there's quantum spirituality and quantum mysticism. Then there's New Age entrepreneur Deepak Chopra, who, in his 1989 book *Quantum Healing,* promises a pathway to curing cancer, reversing aging, and, heavens to Betsy, even immortality.* There's quantum activism, which, as espoused by a New Age physicist in his seminars, "is the idea of changing ourselves and our societies in accordance with the principles of quantum physics." There's "quantum cognition," "spin-mediated consciousness," "quantum neurophysics," and—wait for it—a "Nebulous Cartesian system" of oscillations and quantum dynamics, explaining our freely choosing brains. And as a branch that particularly gets under my skin, there's quantum psychotherapy, a field where one paper proposes that clinical depression is rooted in quantum abnormalities in the fatty acids found in the membranes of platelet cells; gain hope from the knowledge that there are folks pursuing this angle to help you, should you feel suffocatingly sad day after day. Meanwhile, the same journal contains a paper aiming to aid the treatment of schizophrenia sufferers, entitled "Quantum Logic of the Unconscious and Schizophrenia" (in which *quantum* comprises 9.6 percent of the words in the paper's abstract). I'm not gonna lie—I'm not a big fan of folks touting crap like this concerning people in pain.[1]

The nonsense has some consistent themes. There's a notion that if particles can be entangled and communicate with each other instantaneously, there is a unity, a oneness that connects all living things together, including all humans (except for people who are mean to dolphins or elephants). The time travel spookiness of entanglement can be hijacked with the idea that there is no unfortunate event in your past that cannot, in theory, be

*By the way, the quote at the beginning of this section, "Attention and intention are the mechanics of manifestation," was made by someone named Tom Williamson who randomly strings together words from Deepak Chopra's Twitter stream. Two of today's random fictional Chopra quotes at Williamson's site (wisdomofchopra.com) are "A formless void is inside the barrier of facts" and "Intuition reflects your own molecules." The site is discussed in an irresistibly interesting paper by psychologist Gordon Pennycook, entitled "On the Reception and Detection of Pseudo-profound Bullshit."

gone back to and fixed. There's the theme that if you can supposedly collapse a quantum wave just by looking at it, you can achieve nirvana or go into the boss's office and get a raise. According to the same New Age physicist, "The material world around us is nothing but possible movements of consciousness. I am choosing moment by moment my experience." There is also the usual trope that whatever quantum physicists found out with their high-tech gizmos merely confirms what was already known by the Ancients; lotus positions galore. And near-villainous antigrooviness comes from "materialists" with their "classical physics"*—"these elitists who dictate people's experiences of meaning." All this infinite potential is one big blowout salute to the renowned New Age healer Mary Poppins.[†,2]

Some problems here are obvious. These papers, which are typically unvetted and unread by neuroscientists, are published in journals that scientific indexes won't classify as scientific journals (e.g., *NeuroQuantology*) and are written by people not professionally trained to know how the brain works.[3]

But now and then, one's critique of this thinking has to accommodate someone who knew how the brain works, bringing us to the challenging case of the Australian neurophysiologist John Eccles. He wasn't just a good, or even a great, scientist. He was Sir John, Nobel laureate, who pioneered understanding in the 1950s of how synapses work. Thirty years later, in his book *How the Self Controls Its Brain* (Springer-Verlag, 1994), Eccles posited that the "mind" produces "psychons" (i.e., fundamental units of consciousness, a term previously mostly used in cheesy science fiction), which regulate "dendrons" (i.e., functional units of neurons)

---

*That said, some experts, such as philosopher of physics J. T. Ismael of Columbia University, view free will as the product of classical physics.

†In the Broadway musical version (but *not* the movie, I say with inexplicable bitterness), Mary empowers Jane and Michael by singing, "Anything can happen if you let it," a view about exercising free will to prevent exercising unwanted free won't. The song then makes Broadway musical history by rhyming *marvel* ("anything can happen, it's a marvel") with *larval* (Michael: "You can be a butterfly," Jane: "Or just stay larval"). It took decades for Idina Menzel to top this, singing about fractals in "Let It Go."

through quantum tunneling. He didn't merely reject materialism in favor of dualism; he declared himself a "trialist," making room for the category of soul/spirit, which freed the human brain from some of the laws of the physical universe. In his book *Evolution of the Brain: Creation of the Self* (Routledge, 1989), an unironic amalgam of spirituality and paleontology, Eccles tried to pinpoint when this uniqueness first evolved, which hominin ancestor gave birth to the first organism with a soul. He also believed in ESP and psychokinesis, querying new lab members whether they shared these beliefs. By my student days, the mention of Eccles, with his religious mysticism and embrace of the paranormal, elicited nothing but eye-rolling. As a scathing *New York Times* review of *Evolution of the Brain* concluded, Eccles's descent into spirituality invited "Ophelia's lament for Hamlet, 'O! what a noble mind is here o'erthrown.'"[*,4]

Obviously, it's not sufficient for me to reject the idea that quantum indeterminacy is an opening for free will merely by citing the paucity of neuroscientists thinking this way, or by performing the Dirge for Eccles. Time to examine what I see as, collectively, three fatal problems with the idea.

PROBLEM #1: BUBBLING UP

The starting point here is the idea that quantum effects, down there at the level of electrons entangling with each other, will affect "biology." There is precedent for this concerning photosynthesis. In that realm, electrons that have been excited by light are impossibly efficient at finding the fastest way to move from one part of a plant cell to another, seemingly because each electron does this by being in a quantum superposition state, checking out all the possible routes at once.[5]

So that's plants. Trying to pull free will out of electrons in the brain is the immediate challenge—can quantal effects bubble upward, amplify in

*Eccles is usually framed as a sad tale of the ravages of time, a pitiable octogenarian scientist suddenly proclaiming that the brain runs on invisible star stuff. In reality, Eccles was already heading in this direction in his late forties.

their effects, so that they can influence gigantic things, like a single molecule, or a single neuron, or a single person's moral beliefs? Nearly everyone thinking about the subject concludes that it cannot happen because, as we'll soon cover, quantal effects get washed out, cancel each other out in the noise—the waves of superposition "decohere." As summarized nicely by the title of a book by physicist David Lindley, *Where Does the Weirdness Go? Why Quantum Mechanics Is Strange, but Not as Strange as You Think* (Basic Books, 1996).

Nonetheless, people linking quantum indeterminacy with free will argue otherwise. Their challenge is to show how any building block of neuronal function is subject to quantum effects. One possibility is explored by Peter Tse, who considers the neurotransmitter glutamate, where the workings of one of its receptors requires popping a single atom of magnesium out of an ion channel that it blocks. In Tse's view, the location of the magnesium can change in the absence of antecedent causes, because of indeterminate quantal randomness. And these effects bubble up further: "The brain has in fact evolved to *amplify* quantum domain randomness . . . up to a level of neural spike timing randomness" (my emphasis)—i.e., up to the level of individual neurons being indeterminate. And the consequences then ripple upward further into circuits of neurons and beyond.[6]

Other advocates have also focused on quantal effects occurring at a similar level, as captured in one book's title—*Chance in Neurobiology: From Ion Channels to the Question of Free Will.** Psychiatrist Jeffrey Schwartz of UCLA views the level of single ion channels and ions as fair game for quantal effects: "This extreme smallness of the opening in the calcium ion channels has profound quantum mechanical implications." Biophysicist Alipasha Vaziri of Rockefeller University examines the role of "non-classical" physics in determining which *type* of ion flows through a particular channel.[7]

In the views of anesthesiologist Stuart Hameroff and physicist Roger Penrose, consciousness and free will arise from a different part of neurons,

*This is as far as I could get myself to go with Google Translate, as it's in German.

namely microtubules. To review, neurons send axonal and dendritic projections all over the brain. This requires a transport system within these projections to, for example, deliver the building blocks for new copies of neurotransmitter or neurotransmitter receptors. This is accomplished with bundles of transport tubes—microtubules—inside projections (this was briefly touched on in chapter 7). Despite some evidence that they can themselves be informational, microtubules are mostly like the pneumatic tubes in office buildings circa 1900, where someone in accounting could send a note in a cylinder downstairs to the folks in marketing. Hameroff and Penrose (with papers with titles such as "How Quantum Biology Can Rescue Conscious Free Will") focus in on microtubules. Why? In their view, the tightly packed, fairly stable, parallel microtubules are ideal for quantum entanglement effects among them, and it's on to free will from there. This strikes me as akin to hypothesizing that the knowledge contained in a library emanates not from the books but from the little carts used to transport books around for reshelving.[8]

Hameroff and Penrose's ideas have gained particular traction among quantum free-willers, no doubt in part because Penrose won the Nobel Prize in Physics for work concerning black holes and also authored the 1989 bestseller *The Emperor's Mind: Concerning Computers, Minds, and the Laws of Physics* (Oxford University Press). Despite this firepower, neuroscientists, physicists, mathematicians, and philosophers have pilloried these ideas. MIT physicist Max Tegmark showed that the time course of quantum states in microtubules is many, many orders of magnitude shorter-lived than anything biologically meaningful; in terms of the discrepancy in scale, Hameroff and Penrose are suggesting that the movement of a glacier over the course of a century could be significantly influenced by random sneezes among nearby villagers. Others pointed out that the model depends on a key microtubule protein having a conformation that doesn't occur, on types of intercellular connections that don't happen in the adult brain, and on an organelle in neurons being in a place where it isn't.[9]

So, this savaging aside, can quantal effects actually bubble up enough to influence behavior? The indeterminacy that releases magnesium from a single glutamate receptor doesn't enhance excitation across a synapse all that much. And even major excitation of a single synapse is not enough to trigger an action potential in a neuron. And an action potential in one neuron is not enough to make a signal propagate through a network of neurons. Let's put some numbers behind these facts. The dendrite in a single glutamatergic synapse contains approximately 200 glutamate receptors, and remember that we're considering quantal events in a single receptor at a time. A neuron has, conservatively, 10,000–50,000 of those synapses. Just to pick a brain region at random, the hippocampus has approximately 10 million of those neurons. That's 20–100 trillion glutamate receptors (200 x 10,000 x 10,000,000 = 20 trillion, and 200 x 50,000 x 10,000,000 = 100 trillion).* It is possible that an event having no prior deterministic cause could alter the functioning of a single glutamate receptor. But how likely is it that quantum events like these just happen to occur at the same time and in the same direction (i.e., increasing or decreasing receptor activation) in enough of those 20–100 trillion receptors to produce an actual neurobiological event that has no prior deterministic cause?[10]

Apply some similar numbers in the hippocampus to those putative consciousness-producing microtubules: Their basic building block, a protein called tubulin, is 445 amino acids long, and amino acids average out to close to 20 atoms each. Thus, around 9,000 atoms in each molecule of tubulin. Each stretch of microtubule is made up of 13 tubulin molecules. Each stretch of axon contains about 100 bundles of microtubules, each axon helping to make the 10,000–50,000 synapses in each of those 10 million neurons. Again with the zeros.

This is the bubbling-up problem in going from quantum indeterminacy at the subatomic level up to brains producing behavior—you'd need to have a staggeringly large number of such random events occurring at the same time, place, and direction. Instead, most experts conclude that the more likely scenario is that any given quantum event gets lost in the noise

*This is an overestimate, since you're not using every hippocampal neuron at the same time. Still, it's in this ballpark.

of a staggering number of other quantum events occurring at different times and directions. People in this business view the brain not only as "noisy" in this sense but also as "warm" and "wet," the messy sort of living environment that biases against quantum effects persisting. As summarized by one philosopher, "The law of large numbers, combined with the sheer number of quantum events occurring in any macro-level object, assure us that the effects of random quantum-level fluctuations are entirely predictable at the macro level, much the way that the profits of casinos are predictable, even though based on millions of 'purely chance' events." The early-twentieth-century physicist Paul Ehrenfest, in the theorem bearing his name, formalizes how as one considers larger and larger numbers of elements, the nonclassical physics of quantum mechanics merges into old-style, predictable classical physics.* To paraphrase Lindley, this is why the weirdness disappears.[11]

So one glutamate receptor does not a moral philosophy make. The response to this by quantum free-willers is that various features of nonclassical physics can coordinate quantum events among a lot of constituents in the nervous system (and some posit that quantum indeterminacy bubbles up to some extent and meets chaoticism there, piggybacking all the way up to behavior). For Eccles, quantum tunneling across synapses allows for the coupling of networks of neurons in shared quantum states (and note that implicit in this idea and those to follow is that entanglement occurs not just between two particles, but between whole neurons as well). For Schwartz, quantum superposition means that a single ion flowing through a channel is not really singular. Instead, it is a "*quantum cloud of possibilities* associated with the [calcium] ion to *fan out* over an increasing area as it moves away from the tiny channel to the target region where the ion will be absorbed as a whole, or not absorbed at all." In other words, thanks to particle/wave duality, each ion can have coordinated effects far and wide.

*Physicist Sean Carroll emphasizes this dichotomy, noting how in the nonclassical micro world, there is no arrow of time; the only difference between the past and the future is that one is easier to explain and the other is easier to influence, and neither interests the universe. It is only at the macro level of classical physics that our usual sense of time becomes meaningful.

And, Schwartz continues, this process bubbles upward to encompass the whole brain: "In fact, because of uncertainties on timings and locations, what is generated by the physical processes in the brain will be not a single discrete set of non-overlapping physical possibilities but rather a huge *smear* of classically conceived possibilities" now subject to quantum rules. Sultan Tarlaci and Massimo Pregnolato cite similar quantum physics in speculating that a single neurotransmitter molecule has a similar cloud of superposition possibilities, binding to an array of receptors at once and lassoing them into collective action.[*,12]

So the notion that random, indeterministic quantum effects can bubble all the way up to behavior strikes me as a little dubious. Moreover, nearly all the scientists with the appropriate expertise think it is resoundingly dubious.

Somewhere around here it seems useful to approach things on a more empirical level. Do synapses ever actually act randomly? How about entire neurons? Entire networks of neurons?

NEURONAL SPONTANEITY

As a brief reminder: When an action potential occurs in a neuron, it goes hurtling down the axon, eventually reaching all of the thousands of that neuron's axon terminals. As a result, packets of neurotransmitter are released from each terminal.

If you were designing things, maybe each axon terminal's neurotransmitters would be contained in a single bucket, a single large vesicle, which would then be emptied into the synapse. That has a certain logic. Instead, that same amount of neurotransmitter is stored in a bunch of much smaller

[*]For Hameroff, this spatial nonlocality (i.e., how, say, one molecule of neurotransmitter can be interacting with a smear of receptors at once) is accompanied by temporal nonlocality. Back to Libet and chapter 2, where neurons commit to activating muscles before the person consciously believes they have made that decision. But there's an end-around for Hameroff. Quantum phenomena "can cause temporal non-locality, sending quantum information *backward in classical time*, enabling conscious control of behavior" (my italics).

buckets, and all of them are emptied into the synapse in response to an action potential. Your average hippocampal neuron that releases glutamate as its neurotransmitter has about 2.2 million copies of glutamate molecules stored in each of its axon terminals. In theory, each terminal could have all of those copies in our single big bucket vesicle; instead, as noted before, the terminal contains an average of 270 little vesicles, each containing about eight thousand copies of glutamate.

Why has this organization evolved, instead of the single-bucket approach? Probably because it gives you more fine control. For example, it turns out that a large percentage of vesicles are usually mothballed at the back end of the terminal, kept in storage for when needed. Therefore, an action potential doesn't really cause the release of neurotransmitter from *all* the vesicles in each axon terminal. More correctly, it causes releases from all of the vesicles in the "readily releasable pool." And neurons can regulate what percentage of their vesicles are readily releasable versus in storage, a way of changing the strength of the signal across the synapse.

This was the work of Bernard Katz, who got some of his training with Eccles and went on to his own knighthood and Nobel Prize. Katz would isolate a single neuron and, with the use of a particular drug, make it impossible for it to have an action potential. He'd then study what would be happening at a given axon terminal. What he saw was that, amid action potentials being blocked, every now and then, maybe once a minute,* the axon terminal would release a tiny hiccup of excitation, something eventually called a miniature end-plate potential (MEPP). Showing that little bits of neurotransmitter were spontaneously and randomly released.

Katz noted something interesting. The hiccups were all roughly the same size, say, 1.3 smidgens of excitation. Never 1.2 or 1.4. To the limits of measurement, always 1.3. And then, after sitting there recording the occasional 1.3 smidgen-size blip, Katz noticed that much more rarely than that, there'd be a hiccup that was 2.6 smidgens. Whoa. And even more rarely, 3.9 smidgens. What was Katz seeing? 1.3 smidgens was the amount

*Which is glacial from the standpoint of the nervous system—an action potential takes a few thousandths of a second.

of excitation of one single vesicle being spontaneously released; 2.6, the much rarer spontaneous release of two vesicles simultaneously, and so on.* From that came the insight that neurotransmitters were stored in individual vesicular packets, and that every now and then, in a purely probabilistic fashion, an individual vesicle would dump its neurotransmitters—drumroll please—*in the absence of an antecedent cause.*[†,13]

While the field has often viewed the phenomenon as not hugely interesting, often referring to it semisarcastically as "leaky synapses," the notion of there being no antecedent causes turned spontaneous vesicular release of neurotransmitter into an amusement park in which neuroquantologists can gambol. Aha, spontaneous, nondeterministic vesicular neurotransmitter release as the building block for the brain as a cloud of potentials, for being the captain of your fate. Four reasons to be very cautious about this:[14]

> —Not so fast with the no-antecedent-cause part. There's a whole cascade of molecules involved in the process of an action potential causing vesicles to dump their neurotransmitter into the synapse—ion channels open or close, ion-sensitive enzymes are activated, a matrix of proteins holding a vesicle still in its inactive state has to be cleaved, a molecular machete has to cut through more matrix to allow the vesicle to then move toward the neuron's membrane, the vesicle has to now dock to a specific release portal in the membrane. The insights of many fruitful careers in science. Okay, you think you see where I'm going—yeah, yeah, neurotransmitter doesn't just get dumped from out of nowhere, there's this whole complex mechanistic cascade explaining intentional neurotransmitter release, so we'll reframe our free

*"Much rarer." If there was spontaneous release of a single vesicle from an axon terminal an average of once every one hundred seconds, then the probability of two being released simultaneously was once every ten thousand seconds (as in 100 x 100 = 10,000). Three at once? Once every one million seconds. Katz was sitting there for a long time to notice all this.

†I'm forced here to use a term that I have desperately tried to avoid in the main text, because of the confusion it would sow. The phenomenon of neurotransmitter being released in irreducible-size little packets is known as "quantal" release. I'm not going anywhere near why *quantal* and *quantum* have the same roots.

will as when this deterministic cascade happens to be triggered in the absence of an antecedent cause. But no—it's not just when the usual process is triggered randomly, because it turns out that the mechanistic cascade for spontaneous vesicular release is *different* from the cascade for release evoked by an action potential. It's not a random universe hitting a button that normally represents intent. A separate button evolved.[15]

—Moreover, the process of spontaneous vesicular release is *regulated* by factors extrinsic to the axon terminal—other neurotransmitters, hormones, alcohol, having a disease like diabetes, or having a particular visual experience can all alter spontaneous release without having a similar effect on evoked neurotransmitter release. Events in your big toe can change the likelihood of these hiccups happening in the axon terminal of some neuron in the corner of your brain. How would, say, a hormone do this? It sure wouldn't be changing the fundamental nature of quantum mechanics ("Ever since puberty and hormones hit, all I get from her is sullenness and quantum entanglement"). But a hormone can alter the opportunity for quantum events to occur. For example, many hormones change the composition of ion channels, changing how subject they are to quantum effects.[16]

Thus, deterministic neurobiology can make indeterministic randomness more or less likely to occur. It's like you're the director of a show where, at some point, the new king emerges, to much acclaim. And as your direction, you tell the twenty people in the ensemble, "Okay, when the king appears from stage left, shout out stuff like 'Hoorah!' 'Behold, the king!' 'Long life, sire!' 'Huzzah!'—just pick one of those."* And you're pretty much guaranteed to get the mélange of responses you were aiming for. *Determined indeterminacy*. This certainly does not count as randomness being an uncaused cause.[17]

*As noted earlier, my wife is a musical theater director in a school, which is why this scenario comes to mind. And despite expectations, the outcome is never random—in a pattern well known in psychology circles, the ensemble members are most likely to shout out the first or last options in the list, or the one that is most fun to say (e.g., "Yippee!"), particularly loudly. Then there's the rare kid who shouts something like "Elmo!" or "Tofu!" and who is destined for greatness and/or sociopathy.

—Spontaneous vesicular release of neurotransmitters serves a useful purpose. If a synapse has been silent for a while, the likelihood of spontaneous release increases—the synapse gets up and stretches a bit. It's like, during a long period at home, running the car occasionally to keep the battery from dying.* In addition, spontaneous neurotransmitter release plays a large role in the developing brain—it's a good idea to excite a newly wired synapse a bit, make sure everything is working right, before putting it in charge of, say, breathing.[18]

—Finally, there's still the bubbling-up problem.

The bubbling issue brings us to our next level. So individual vesicles randomly dump their contents now and then, ignoring for the moment the issues of its involving unique machinery, being intentionally regulated, and being purposeful. Do enough vesicles ever get dumped all at once to make a major burst of excitation in a single synapse? Unlikely; an action potential evokes about forty times the excitation as does the spontaneous dump of a single vesicle.† You'd need a *lot* of those hiccups at once to produce this.

Scaling up one step higher, do neurons ever just randomly have action potentials, dumping vesicles in all ten thousand to fifty thousand axon terminals, seemingly in the absence of an antecedent cause?

Now and then. Have we now leapfrogged up to a more integrated level of brain function that could be subject to quantum effects? The same caution is called for again. Such action potentials have their own mechanistic antecedent causes, are regulated extrinsically, and serve a purpose. As an example of the last point, neurons that send their axon terminals into muscles, stimulating muscle movement, will have spontaneous action potentials. It turns out that when the muscle has been quiet for a while, a part of it (called the muscle spindle) can make the neurons more likely to have spontaneous action potentials—when you've been still for a long while,

*Yeah, it's mid-2020, and we just discovered that the car's battery is dead, three months now into the pandemic lockdown.

†If you insist: about twenty millivolts for the former, half a millivolt for the latter.

your muscles get twitchy, just so the battery doesn't run down.* Another case where a mechanistic, deterministic regulatory loop can make indeterministic events more likely. Again, we'll get to what to make of such determined indeterminacy.

One level higher—do entire networks, circuits of neurons, ever activate randomly? People used to think so. Suppose you're interested in what areas of the brain respond to a particular stimulus. Stick someone in a brain scanner and expose them to that stimulus, and see what brain regions activate (for example, the amygdala tends to activate in response to seeing pictures of scary faces, implicating that brain region in fear and anxiety). And in analyzing the data, you would always have to subtract out the background level of noisy activity in each brain region, in order to identify what was explicitly activated by the stimulus. *Background noise.* Interesting term. In other words, when you're just lying there, doing nothing, there's all sorts of random burbling going on throughout the brain, once again begging for an indeterminacy interpretation.

Until some mavericks, principally Marcus Raichle of Washington University School of Medicine, decided to study the boring background noise. Which, of course, turns out to be anything but that—there's no such thing as the brain doing "nothing"—and is now known as the "default mode network." And, no surprise by now, it has its own underlying mechanisms, is subject to all sorts of regulation, serves a purpose. One such purpose is really interesting because of its counterintuitive punch line. Ask subjects in a brain scanner what they were thinking at a particular moment, and the default network is very active when they are daydreaming, aka "mind-wandering." The network is most heavily regulated by the dlPFC. The obvious prediction now would be that the uptight dlPFC inhibits the default network, gets you back to work when you're spacing out thinking about your next vacation. Instead, if you stimulate someone's dlPFC, you *increase* activity of the default network. An idle mind isn't the Devil's playground. It's a state that the most superego-ish part of your brain *asks for*

*And now we can't find our AAA card for when the tow truck gets here.

now and then. Why? Speculation is that it's to take advantage of the creative problem solving that we do when mind-wandering.[19]

What is to be made of these instances of neurons acting spontaneously? Back, once again, to the show-me scenario—if free will exists, show me a neuron(s) that just caused a behavior to occur in the complete absence of any influences coming from other neurons, from the neuron's energy state, from hormones, from any environmental events stretching back through fetal life, from genes. On and on. And none of the versions of ostensibly spontaneous activation of a single vesicle, synapse, neuron, or neuronal network constitutes an example of this. None are truly random events that could be directly rooted in quantum effects; instead, they are all circumstances where something very mechanistic in the brain has determined that it's time to be indeterministic. Whatever quantum effects there are in the nervous system, none bubble up to the level of telling us anything about someone pulling a trigger heartlessly or heroically.

PROBLEM #2: IS YOUR FREE WILL A SMEAR?

Which brings us to the second big problem with the idea that quantum mechanics means that our macroscopic world cannot actually be deterministic and free will is alive and well. Rather than the technicalities of leaky synapses, muscle spindles, and quantumly entangled vesicles, this problem is simple. And, in my opinion, devastating.

Suppose there were no issues with bubbling—indeterminacy at the quantum level was not canceled out in the noise and instead shaped macroscopic events dozens of orders of magnitude larger in size. Suppose the functioning of every part of your brain as well as your behavior could most effectively be understood on the quantum level.

It's difficult to imagine what that would look like. Would we each be a cloud of superimposition, believing in fifty mutually contradictory moral systems at the same time? Would we simultaneously pull the trigger and not pull the trigger during the liquor store stickup, and only when the police arrive would the macro-wave function collapse and the clerk be either dead or not?

This raises a fundamental problem that screams out, one that every stripe of scholar thinking about this topic typically wrestles with. If our behavior were rooted in quantum indeterminacy, it would be random. In his influential 2001 essay "Free Will as a Problem in Neurobiology," philosopher John Searle wrote, "Quantum indeterminism gives us no help with the free will problem because that indeterminism introduces randomness into the basic structure of the universe, and the hypothesis that some of our acts occur freely is not at all the same as the hypothesis that some of our acts occur at random. . . . How do we get from randomness to rationality?"* Or as often pointed out by Sam Harris, if quantum mechanics actually played a role in supposed free will, "every thought and action would seem to merit the statement 'I don't know what came over me.'" Except, I'd add, you wouldn't actually be able to make that statement, since you'd just be making gargly sounds because the muscles in your tongue would be doing all sorts of random things. As emphasized by Michael Shadlen and Adina Roskies, whether you believe that free will is compatible with determinism, it isn't compatible with indeterminism.† Or

*Searle, a particularly clear thinker and writer, attacks the implausibility of a dualism that separates self, mind, consciousness, from the underlying biology, sarcastically asking whether, in a restaurant, it would make sense to say to the waiter, "Look, I am a determinist—que será será, I'll just wait and see what I order." What is the problem of free will in neurobiology? According to Searle, it's not whether it exists, independent of underlying biology—it doesn't. For him, the philosophical "solution kicks the problem upstairs to neurobiology." For him, the problem is why we have such strong illusions of free will, and whether that is a good thing. Definitely not, but we'll get to that near the end of the book.

†In addition to randomness being a pretty implausible building block for free will, it turns out that it is extremely hard for people to actually produce randomness. Ask people to randomly generate a sequence of ones and zeros, and inevitably, a significant degree of patterning slips in.

in the really elegant words of one philosopher, "Chance is as relentless as necessity."[20]

When we argue about whether our behavior is the product of our agency, we're not interested in random behavior, why there might have been that one time in Stockholm where Mother Teresa pulled a knife on some guy and stole his wallet. We're interested in the *consistency* of behavior that constitutes our moral character. And in the consistent ways in which we try to reconcile our multifaceted inconsistencies.* We're trying to understand how Martin Luther would stick to his guns and say, "Here I stand, I can do no other," when ordered to renounce his views by ecumenical thugs who burned people at the stake as a hobby. We're trying to understand that lost-cause person who is trying to straighten out their life yet makes self-destructive, impulsive decisions again and again. It's why funerals so often include a eulogy from that person's oldest friend, a historical witness to consistency: "Even when we were in grade school, she already was the sort of person who . . ."

Even if quantum effects bubbled up enough to make our macro world as indeterministic as our micro one is, this would not be a mechanism for free will worth wanting. That is, unless you figure out a way where we can supposedly harness the randomness of quantum indeterminacy to direct the consistencies of who we are.

---

*As an aside that might just be mighty relevant to a book about behavior and responsibility, Searle presents an example of those challenges of integrating dramatic inconsistencies into a coherent whole. He was a renowned philosopher at UC Berkeley, with honorary degrees out the wazoo and a philosophy center named for him. Sociopolitically, he was on the side of angels—as an undergrad at University of Wisconsin in the 1950s, he organized student protests against Wisconsin senator Joe McCarthy, and in the 1960s, he was the first tenured Berkeley professor to join the Free Speech Movement. Admittedly, in his later years, his progressive politics gave way to neoconservatism, but that's the trajectory of many an aging ex-leftist. But most important, in 2017, the then-eighty-four-year-old Searle, with so much to say about moral philosophy, was accused of sexual assault by a research assistant, and following that, a career's worth of allegations of harassment, assault, and sexual quid pro quos with students and staff came to light. Allegations that the university concluded were credible. Thus, moral philosophizing and moral behavior aren't synonymous.

PROBLEM #3: HARNESSING THE RANDOMNESS OF QUANTUM INDETERMINACY TO DIRECT THE CONSISTENCIES OF WHO WE ARE

Which is precisely what is argued by some free-will believers leaning on quantum indeterminacy. In the words of Daniel Dennett in describing this view, "Whatever you are, you can't influence the undetermined event—the whole point of quantum indeterminacy is that such quantum events are not influenced by anything—so you will somehow have to *co-opt it or join forces with it, putting it to use* in some intimate way" (my italics). Or in the words of Peter Tse, your brain "would have to be able to harness this randomness to fulfill information processing aims."[21]

I see two broad ways of thinking about how we might harness, co-opt, and join forces with randomness for moral consistency. In a "filtering" model, randomness is generated indeterministically, the usual, but the agentic "you" installs a filter up top that allows only some of the randomness that has bubbled up to pass through and drive behavior. In contrast, in a "messing with" model, your agentic self reaches all the way down and messes with the quantum indeterminacy itself in a way that produces the behavior supposedly chosen.

Filtering

Biology provides at least two fantastic examples of this sort of filtering. The first is evolution—the random physical chemistry of mutations occurring in DNA provides genotypic variety, and natural selection is then the filter choosing which mutations get through and become more common in a gene pool. The other example concerns the immune system. Suppose you get infected with a virus that your body has never seen before; thus, there's no antibody against it in your body's medicine cabinet. The immune system now shuffles some genes to randomly generate an enormous array of different antibodies. At which point filtering begins.

Each new type of antibody is presented with a piece of the virus, to see how well the former reacts to the latter. It's a Hail Mary pass, hoping that some of these randomly generated antibodies happen to target the virus. Identify them, and then destroy the rest of the antibodies, a process termed positive selection. Now check each remaining antibody type and make sure it doesn't happen to do something dangerous as well, namely targeting a piece of you that happens to be similar to the viral fragment that was presented. Check each candidate antibody against a "self" fragment; find any that attack it and get rid of them and the cells that made them—negative selection. You now have a handful of antibodies that target the novel virus without inadvertently targeting you.[22]

As such, this is a three-step process. One—the immune system determines it's time to induce some indeterministic randomness. Two—the random gene shuffling occurs. Three—your immune system determines which random outcomes fit the bill, filtering out the rest. Deterministically inducing a randomization process; being random; using predetermined criteria for filtering out the unuseful randomness. In the jargon of that field, this is "harnessing the stochasticity of hypermutation."

Which is what supposedly goes on in the filtering version of quantum effects generating free will. In Dennett's words:

> The model of decision making I am proposing has the following feature: when we are faced with an important decision, a consideration-generator whose output is to some degree undetermined, produces a series of considerations, some of which may of course be immediately rejected as irrelevant by the agent (consciously or unconsciously). Those considerations that are selected by the agent as having a more than negligible bearing on the decision then figure in a reasoning process, and if the agent is in the main reasonable, those considerations ultimately serve as predictors and explicators of the agent's final decision.[23]

As such, determining that you are at a decision-making juncture activates an indeterministic generator, and you then reason through which consideration is chosen.* As noted, Roskies does not equate the random noise of nervous systems (rooted in quantum indeterminacy or otherwise) with the headwaters of free will; instead, for Roskies, writing with Michael Shadlen, free will is what's happening when you filter out the chaff from the wheat: "Noise puts a limit on an agent's capacities and control, but invites the agent to compensate for these limitations by high-level decisions or policies† that may be (a) consciously accessible; (b) voluntarily malleable; and (c) indicative of character." Filtering, picking, choosing as an act of sufficient free will and character that, as they state, this "can provide a basis for accountability and responsibility."[24]

Such a harnessing scenario has at least three limitations, of increasing significance:

> —A child has fallen into an icy river, and your consideration generator produces three possibilities to choose among: leap in and save the child; shout for help; pretend you didn't see and scurry away. Choose. But since we're dealing with quantum indeterminacy, what if the first three possibilities are: tango in the absence of a partner; confess to cheating on your taxes; make squawking sounds while jumping backward like the dolphins at Sea World? Perfectly plausible, if superpositioned electron waves are the wellsprings from which your moral decisions flow.

> —To avoid having only tangoing, confessing, and dolphining as options, determine that you need to indeterminately generate *every* random possibility. But now you have to spend a lifetime evaluating and comparing each before choosing which is best. You need to have an impossibly efficient search algorithm.‡,[25]

*Dennett is not necessarily tying his wagon to quantum indeterminacy in this scenario; this is merely a clear description of what harnessing random indeterminacy might look like.

†With Roskies and Shadlen defining "policies" as meaning "constitution, temperament, values, interests, passions, capacities, and so forth."

‡People often frame this in the context of the infinite monkey theorem, the thought experiment where an infinite number of monkeys typing for an infinite length of time eventually produce all of Shakespeare. A feature of the thought experiment explored by many computer scientists is how to most efficiently check which of the infinitely large number of massive manuscripts generated fits the Bard down to each comma. This is hard work because among the manuscripts

—So, phew, generate enough options so that they aren't all silly, figure out how to efficiently evaluate them all, and then use your criteria to filter out all but the winner. But where does that filter, reflecting your values, ethics, and character, come from? It's chapter 3. And where does intent come from? How is it that one person's filter filters out every random possibility other than "Rob the bank," while another's goes for "Wish the bank teller a good day"? And where do the values and criteria come from in even first deciding whether some circumstance merits activating Dennett's random consideration generator? One person might do so when considering whether to commence an act of civil disobedience at great personal cost, while another would when making a fashion decision. Likewise, where do the differences come from as to which search algorithm is used and for how long? Where do all of those come from? From the events, outside the person's control, occurring one second before, one minute before, one hour before, and so on. Filtering out nonsense might prevent quantum indeterminacy from generating random behavior, but it sure isn't a manifestation of free will.

produced will be a zillion that perfectly reproduce Shakespeare until the last page of his final play, until veering off into unique gibberish. One experiment used virtual monkeys typing; after over a billion monkey years (how long is a monkey year of typing?), one monkey typed, "VALENTINE. Cease toIdor:eFLP0FRjWK78aXzVOwm)-';8.t . . . ;." The first nineteen letters occur in *The Two Gentlemen of Verona*; this holds the record for the longest Shakespearean quote by a virtual monkey. Finding algorithms that efficiently filter out the non-Shakespeare from the Shakespeare is often called Dawkins' weasel (after Richard Dawkins [author of *The Blind Watchmaker*], who proposed sorting algorithms in the context of the generation of random variation in evolution. This name represents a merciful reduction in the task for the monkeys, who now merely must type one sentence from *Hamlet*. Hamlet points out a cloud to Polonius that is shaped like a camel. "Yeah, looks like a camel to me," says Polonius. "Methinks it is like a weasel," opines Hamlet, questioning the notion of shared reality while throwing down the gauntlet to the monkey typists.

Footnote about a footnote: Killjoys have suggested that even if a monkey typed all of *Hamlet*, it wouldn't be *Hamlet*, because the monkey hadn't intended to type *Hamlet*, didn't understand Elizabethan culture, and so on. This seems immensely cool to think about, with relevance to Turing machines and artificial intelligence. Borges wrote a wonderful story, "Pierre Menard, Author of the Quixote," about a twentieth-century writer who attempts to so completely immerse himself in seventeenth-century Spanish life that when he re-creates the manuscript of *Don Quixote*, generates it on his own, it will not be a plagiarized copy of Cervantes's *Don Quixote*. Instead, despite the word-for-word similarity, it will actually be Menard's *Don Quixote*. The story is funny as hell and illustrates why there will never be a Chim-Chim's *The Tragedy of Hamlet, Prince of Denmark*.

Okay, another footnote about a footnote: If you search "infinite monkey theorem" on Google Images, about 90 percent of the primates pictured are chimps, who are *apes*, not monkeys. Pisses me off. Some good cartoons, though, about "monkeys" typing sonnets about bananas.

Messing With

To reiterate, in a messing-with model, you don't merely pick and choose among the random quantum effects generated. Instead, you reach down and alter the process. As discussed in the last chapter, downward causation is perfectly valid; the metaphor often used is that when a wheel is rolling, its high-level wheel-ness is causing its constituent parts to do forward rolls. And when you choose to pull a trigger, all of your index finger's cells, organelles, molecules, atoms, and quarks move about an inch.

Thus, supposedly, some high-level "me" reaches down, does some downward causation such that subatomic events produce free will. In the words of Irish neuroscientist Kevin Mitchell, "indeterminacy creates some elbow room. . . . What randomness does, it is posited, is to introduce some room, some causal slack in the system, for higher-order factors to exert a *causal* influence" (my emphasis).[26]

As a first problem, the "controlled randomness" implicit in reaching down and messing with quantum events is as much of an oxymoron as "determined indeterminacy." And where do the criteria come from as to how you're going to mess with your electrons? Amid these issues, the biggest challenge I have in evaluating this idea is that it is truly difficult to understand what exactly is being suggested.

One picture of downward causation changing the ability of quantum events to influence our behavior is offered by libertarian philosopher Robert Kane, who, it will be recalled from chapter 4, suggests that at times of life when we are at a major crossroads of decision-making, the consistent character at play when we choose was formed in the past out of free will (i.e., his idea of "Self-Forming Actions"). But how does that self-formed self actually bring about that decision? At such consequential crossroads, "there is tension and uncertainty in our minds about what to do, I suggest, that is reflected in appropriate regions of our brains by movement away from thermodynamic equilibrium—in short, a kind of stirring up of chaos in the brain that makes it sensitive to microindeterminacies at the neuronal level." In this view, your conscious self uses downward causation to induce neuronal chaoticism in a way that allows quantum indeterminacy to bubble all the way up in exactly the way you've chosen.[27]

Similar messing-with comes from Peter Tse, who, as quoted earlier, argues that "the brain has in fact evolved to amplify quantum domain randomness" (and then speculates that animals that had brains that could do this "procreate better than those that did not"). For him, the brain reaches down and messes with fundamental indeterminacy: "This permits information to be downwardly causal regarding which indeterministic events at the root-most level will be realized."[*,28]

I am nontrivially unsure how Tse proposes this happens. He wisely emphasizes how cause and effect in the nervous system can be conceptualized as the flow of "information." But then a cloud of dualism comes in. For him, downwardly causal information is not materially real, which runs counter to the fact that in the brain, "information" is comprised of real, material things, like neurotransmitter, receptor, and ion channel molecules. Neurotransmitters bind to particular receptors for particular durations; chains of proteins change conformations such that channels open or close like the locks in the Panama Canal; ions flow like tsunamis into or out of cells. But despite that, "information cannot be anything like an energy that imposes forces." However, such information, which is not causal, can allow information that *is* causal: "Information is not causal as a force. Rather, it is causal by allowing those physical causal chains that are *also* informational causal chains . . . to become real." And while informational "patterns" are not material, there are "physically realized pattern detectors." In other words, while information might be made of immaterial dust, the brain's immaterial dust detectors are made of reinforced concrete, steel rebar, and, if you're on the old side, asbestos.

My problem with Kane's and Tse's views, and the similar ones of other philosophers, is that, for the life of me, I can't figure out how such reaching down and messing with microscopic indeterminacy in the brain is supposed to work. I can't get past information being both a force and not

*Note that he is using the less common meaning of *realized*, as something coming into being.

without sensing cake being both had and eaten. When Kane writes, "There is tension and uncertainty in our minds about what to do, I suggest, that is reflected in appropriate regions of our brains by movement away from thermodynamic equilibrium,"[29] I am unclear whether "reflected" is meant to be causal or correlative. Moreover, I know of no biology that explains how having to make a tough decision causes thermodynamic disequilibrium in the brain; how chaoticism can be "stirred up" in synapses; how chaotic and nonchaotic determinism differ in their sensitivity to quantum indeterminacy occurring at a scale many, many orders of magnitude smaller; whether downward causality causing quantum randomness to fuel the consistency of one's choices in life does so by changing *which* electrons entangle with each other, how much nonlocality of time and backward time travel is occurring, or whether the spread of clouds of superpositioned possibilities can be expanded far enough so that, in principle, your olfactory cortex, rather than your motor cortex, sometimes makes you sign a check. It is no longer the challenge I keep raising—"show me a neuron that initiates a complete, coherent behavior for no reason whatsoever, and we can talk seriously about free will." Instead, it's "show me how a neuron accomplishes this for the sorts of reasons offered by these scholars." What we have is a murky version of highly unlikely strong downward causality.

Please believe me—I am so trying to not sound snarky, and to instead seem respectful. I'd certainly come up with bigger cock-ups if I hypothesized about philosophy topics such as agnotology, mereology, or the philosophy of mathematical antirealism. Nevertheless, it seems to me that these free-will advocates are indignantly saying, "We're not claiming that quantum indeterminacy generates our freely chosen decisions *for no reason*. We're saying that quantum indeterminacy does so for magical reasons."*

*Searle gives a particularly clear explication of why the idea of top-down harnessing of randomness to create free will is silly. J. Searle, "Philosophy of Free Will," Closer to Truth, September 19, 2020, YouTube video, 10:58, youtube.com/watch?v=973akk1q5Ws&list=PLFJr3pJl27pIqOCeXUnhSXsPTcnzJMAbT&index=14.

SOME CONCLUSIONS

When people are suggesting that fundamental indeterminacies in how the universe works can be the bases of free will, responsibility, and our sacred sense of agency, only weirdos are referring to Brownian motion of dust particles.

Quantum indeterminacy is beyond strange, and in the legendary words of physics god Richard Feynman, "If you think you understand quantum mechanics, you don't understand quantum mechanics."*

It is perfectly plausible, maybe even inevitable, that there will be quantum effects on how things like ions interact with the likes of ion channels or receptors in the nervous system.

However, there is no evidence that those sorts of quantum effects bubble up enough to alter behavior, and most experts think that it is actually impossible—quantum strangeness is not *that* strange, and quantum effects are washed away amid the decohering warm, wet noise of the brain as one scales up.

Even if quantum indeterminacy did bubble all the way up to behavior, there is the fatal problem that all it would produce is randomness. Do you really want to claim that the free will for which you'd deserve punishment or reward is based on randomness?

The supposed ways by which we can harness, filter, stir up, or mess with the randomness enough to produce free will seem pretty unconvincing. If determined indeterminism is a valid building block for free will, then taking an improv acting class is a valid building block for, à la Sartre, believing that we are condemned to be free.

---

*"Legendary," as in everyone attributes that to Feynman, but I couldn't find an exact source, other than "in one of his [famed] lectures."

AND SOME CONCLUSIONS ABOUT THE LAST SIX CHAPTERS

Reductionism is great. It's a whole lot better to take on a pandemic by sequencing the gene for a viral coat protein than by trying to appease a vengeful deity with sacrificial offerings of goat intestines. Nonetheless, it has its limits, and what the revolutions of chaoticism, emergent complexity, and quantum indeterminism show is that some of the most interesting things about us defy pure reductionism.

This rejection of reductionism carries all sorts of subversive, liberating implications. That bottom-up collectivity built on neighbor-neighbor interactions and random encounters can potentially crush top-down authoritarian control. That in such circumstances, generalists, rather than specialists, are most valuable. That what appears to be a norm, on closer examination, is never actually reached; instead, it is reality oscillating strangely, aperiodically, around a Platonic ideal. That this business about norms applies to being normal, no matter what the cool kids say; there are no actual forms of perfection that we fail to reach—*normal* is a not-quite-accurate descriptor, certainly not a prescription. And that, as a point I emphasize to my students with ham-hocked unsubtlety, if you can explain something of breathtaking complexity, adaptiveness, and even beauty without invoking a blueprint, you don't have to invoke a blueprint maker either.[30]

But despite the moving power in these nonreductive revolutions, they aren't mother's milk that nurtures free will. Nonreductionism doesn't mean that there are no component parts. Or that component parts work differently once there are lots of them, or that complex things can fly away untethered from their component parts. A system being unpredictable doesn't mean that it is enchanted, and magical explanations for things aren't really explanations.

10.5

Interlude

Why did that behavior—dastardly, noble, or ambiguously in between—just occur? Because of what happened a second before, and a minute before, and a . . . The easy takeaway from the first half of this book is that the biological determinants of our behavior stretch widely over space and time—responding to events in front of you this instant but also to events on the other side of the planet or that shaped your ancestors centuries back. And those influences are deep and subterranean, and our ignorance of the shaping forces beneath the surface leads us to fill in the vacuum with stories of agency. Just to restate that irritatingly-familiar-by-now notion, we are nothing more or less than the sum of that which we could not control—our biology, our environments, their interactions.

The most important message was that these are not all separate -ology fields producing behavior. They all merge into one—evolution produces genes marked by the epigenetics of early environment, which produce proteins that, facilitated by hormones in a particular context, work in the brain to produce you. A seamless continuum leaving no cracks between the disciplines into which to slip some free will.

Because of this, as covered in chapter 2, it doesn't really matter what Libet-style experiments do or don't show; it doesn't really matter when intent occurred. All that matters is how that intent came to be. We can't successfully wish to not wish for what we wish for; we can't announce that good and bad luck even out over time, since they're far more likely to progressively diverge. Someone's history can't be ignored, because all we are is our history.

Moreover, as the point of chapter 4, it's biological turtles all the way down with respect to *all* of who we are, not just some parts. It's not the case that while our natural attributes and aptitudes are made of sciencey stuff, our character, resilience, and backbone come packaged in a soul. Everything is turtles all the way down, and when you come to a juncture where you must choose between the easy way and the harder but better way, your frontal cortex's actions are the result of the exact same one-second-before-one-minute-before as everything else in your brain. It is the reason that, try as we might, we can't will ourselves to have more willpower.

Moreover, this seamless continuum of biology and environment forming us doesn't leave room for novel portals of free will by way of the revolutions of chapters 5–10. Yes, all the interesting things in the world can be shot through with chaoticism, including a cell, an organ, an organism, a society. And as a result, there are really important things that can't be predicted, that can never be predicted. But nonetheless, every step in the progression of a chaotic system is made of determinism, not whim. And yes, take a huge number of simple component parts that interact in simple ways, let them interact, and stunningly adaptive complexity emerges. But the component parts remain precisely as simple, and they can't transcend their biological constraints to contain magical things like free will—a brick may want to be something elegant and glamorous, but it will always remain a brick. And yes, truly indeterministic things seem to happen way down at the subatomic level. Nonetheless, it's not possible for that level of weirdness to percolate all the way up to influence behavior, and besides, if you base your notion of being a free, willful agent on randomness, you got problems. As do the people stuck around you; it can be very unsettling when a sentence doesn't end in the way that you potato. Likewise when behavior is random.

As shown in everyday life, in jury boxes, schoolrooms, award ceremonies, eulogies, and the work of experimental philosophers, people hold on to the notion of free will with ferocious tenacity. The pull toward attribution and judgment, whether of others or of ourselves, is enormous and is demonstrable (to varying extents) in cultures all over the world. Heck, even chimps believe in free will.[*,1]

Given that, my goal hasn't been to convince every reader that there is no free will whatsoever. I recognize that I'm on the fringe here, fellow traveling with only a handful of scholars (e.g., Gregg Caruso, Sam Harris, Derk Pereboom, Peter Strawson). I'll settle for merely significantly challenging someone's free-will faith. Sufficiently so that they will reframe their thinking about both our everyday lives and our most consequential moments. Hopefully, you've reached that point.

Nonetheless, we have a big problem, which is that amid all this science and determinism and mechanism, we're still not very adept at predicting behavior. Take someone with extensive frontal cortical damage, and you're on solid ground predicting that their social behavior will be inappropriate, but good luck predicting whether they'll become an impulsive murderer or someone who is rude to a dinner host. Take someone raised in a hellhole of adversity and deprivation, and you're pretty safe predicting that the outcome won't be good, but not much beyond that.

In addition to the unpredictable versions of predictable outcomes, there are a world's worth of exceptions, of thoroughly unpredictable outcomes. Every so often, two rich, brilliant law students murder a fourteen-year-old

---

*Both monkeys and chimps interact differently with a person who is unable to give them food, versus one who is able but unwilling; they don't want to be around the latter: "What a mean hairless primate—they could have given me food but *chose* not to." Particularly interesting work, by psychologist Laurie Santos of Yale, has shown that other primates have their own sense of agency. A human test subject rates their preferences for an array of household items. Find two that are rated equivalently, and force the person to choose one over the other; thereafter, they show a preference for that item: "Hmm, I'm a rational agent of free will, and if I chose this one over that one, it must have been for a good reason." Do the same thing with capuchin monkeys—force them to choose one of two different colored M&M's, have them *believe* that they made a choice (even under circumstances where, unknown to them, their choice is actually forced)—and they show a preference for that color thereafter. If a human chooses for them, no preference emerges.

as a test of their addled philosophy.* Or a Crips gang member facing his second stint in jail has his mug shot go viral and winds up as an international fashion model and brand ambassador for a Swiss fragrance line, squiring around the daughter of a knighted Brit business mogul.† Maybe Laurey, out among the waving wheat in Oklahoma, realizes that Curley's a dull pretty boy, and shacks up with Jud Fry.[2]

Will we ever get to the point where our behavior is entirely predictable, given the deterministic gears grinding underneath? Never—that's one of the points of chaoticism. But the rate at which we are accruing new insights into those gears is boggling—nearly every fact in this book was discovered in the last fifty years, probably half in the last five. The Society for Neuroscience, the world's premier professional organization for brain scientists, grew from five hundred founding members to twenty-five thousand in its first quarter century. In the time it has taken you to read this paragraph, two different scientists have discovered the function in the brain of some gene and are already squabbling about who did it first. Unless the process of discovery in science grinds to a halt tonight at midnight, the vacuum of ignorance that we try to fill with a sense of agency will just keep shrinking. Which raises the question that motivates the second half of this book.[3]

I'm sitting at my desk during afternoon office hours; two students from my class are asking questions about topics from lectures; we wander into biological determinism, free will, the whole shebang, which is what the course is ultimately about. One of the students is dubious about the extent to which we lack free will: "Sure, if there's major damage to this part of the brain, if you have a mutation in this or that gene, free will is diminished, but it just seems so hard to accept that it applies to everyday, normal behavior." I've been at this juncture in this discussion many times, and I've come to recognize that there is a significant likelihood that this student

*Leopold and Loeb. Not to be confused with Lerner and Loewe.
†Jeremy Meeks, the famed "hot felon."

will now carry out a particular behavior—they will lean forward, pick up a pen on my desk, hold it up in the air and say to me, with great emphasis, "There, I just decided to pick up this pen—are you telling me that was completely out of my control?"

I don't have the data to prove it, but I think I can predict above the chance level which of any given pair of students will be the one who picks up the pen. It's more likely to be the student who skipped lunch and is hungry. It's more likely to be the male, if it is a mixed-sex pair. It is especially more likely if it is a heterosexual male and the female is someone he wants to impress. It's more likely to be the extrovert. It's more likely to be the student who got way too little sleep last night and it's now late afternoon. Or whose circulating androgen levels are higher than typical for them (independent of their sex). It's more likely to be the student who, over the months of the class, has decided that I'm an irritating blowhard, just like their father.

Marching further back, it's more likely to be the one of the pair who is from a wealthy family, rather than on a full scholarship, who is the umpteenth generation of their family to attend a prestigious university, rather than the first member of their immigrant family to finish high school. It's more likely if they're not a firstborn son. It's more likely if their immigrant parents chose to come to the U.S. for economic gain as opposed to having fled their native land as refugees from persecution, more likely if their ancestry is from an individualist culture rather than a collectivist one.

It's the first half of this book, providing an answer to their question, "There, I just decided to pick up this pen—are you telling me that was completely out of my control?" Yes, I am.

By now, easy. But I'm really cornered if instead, the student asks something different: "What if everyone started believing that there is no free will? How are we supposed to function? Why would we bother getting up in the morning if we're just machines?" Hey, don't ask me that; that's too difficult to answer. The second half of this book is an attempt to provide some answers.

11

Will We Run Amok?

The notion of running amok has a certain appeal. Rampaging like a frenzied, headless chicken can let off steam. It's often a way to meet new, interesting people, plus it can be pretty aerobic. Despite those clear pluses, I haven't been seriously tempted to run amok very often. It seems kind of tiring and you get all sweaty. And I worry that I'll just seem insufficiently committed to the venture and wind up looking silly.

Nevertheless, there has been no shortage of people who have been delighted to run amok—spittle-flecked, gibbering, and hell-bent on wreaking havoc. While it can break out at any time, certain circumstances predispose people to run amok, particularly ones that promise being spared punishment. Anonymity helps. During what was officially labeled as a "police riot" at the 1968 Democratic National Convention, cops notoriously removed their ID badges before running amok, beating both peaceful protesters and bystanders and destroying film crews' cameras. In a similar vein, across various traditional cultures, when warriors are anonymous (for example, because of wearing masks), the odds increase that they will mutilate the corpses of their enemies. Related to the shield of anonymity, there's "but everyone else was running amok," clearly a variant of amoking because you won't get caught.[1]

The last century brought us a subtler path to feeling like you can run amok with impunity, even if you do so in the glare of the noonday sun. The excuse given was front and center during the Nuremberg trials, as well as among the World War II generation of Germans trying to explain themselves to their sickened descendants. The core of "I was just following orders" when genocidally running amok presupposes a lack of responsibility, culpability, or volition.

The direction this is going should be clear this far into the book, namely the opposite of all those French philosophers contemplating murdering strangers to proclaim their existentialist freedom to choose. If free will is a myth, and our actions are the mere amoral outcome of biological luck for which we are not responsible, why not just run amok?

The recognition that whatever dreadful thing you do is *not your fault* is at the core of the original running amok. *Meng-âmuk*, the Malaysian/Indonesian word that spawned the *amok* of English, refers to the occasional circumstance of some peaceful milquetoast suddenly exploding into inexplicable, indiscriminate, raging violence. The traditional interpretation is one that deftly sidesteps free will—through no fault of their own, the person is believed to be possessed by an evil spirit and is not held accountable for their actions.[2]

"Don't blame me; I was possessed by Hantu Belian, the evil tiger spirit of the forest" is just a hop, skip, and a jump away from "Don't blame me; we are just biological machines."

So if people accept that there is no free will, will everyone just run amok? Some research appears to suggest exactly that.

HARD DETERMINISTS CAREENING THROUGH THE STREETS

To test this, the experimental approach is simple—prime people to decrease their belief in free will, see if they now become jerks. How to make test subjects doubt free will? One effective technique is to have them spend twenty years studying neuroscience, with some behavior genetics, evolutionary theory, and ethology thrown in for good measure. Impractical. Instead, the most common alternative in these studies is for subjects to read a cogent discussion about our lack of free will. Studies have often used a passage from Francis Crick's 1994 book, *The Astonishing Hypothesis: The Scientific Search for the Soul* (Scribner). Crick, of the Watson-and-Crick duo who identified the structure of DNA, grew fascinated with the brain and consciousness in his later years. A hard determinist as well as an elegant, clear writer, Crick summarizes the scientific argument for our being merely the sum of our biological components. "Who you are is nothing but a pack of neurons," he concludes.[3]

Have subjects read that passage by Crick. Control subjects read a doctored version arguing the opposite (e.g., "Who you are is much more than just a pack of neurons") or an excerpt about something dull and unprovocative.* Subjects then fill out a questionnaire about free-will belief (e.g., "How much do you agree with the statement that people must take full responsibility for any bad choices that they make?"); this is to make sure the manipulation actually manipulated subjects effectively.[4]

What happens in the brain when you experimentally diminish people's belief in free will? For one thing, there is a lessening of what is probably best described as the intentionality or effort that people put into their actions. This is shown with using electroencephalography (EEG) to monitor brain waves. Back to the Libet experiment. When a test subject decides to move her finger, there is a characteristic wave pattern, most probably emanating from the motor cortex, about a half second before. But the first sign of the impending behavior is detectable as a wave a few seconds earlier, termed the "early readiness potential." This seems to arise in the presupplementary motor area, one step earlier in the circuit leading to movement and is interpreted as a signal of the intentionality that is going into the

*Variants on the manipulations: Reading single sentences saying things like "Scientists believe that free will is . . ." versus "Scientists believe that free will is not . . ." Having to write a summary of the Crickian (or control) reading. Being asked to recount a time when they exercised a great deal of free will or when they had none.

subsequent movement (and recall that as the centerpiece of chapter 2, Libet reported that the early readiness potential occurred *before* people became consciously aware that they intended to do something; the endless debates ensue). When people are made to feel helpless and with less agency by being stymied by an unsolvable puzzle, the size of their early readiness potentials decreases. And when people are prompted to believe less in free will, the same occurs, with less belief predicting a greater blunting of the wave (without changing the size of the subsequent wave in the motor cortex itself)—people seem to not be trying as hard, focusing as hard on the task.[5]

Another characteristic EEG wave, termed the "error-related negativity" (ERN) signal, occurs when we realize we have made a mistake. This is shown in a "go/no-go" task where a computer screen displays one of two stimuli (say, a red or a green dot), and you have to quickly push a button for one color and inhibit yourself from pushing for the other. The task goes crazy fast, and when people make a mistake, there's an ERN signal from the prefrontal cortex—"Aii, I messed up"—and a slight delay in responding afterward, as people put more effort and attention into getting the right response—"Come on, I can do better than that." First induce a sense of helplessness and inefficacy in subjects, and they then show less of an ERN wave and less post-error slowing (without a change in the actual error rate). Prompt people to believe less in free will and you see the same. Collectively, these EEG studies show that when people believe less in free will, they put less intentionality and effort into their actions, monitor their errors less closely, and are less invested in the outcomes of a task.[6]

Once you're sure that you've induced some free-will skepticism in your subjects, whether assessed by questionnaire or EEG, time to let them loose on the unsuspecting world. Do they run amok? Seemingly.

A series of studies initiated by behavioral economist Katherine Vohs of the University of Minnesota show that free-will skeptics become more antisocial in their behaviors. In experiments, they are more likely to cheat on a test and to take more than their fair share of money from a common pot. They become less likely to help a stranger in need and more aggressive (after being rebuffed by someone, the subject gets to take revenge by determining how much hot sauce the person will have to consume—make someone a free-will skeptic and they nearly double the amount of retributive hot sauce). Less free-will belief and subjects feel less grateful to someone who has done them a favor—why feel gratitude for an act that was someone's mere biological imperative? And just in case it seems like these skeptics are now having too much nihilistic fun by getting to take revenge with a dish served spicily, the manipulation also makes people feel less meaning in their lives and less of a sense of belonging to other humans. Moreover, lessened free-will belief leads to people feeling like they have less self-knowledge, and to feeling alienated from their "true selves" when making a moral decision. This is hardly surprising, whether because the main thing that free-will skepticism does is make you accept that the vast majority of your actions arise from subterranean biological forces that you're completely unaware of, or because of the more global challenge of trying to imagine where the "me" is inside the machine.[*,7]

But there's more. Lessening people's belief in free will lessens their sense of agency, as shown with the clever phenomenon of "intentional binding." Subjects view a hand sweeping around a clock face (at a rate of one rotation every three seconds). Whenever they wish, they press a button, and then estimate where the hand was on the clock at the time. Alternatively, a tone is played at random, and subjects estimate where the hand was when that occurred. Then couple the two—the subject presses the button and the tone comes a fraction of a second later. And people see agency there, unconsciously perceive the tone as being *caused* by their button press, perceive the two events as bound by intentionality, and thus minutely underestimate the time delay between the two.[†] Lessen people's belief in free will and you lessen this binding effect.[8]

Lessening people's belief in free will probably even has bad impli-

---

[*]Vohs's work has been extremely influential and widely cited.

[†]The implicit binding phenomenon has some elaborations. In one study, the button was pressed by another individual; subjects typically underestimated the interval between the pressing and the subsequent tone, showing that they were projecting agency onto the other person . . . unless they thought the timing of the button press was determined by a computer rather than a human.

cations for battling addiction. No, this is not an experiment where volunteers are turned into crackheads and we then see if it gets harder for them to kick the habit if they've been reading Francis Crick. Instead, one can infer this. People generally perceive addiction as involving a loss of free will; moreover, many addiction experts believe that addicts often adopt a deterministic view of addiction as a destructive attribution that allows them to make excuses for themselves. This is a fine line being negotiated. If the choice is between labeling addiction as a biological disease and labeling it as a weak soul pickling in bathtub gin, the former is a vast, humane advance in thinking. But as a step further, if the choice is between labeling addiction as a biological disease that is incompatible with free will and labeling it as one that is, most clinicians would view the latter label as one more likely to help the addict stop. Note, though, that the assumption is that viewing addiction as incompatible with free will is the same thing as its being incompatible with change. That is not remotely the case—wait for chapter 13.[9]

Thus, undermine someone's belief in free will and they feel less of a sense of agency, meaning, or self-knowledge, less gratitude for other people's kindness. And most important for our purposes, they become less ethical in their behavior, less helpful, and more aggressive. Burn this book before anyone else stumbles upon it and has their moral compass unmoored.

Naturally, things are more complicated. For starters, the effects on behavior in these studies are quite small; reading Crick doesn't make subjects more likely both to cheat at some task and to steal the researcher's laptop on the way out. The outcomes were more amok-ish than amok. Reflecting this is the important fact that you don't typically destroy someone's belief in free will with a dose of Crick. Instead, you just make them a bit less ardent in their belief (without changing the extent to which they value their free will).* This is hardly surprising—how likely is it that reading a passage from a book, being informed that "scientists now

*Which, it should be noted, suggests that even if you decrease free-will belief a smidgen, people who nonetheless still believe overall in free will become more amok-ish. Not great news.

question . . . ," or even being prompted to recall a time when you had less free will than you thought, will have much of an effect on your fundamental feelings about how much agency you have in life? A belief in free will is generally ingrained in us by the time we learn about the sins of gluttony from *The Very Hungry Caterpillar*.[10]

Most important, the bulk of studies have failed to replicate the basic finding that people become less ethical in their behavior when their free-will belief is weakened. Importantly, some of these studies had much larger sample sizes than the original ones that generated the "we'll all run amok" conclusions. A 2022 meta-analysis of the entire literature (consisting of 145 experiments, with 95 unpublished) shows that Crickian manipulations do indeed mildly lessen free-will belief and increase belief in determinism . . . without any consistent effects on ethical behavior.[*,11]

Thus, the literature shows that it is virtually impossible to use a brief experimental manipulation to make someone into a true free-will skeptic; furthermore, even if you merely lessen someone's overall acceptance of free will, there isn't actually the consistent effect of compromising their ethical behavior in laboratory settings.

These conclusions have to be a bit tentative because, all things considered, there hasn't been a huge amount of research in this area. However, "Don't blame me for stealing that child's candy; there's no free will" has a close cousin that has been studied in great depth indeed, and the findings are immensely interesting and teach us a ton.

AN IDEAL MODEL SYSTEM

Thus, we consider the parallel of there's-no-free-will amok-ness: Do people behave immorally when they conclude that they will not ultimately be held responsible for their actions because there is no Omnipotent Someone

[*] On a related note, throwing some Crick at judges lessens their free-will belief . . . without changing their judgments. Why am I bothering writing this book?

doling out the consequences? As per Dostoyevsky, if there is no God, then everything is permitted.

Even before considering atheists, it's worth appreciating something about gods who judge and punish—they are far from universal or ancient. Fascinating work by psychologist Ara Norenzayan of the University of British Columbia shows that such "moralizing gods" are relatively new cultural inventions. Hunter-gatherers, whose lifestyle has dominated 99 percent of human history, do not invent moralizing gods. Sure, their gods might demand a top-of-the-line sacrifice now and then, but they have no interest in whether humans are nice to each other. Everything about the evolution of cooperation and prosociality is facilitated by stable, transparent relationships built on familiarity and the potential for reciprocity; these are precisely the conditions that would make for moral constraint in small hunter-gatherer bands, obviating the need for some god eavesdropping. It was not until humans started living in larger communities that religions with moralizing gods started to pop up. As humans transitioned to villages, cities, and then protostates, for the first time, human sociality included frequent transient and anonymous encounters with strangers. Which generated the need to invent all-seeing eyes in the sky, the moralizing gods who dominate the world religions.[12]

Thus, if belief in a moralizing god(s) is what keeps us in line, it's obvious where lack of belief should take you. This generates the inevitable exchange that every atheist has to endure at some point:

Theist: How can we trust you atheists to be moral if you don't think that God holds you responsible for your actions?

Atheist: Well, what does that say about you religious people, if you only act morally because otherwise you'll burn in hell?

Theist: At least we have morals.

Etc.

How are we supposed to function if no one believes in free will? Much can be gleaned by seeing how people function when they don't believe in a moralizing god.

(Not: Amid the common picture, one's attitudes about religion and about the existence of free will are not inevitably connected. We're just looking at atheism in depth as a warm-up to returning to the challenges of rejecting the notion of free will.)

ATHEISTS GONE WILD

Do atheists run amok? Most people sure believe that, and antiatheist prejudice runs wide and deep. There are fifty-two countries in which atheism is punishable by death or prison. Most Americans have negative perceptions of atheists, and antiatheist prejudice is more prevalent than antipathy toward Muslims (which comes in in second place), African Americans, LGBQT individuals, Jews, or Mormons. Such negativity is permeating in its consequences. Mock juries give atheists longer jail sentences; defense attorneys increase their likelihood of success in emphasizing their client's theism; people put the supposed atheist's name further down on a hypothetical list for an organ transplant; custody of a child has been denied to parents because of their atheism. Some states still have laws on the books barring atheists from holding public office; in more enlightened cantons, voters are less likely to elect people because of their atheism. In the U.S., atheists have higher rates of clinical depression than do the religious, and some of this likely reflects atheists' marginalized, minority status (approximately 5 percent of Americans, according to surveys).[*,13]

Here's how unlikely a place antiatheist prejudice can pop up. Psychologists Will Gervais and Maxine Najle of the University of Kentucky recount the story of a shoe company in Germany that was getting a lot of

* Is this because of that depressogenic void left by a lack of a god? Perhaps in part, but the minority status probably plays a role as well—in markedly secular Scandinavian countries, it is the minority who are highly religious who have higher rates of depression.

complaints from Americans—shoes bought online were greatly delayed or never delivered. The name of the company? Atheist Shoes. The owner did an experiment where half the shipments to America were sent without the company's name on the label, half with. The former were delivered promptly; the latter were frequently delayed or lost. U.S. postal workers were taking a stand against the presumed immorality of those atheistic shoemakers, making sure no God-fearin' American might inadvertently walk a mile in *those* shoes. No such phenomenon was observed with shoes sent within Europe.[14]

Why the bias against atheists? It's not because they are viewed as less warm or competent than religious people. Instead, it is always about morality—the widespread belief that believing in a god is essential for morality, held by the majority of Americans and more than 90 percent of people in places like Bangladesh, Senegal, Jordan, Indonesia, and Egypt. People in most countries surveyed associate atheism with moral norm violations, such as serial murder, incest, or necrobestiality.* In one study, religious Christians reported a sense of visceral disgust when reading an atheistic tract. Even atheists associate atheism with norm violations, which is pretty pathetic; behold, the self-hating atheist.[†,15]

Thus, the expectation that atheists might run amok at any moment is deeply entrenched (just to soften the sting a bit, religious people have similar, if lesser, biases against those "spiritual but not religious" fellow travelers). But on to the key question: Do atheists actually show fewer prosocial behaviors and more antisocial ones than do religious people?[16]

Right off the bat, there's a huge impediment to getting a clear answer

*Necrophilia *and* bestiality—come on, really? This atheist is finally getting a bit fed up with this.

†This antiatheist bias runs alongside a widespread belief that being a scientist precludes being moral (amid scientists generally being respected and viewed as "normal" in degree of caring, trustworthiness, or valuing fairness, and not particularly prone toward atheism). Instead, scientists are viewed as being immoral in realms of loyalty, purity, and obedience to authority. One reason makes sense to me, amid its nearly always being wrong—that in the pursuit of scientific findings, scientists would not hesitate to do things that would be considered immoral by some people (e.g., vivisection, human experimentation, fetal tissue research). The second reason kind of floors me—that scientists would be willing to undermine moral norms by promulgating something, just because it happens to be . . . true.

to a question like this. Suppose you wonder if some new drug can protect against some disease. What do you do? You get two groups of volunteers, matched for age, sex, medical history, and so on; a randomly selected half get the drug, half a placebo (without subjects knowing which they got). But you can't do that with studies of things like religiosity. You don't take two groups of blank-slate volunteers, command half to embrace religion and half to reject it, and then see who is nice out in the world.* It's not random who winds up being religious or atheist—as one example that we will return to, men are more than twice as likely as women to be atheists. Similarly, free-will believers and skeptics don't get to those stances by a coin toss.

Another complication in these theist/atheist studies is obvious to anyone who has been stranded on a desert island with a Unitarian and an evangelical Southern Baptist—religion and religiosity are crazy heterogeneous. Which religion? Is someone a lifelong adherent or a recent convert? Is the person's religiosity mostly about their personal relationship with their deity, with their coreligionists, with humans in general? Is their god all about love or smiting? Do they typically pray alone or in a group? Is their religiosity more about thoughts, emotions, or ritualism?[†,17]

Nonetheless, the bulk of studies in this large literature support the

*A similar challenge hobbles the literature showing that religious belief seems to have some health benefits: "You, yeah you, you start believing. You, over there, you don't. Let's meet in twenty years and check your cholesterol levels."

†There is, of course, similar if less studied heterogeneity to styles of atheism—people who mostly arrived at their stance analytically versus emotionally, people raised with belief who seceded versus those who were never believers, people whose stance is an active versus a passive one (stay tuned for the end of this chapter), gradually acquired versus arising from a non-Zeusian bolt of lightning. Amid that heterogeneity, though, most atheists seem to have gotten to where they are by an analytic route (not me, though), and when people are experimentally prompted to think more analytically, they also then report less religious belief. And then there are atheists who, nonetheless, embrace some religion's culture and rituals or embrace the stable supportiveness of a humanistic community of nonbelievers, versus those doing their atheism in a solitary way. All this brings to mind the argument in *Catch-22* between Yossarian and Mrs. Scheisskopf, both atheists, about the nature of the God they don't believe in. The bitter Yossarian wishes there were a God so that he could express the violence and hatred he feels toward Him for His divine cruelty; Mrs. Scheisskopf is horrified by this blasphemy, insisting that the God she doesn't believe in is warm, loving, and benevolent.

notion that deciding that there is no god to monitor you makes for rottener people. As compared with religious people, atheists are less honest and trustworthy, are less charitable in both experimental settings and out in the real world, volunteer less of their time. Case closed. The only question now is whether people who don't believe in a god or people who don't believe in free will run amok faster.

What we now need to do is to deconstruct this general finding. Because, naturally, the actual picture is way different and very pertinent to free-will skepticism.

SAYING VERSUS DOING

The first issue to deal with should be a no-brainer. If you are interested in these issues, do you observe how charitable study subjects are, or do you just ask them how often they give to charity? Asking someone just tells you how charitable they want to appear. A large percentage of the relevant literature is based on "self-reporting" rather than empirical data, and it turns out that religious people are more concerned than atheists with maintaining a moral reputation, arising from the more common personality trait of being concerned about being socially desirable.[18] This no doubt reflects the fact that theists are more likely than atheists to live their moral lives in the context of a cohesive social group. Moreover, concern of religious people with social desirability is greater in more religious countries.[19]

Once you actually observe what people do, rather than listen to what they say, there's no difference between theists and atheists in rates of blood donations, amount of tipping, or compliance with "honor system" payments; ditto for a lack of difference in being altruistic, forgiving, or evincing gratitude. Furthermore, there's no difference in being aggressive or vengeful in experimental settings where subjects can retaliate against a norm violation (for example, by administering what they believe to be a shock to someone).[20]

Thus, observe what people do rather than what they say, and the differences in prosociality between theists and atheists mostly disappears. The lesson for studying free-will believers versus skeptics is obvious. Collectively, the studies examining what people actually do in an experimental setting show no difference in ethical behavior between the two groups.

OLD, RICH, SOCIALIZED WOMEN VERSUS YOUNG, POOR, SOLITARY GUYS

Back to that self-selection challenge: when compared with atheists, religious people are more likely to be female, older, married, and of higher socioeconomic status, and to have a larger and more stable social network. And this is a minefield of confounds because, independent of religiosity, these are all traits associated with higher levels of prosocial behavior.[21]

Being in a stable social network seems to be really important. For example, the increased charitability and volunteering found in religious people is not a function of how often they pray but, instead, how often they attend their house of worship, and atheists who show the same degree of involvement in a close-knit community show the same degree of good-neighborliness (in a similar vein, controlling for involvement in a social community significantly lessens the difference in rates of depression among theists versus atheists). Once you control for sex, age, socioeconomic status, marital status, and sociality, most of the differences between theists and atheists disappear.[22]

The relevance of this point to free-will issues is clear; the extent to which someone does or doesn't believe in free will, and how readily that view can be altered experimentally, is probably closely related to variables about age, sex, education, and so on, and these might actually be more important predictors of running-amok-ness.

WHEN YOU'RE PRIMED TO BE GOOD FOR GOODNESS' SAKE

Where religious people tend to become more prosocial than atheists is when you remind the former of their religiosity. This can be done explicitly: "Do you consider yourself to be religious?" More interesting is when religious people become more prosocial after being implicitly primed about their religiosity—for example, ask someone to unscramble a list of words that includes religious terms (versus no such words listed) or to list the Ten Commandments (versus ten books they read in high school). Other approaches include having a subject walk down a block that does or doesn't contain a church, or playing religious versus secular music in the background in the test room.[23]

Collectively, these studies show that religious primes bring out the best in religious people, making them more charitable, generous, and honest, more resistant to temptation, and more capable of exerting self-control. In such studies, some of the most effective implicit primes bring divine reward and punishment to mind (raising the interesting question of whether better behavior is evoked by unscrambling "lehl" versus "neehav").[24]

Now we're getting somewhere. When religious people are not thinking about their religious principles, they sink into the same immoral muck as atheists. But remind them of what really matters, and the halo comes out.

Two big complications: The first is that in a lot of these studies, implicit religious primes make atheists more prosocial as well. After all, you don't have to be a Christian to decide that the Sermon on the Mount has good parts. But as a more informative complication, while prosociality in religious people is boosted by religious primes, prosociality in atheists is boosted *just as much* by the right kinds of secular primes. "I'd better be good or else I'll get into trouble" can certainly be primed by "alij" or "eocpli." Prosociality in atheists is also prompted by loftier secular concepts, like "civic," "duty," "liberty," and "equality."[*,25]

*An interesting parallel occurs with the notion that during times of trouble, atheists lack the larger structures of comfort available to theists. In reality, at such times, many atheists resort to and gain

In other words, reminders, including implicit ones, of one's ethical stances, moral principles, and values bring forth the same degree of decency in theists and atheists. It's just that the prosociality of the two groups is moored in different values and principles, and thus primed in different contexts.

Obviously, then, what counts as moral behavior is crucial. Work by psychologist Jonathan Haidt of New York University groups moral concerns into five domains—those related to obedience, loyalty, purity, fairness, and harm avoidance. His influential work has shown that political conservatives and highly religious people tilt in the direction of particularly valuing obedience, loyalty, and purity. The Left and the irreligious, in contrast, are more concerned with fairness and harm avoidance. This can be framed with highfalutin philosophy. One can approach a moral quandary as a deontologist, believing that the morality of an action should be evaluated independently of its consequence ("I don't care how many lives it saves, it's never okay to . . ."). This contrasts with being a consequentialist ("Well, I'm normally opposed to X, but the good that it will accomplish in this case outweighs . . ."). So who are the deontologists, theists or atheists? It depends. Religious people tend toward deontology about obedience, loyalty, and purity—it is never okay to disobey an order, turn on your group, or desecrate the sacred. However, when it comes to issues of fairness and harm avoidance, atheists tend to be as deontological as the religious.[26]

Differences in values show up in an additional way. The highly religious tend to view good works more in a personal, private context, helping to explain why religious Americans donate more of their income to charity than do the secular. In contrast, atheists are more likely to view good works as a collective responsibility, helping to explain why they are the ones who are more likely to support candidates advocating wealth redistribution to decrease inequality. Thus, if you're trying to decide who is more likely to run amok with antisocial behaviors, atheists will look bad if the question is "How much of your money would you give to charity for the

comfort from their belief in science.

poor?" But if the question is "How much of your money would you pay in higher taxes for more social services for the poor?" you'll reach a different conclusion.[27]

The relevance to free-will believers versus skeptics? Obvious—it depends on what the prime is and what value is being evoked. This generates a simple prediction: implicitly prime someone by asking them to spot the misspellings in "Captaim of yeur gate," and free-will believers will be more influenced, and in the direction of showing more self-control. In contrast, try "Victin of vircumsrance," and it is free-will skeptics who will become less punitive and more forgiving.

ONE ATHEIST AT A TIME VERSUS AN INFESTATION OF THEM

The preceding sections suggest that deciding that there is no omnipotent being to punish transgressions doesn't send atheists into a downward moral spiral. It should be noted, however, that a huge percentage of the research discussed has been with American subjects, from a country where only roughly 5 percent of people say they are atheists. We saw that prosociality can even be enhanced in atheists by religious primes. Maybe the relative morality of atheists is due to being surrounded by the morality of all those theists rubbing off on them. What would happen if most people became atheist or irreligious—what sort of society would they construct, when everyone is freed from being nice due to the fear of God?

A moral and humane one, and this conclusion is not based on a thought experiment. What I'm referring to are those ever reliably utopian Scandinavians. Religiosity throughout the region plummeted through the twentieth century, and Scandinavian countries are the most secular in the world. How do they stack up when compared with a highly religious country such as the U.S.? Studies of quality of life and of health show that Scandinavians fare better (on measures such as happiness and well-being, life expectancy, infant mortality rates, and rates of death in childbirth); moreover, poverty rates are lower, and income inequality is tiny in comparison. And measures of the prevalence of antisocial behavior, crime rates, and rates of violence and damaging aggression—from warfare to criminal violence to school bullying to corporal punishment—are lower. And as for some indices of prosociality, Scandinavian countries' per-capita expenditures on social services for their own citizens* and as aid for poor countries are greater.[28]

Furthermore, these differences are not merely Scandinavians feasting on lutefisk versus sweaty, capitalistic Americans. Across a broad range of countries, lower average rates of religiosity predict higher rates of all these salubrious outcomes. Moreover, cross-nationally, lower average rates of religiosity in a country predict lower levels of corruption, more tolerance of racial and ethnic minorities, higher literacy rates, lower rates of overall crime and of homicide, and less frequent warfare.[29]

Correlative studies like these always have the major problem of not telling anything about cause or effect. For example, do lower rates of religiosity result in governments spending more on social services for the poor, or does governments spending more on such services result in lower rates of religiosity (or do both arise from a third factor)? It's hard to answer, even with the well-documented Scandinavian profile, since the decline in religiosity and the Scandinavian social welfare model emerged in parallel. It's probably some of both. The preference of atheists for collective responsibility for good works would certainly help foster Scandinavian models. And as societies become more economically stable and safer, rates of religious belief decline.[30]

Separate from these complex chicken-and-egg issues, we have a clear

---

*It should be noted that while Scandinavian governments expend more money on the poor than does the U.S., Scandinavian people give individually to charities at a lower rate than Americans; however, the higher rates of governmental social services in Scandinavia more than offset the higher rates of charitability in the U.S. The distinctive cultural responses to tragedy in one Scandinavian country will be explored in chapter 14.

answer to our question of whether less religious countries swarm with citizens running amok. Not at all. In fact, they're downright Edenic.*

Thus, it is not the case that atheists match theists in their morality simply because, thank god(s), the former are constrained by the abundance of the latter. My guess is that similarly, ethical behavior by free-will skeptics is not a function of their being a minority surrounded by go-getters brimming with a sense of agency.

This brings us to what is probably the most important point in evaluating whether religious people are more prosocial than atheists, as well as considering the prospects of free-will believers being more prosocial than free-will skeptics.

WHO NEEDS THE HELP?

Even after controlling for factors like self-reporting or demographic correlates of religiosity, and after considering broader definitions of prosociality, religious people still come through as being more prosocial than atheists in some experimental as well as real-world settings. Which leads us to a really crucial point: *religious prosociality is mostly about religious people being nice to people like themselves.* It's mostly in-group. In economic games, for example, the enhanced honesty of religious subjects extends only to other players described to them as coreligionists, something made more extreme by religious primes. Moreover, the greater charitability of religious people in studies is accounted for by their contributing more to coreligionists, and the bulk of the charitability of highly religious people in the real world consists of charity to their own group.[31]

But being on guard for confounds, maybe this is a spurious relationship—maybe religious people are kinder to coreligionists because they are likely

*Okay, despite my obvious enthusiasm, it is crucial to point out how Scandinavian countries have gotten a ton of egalitarian mileage out of being small, ethnically/linguistically homogeneous countries, and more American-esque problems are emerging as they become less so. And then there's ABBA.

to be living among them. Thus, maybe the kindness is driven not by the religiosity but by the familiarity. This is probably not the case, though. For example, one cross-cultural study of fifteen different societies showed that in-group favoritism of religious people extended to distant coreligionists they had never met.[32]

Thus, despite claims toward universal niceness, theistic niceness tends to be in-group. Moreover, this is particularly pronounced in religious groups characterized by fundamentalist belief and authoritarianism.[33]

How about when it comes to out-group members? In those circumstances, it is atheists who are more prosocial, including more accepting of and extending protection to Thems. Moreover, religious primes can make religious people more prejudiced against out-group members, including increasing vengefulness and willingness to punish their transgressions. In one classic study, religious schoolchildren viewed it as unacceptable for a population of innocent people to be destroyed . . . unless it was presented as Joshua's destruction of the innocent population of Jericho in the Old Testament. In another, religious primes resulted in fundamentalist West Bank Jewish settlers expressing more admiration for a Jewish terrorist who had killed Palestinians. In one study, merely walking past a church resulted in religious Christians expressing more negative feelings about atheists, ethnic minorities, and LGBTQ individuals. In another, priming Christian subjects with the Christian version of the Golden Rule did not reduce homophobia; however, priming them with what they were told was the Buddhist equivalent of the Golden Rule *increased* homophobia. Finally, some oft-cited studies looked at how aggressive subjects would be to an opponent in a game (e.g., the volume of loud noise they would choose to blast the other player with). Such aggressiveness was increased when subjects had first read a passage mentioning God or the Bible, relative to passages without that; aggressiveness was increased even more when subjects read a passage about biblical vengefulness sanctioned by God versus the same description of vengefulness without the divine sanction.[34]

Thus, a variety of studies shows that when it comes to theists versus nontheists being kind to someone, it really depends on who that someone is. And the majority of experimental studies examining these issues have involved subjects thinking about in-group members. Just imagine—a professor who studies the subject recruits a bunch of Psych 101 students to participate in a study of how generous and trustworthy they are. As part of it, they play an online economic game, supposedly against someone in the next room. Who do you imagine the students implicitly assume is in that next room—a fellow classmate or a yak herder from Bhutan? Experimental designs like these implicitly prompt subjects to think of other participants, hypothetical or otherwise, as in-group members, thus disproportionately priming for more prosociality from theists than from atheists.

How would the issue of who is being helped play out in comparing free-will believers with skeptics? I would imagine that free-will believers will feel more of a moral imperative (versus an instrumental strategy) to help someone who is making an extra effort at something, while free-will skeptics will feel more of an imperative to understand the actions of someone very different from them.

We return to the broad question in this section: Does disbelief that one's actions are judged by an omnipotent force degrade morality? Seemingly so. That is, as long as you are asking people to say how moral they are rather than to demonstrate it, or you prime them with religious cues rather than secular ones of equivalent symbolic power. And as long as "good acts" are individualistic rather than collective, and are directed at people who look like them. Skepticism about the existence of a moralizing god(s) doesn't particularly generate immoral behavior; this is the case for underlying reasons that help explain why being skeptical about free will doesn't either.

Now to the most important point about the menace of skeptics running amok. Asking about differences between free-will believers and skeptics is the wrong question.

INTO THE VALLEY OF THE INDIFFERENT

Consider this U-shaped curve:

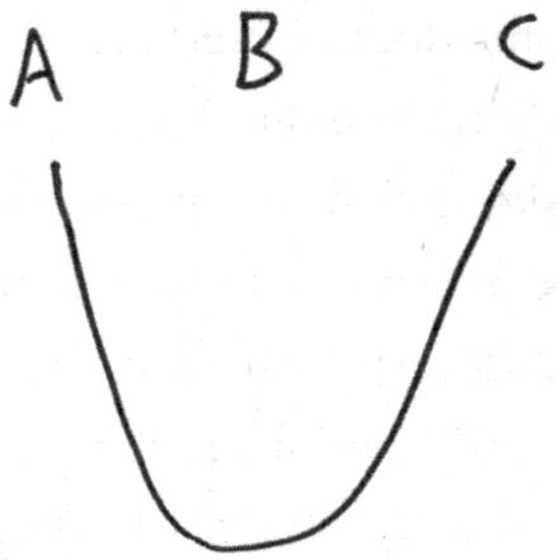

On the left (A) are people who firmly believe that there is no free will, period; in the trough (B) are those whose belief in free will is a bit malleable, while on the right (C) are those whose belief in free will is unshakable.

Back to the Temptation of Crick. The collection of volunteer subjects in the studies reviewed was almost certainly comprised of people in category B or C, given the rarity with which free will is completely rejected. What, collectively, do those studies show?

> —First, when free-will believers read about how there is no free will, on the average, there is a small decrease in belief in free will, and with a lot of variability, reflecting the fact that some of the people are unmoved by arguments against free will. As such, subjects whose belief shifts can be thought of as category B, those who are unshakable, category C.
>
> —The more a subject's faith in free will is shifted, the more likely they are to act unethically in the experiment.

In other words, when it comes to beliefs about the nature of human agency and responsibility, it's category B people who run amok, not those in category C. This entire literature bypasses the thing we're really interested in, which is whether categories A and C differ in their moral uprightness.

To my knowledge, only one study has examined this explicit question, carried out by psychologist Damien Crone, then at the University of Melbourne in Australia, and philosopher Neil Levy, whose ideas have already been discussed. Subjects stoutly believed in free will, or were those who identified their free-will skepticism as long-standing. The really excellent study even examined the reasons why particular subjects rejected free will, contrasting scientific determinists (endorsing statements like "Your genes determine your future"*), with fatalistic determinists ("The future has already been determined by fate"). In other words, these were free-will skeptics who had arrived at their stances through different emotional and cognitive routes. The commonality was that they had rejected belief in free will long ago.[35]

The results? Free-will skeptics (of whatever stripe) and free-will believers were identical in their ethical behavior. And as a finding that ultimately tells the whole story, people who most defined themselves by their moral identity were the most honest and generous, regardless of their stance about free will.[36]

The identical pattern holds when considering religious belief and morality. Category A are atheists whose paths to that view are scarred with craters—"Losing my religion was the loneliest moment of my life" or "It would have been so easy to continue after all those years, but that's when I left my seminary." Category C? People for whom their belief is daily bread rather than cake on Sunday,† informing their every action, who know who they are and what God expects them to do.‡ And then there is category B, covering the range from apatheists, for whom saying that they don't believe in God is like saying that they don't ski,§ as well as those

*Just as an important reminder from chapter 3, genes don't determine your future; instead, they work in different ways in different environments. Nonetheless, a stance of "It's all genetic" is an acceptable stand-in in this case for "It's all biological."

†To paraphrase Henry Ward Beecher.

‡To paraphrase Tevye.

§To paraphrase comedian Ricky Gervais (as cited by, hmm, psychologist Will Gervais).

whose religiosity is out of habit, convention, nostalgia, an example for the kids—of the 90 percent of Americans who are theists, probably half fall into this category, given that approximately half don't go to religious services regularly. As the immensely important point, when it comes to ethical behavior, daily-bread theists and daily-bread atheists resemble each other more than they resemble those in category B.[37]

For example, highly religious and highly secular people score the same on tests of conscientiousness, coming out higher than those in the third group. In experimental studies of obedience (usually variants on the classic research of Stanley Milgram examining how willing subjects are to obey an order to shock someone), the greatest rates of compliance came from religious "moderates," whereas "extreme believers" and "extreme nonbelievers" were equally resistant. In another study, doctors who had chosen to care for the underserved at the cost of personal income were disproportionately highly religious or highly irreligious. Moreover, classic studies of the people who risked their lives to save Jews during the Holocaust documented that these people who could not look the other way were disproportionately likely to be either highly religious or highly irreligious.[38]

Here is our vitally important reason for optimism, about how the sky won't necessarily fall if people *come to* stop believing in free will. There are people who have thought long and hard about, say, what early-life privilege or adversity does to the development of the frontal cortex, and have concluded, "There's no free will and here's why." They are a mirror of the people who have thought long and hard about the same and concluded, "There's still free will and here's why." The similarities between the two are ultimately greater than the differences, and the real contrast is between them and those whose reaction to questions about the roots of our moral decency is "Whatever."

12

The Ancient Gears within Us: How Does Change Happen?

This book has a goal—to get people to think differently about moral responsibility, blame and praise, and the notion of our being free agents. And to feel differently about those issues as well. And most of all, to *change* fundamental aspects of how we behave.

This is the goal of many of the things we are exposed to: to change our behavior. That's certainly what is going on with most speeches, lectures, books—e.g., to change whom you vote for, what you believe the first seven days of the universe were like, or your commitment to the workers of the world uniting and losing their chains. The same for lots of our interpersonal interactions—to persuade, convince, recruit, compel, repel, induce, seduce. And of course, there are the efforts to get you to change your behavior in a way that will make every remaining moment of your life *so* much happier if only you buy the object being advertised.

All these ways to make you and everyone else change their behavior.

Which raises a gigantic question. Last chapter's question was "If people stopped *believing* in free will, would there be amoral chaos?" This chapter's question is "If there *is* no free will, how does anything ever change?" How do you decide shortly after this sentence to change your behavior and grab a brownie? If the world is deterministic on the level that matters, isn't everything thus *already* determined?

The answer is that we don't *change* our minds. Our minds, which are the end products of all the biological moments that came before, are *changed* by circumstances around us. Which seems like a thoroughly unsatisfying response that is incompatible with your intuitions about how you function.

As such, the goal of this chapter is to reconcile an absence of free will with the fact that change occurs. To do so, we're going to look at how behavior changes in organisms far simpler than humans, down on the level of molecules and genes. This will segue to considering behavioral change in us. Hopefully, this will make clear an immensely important point: When our behavior changes, it doesn't involve biology with some themes and motifs similar to ones seen in these simpler organisms. It involves the *same* molecules, genes, and mechanisms of neuronal function. When you begin to be biased against some alien group of people because their customs differ from your own, the biology underlying your change in behavior is the same as when a sea slug learns to avoid a shock administered by a researcher. And that sea slug sure isn't displaying free will when that change occurs. Remarkably and probably most important, the antiquity and ubiquity of these biological gears explaining behavioral change wind up being grounds for optimism.

PROTECTING YOUR GILL

We start with a sea slug, specifically *Aplysia californica*, the California sea hare, a gigantic slug that can be more than two feet long. Neuroscientists love this species, write operas about it, all because one of the most important, beautiful, inspiring pieces of neuroscience research in the twentieth century was done with it.

On the surface of an *Aplysia* is its gill, which is majorly important to an *Aplysia* surviving. If you lightly touch the area surrounding the gill, called the siphon, the *Aplysia* protectively retracts its gill inward for a while:

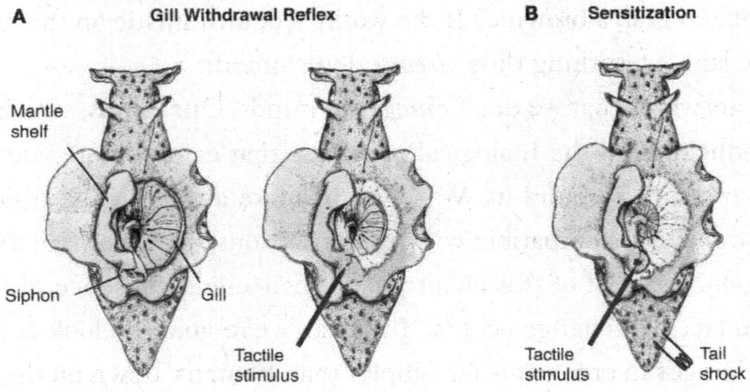

The circuitry underlying this is straightforward: throughout the siphon are sensory neurons (SNs), which have action potentials if anything touches the siphon. Once activated, the SNs activate motor neurons (MNs), which retract the gill:

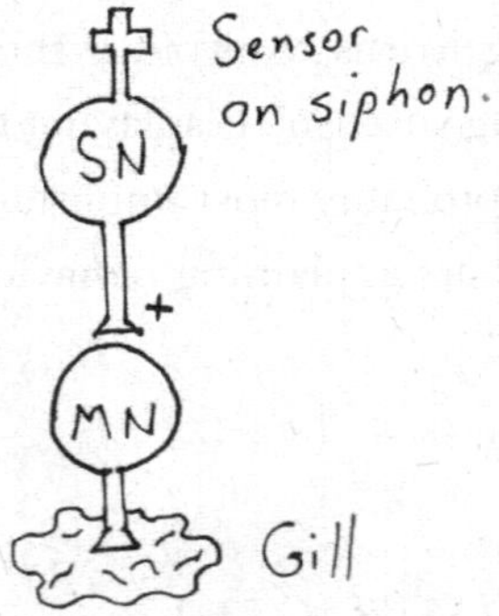

The gill is essential for survival, and *Aplysia* have evolved a backup pathway in case the SN-MN connection fails. It turns out that the SN also sends a projection to a little local excitatory node (Exc). Now, when the siphon is touched, the SN activates both the MNs and this Exc node; the latter sends a projection on to the MN, activating it. Thus, if the SN-MN connection fails, there's still the SN-Exc-MN route available:*

*The SN-Exc-MN route works a little slower than the SN-MN, since the SN-MN signal needs to traverse only one synapse, while SN-Exc-MN involves two.

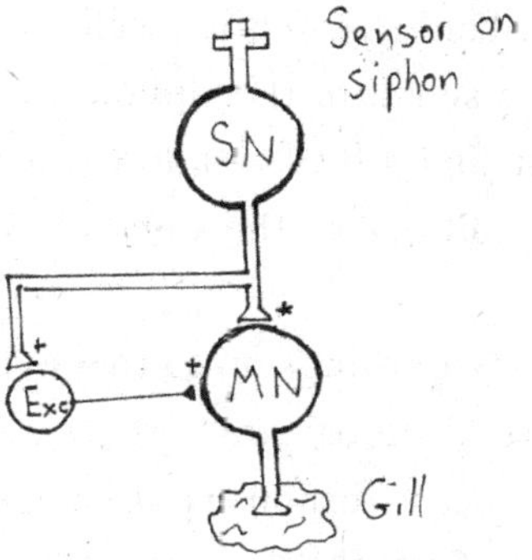

The gill can't remain retracted forever, as it needs to be on the surface to function. Thus, after a bit of time, retraction has to be halted; an off switch has evolved to do this. When the SN is activated, not only does it activate MN and Exc but, after a delay, it also activates a small inhibitory node (Inh). This node then inhibits the Exc branch (which, remember, is the delayed route from SN to MN, so it's the one to target with this delayed inhibition). Result: the MN is no longer being activated, so the gill defaults back to the surface:

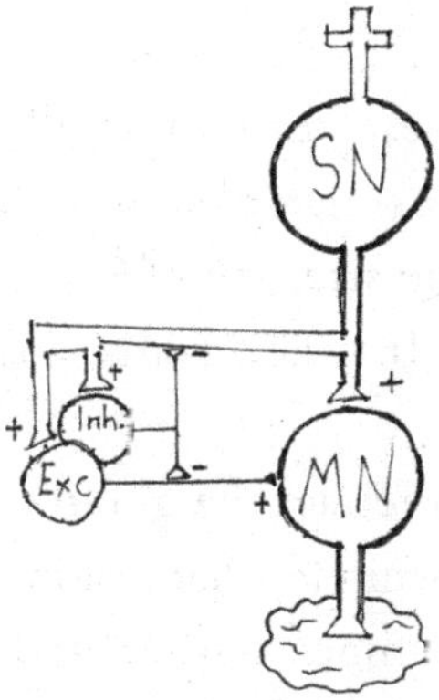

This SN/MN/Exc/Inh circuitry is not a world unto itself; the way it works can be altered by what's happening throughout the rest of the

Aplysia. At the tail end of an *Aplysia* is its, well, tail. If you shock the tail, it basically sends an alarm signal to the siphon; as a result, if the siphon is touched soon after that, the gill is withdrawn for twice as long as usual. Worrisome news at the tail makes the siphon more responsive to its own worrisome news.

How are we going to wire things up so that events in the tail make gill withdrawal more sensitive? Pretty straightforward. There has to be a tail sensory neuron (TSN) that is responsive to shock, and it has to have the means to then talk to the SN/MN/Exc/Inh circuit. When the TSN is activated, it makes both the SN and the Exc more excitable:

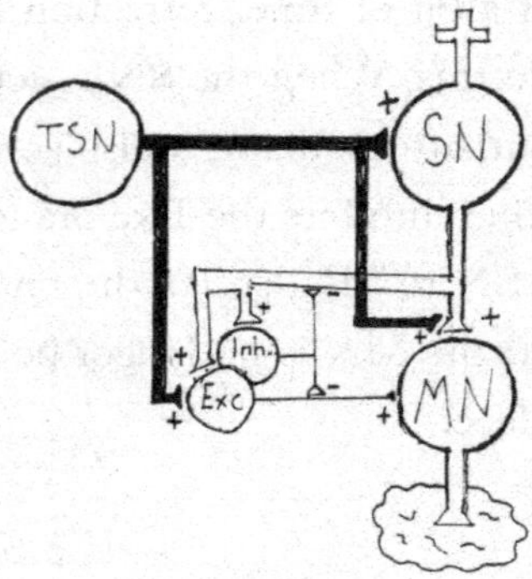

Note that a tail shock doesn't cause the gill to be retracted—the excitation from TSN isn't strong enough to activate MN on its own. Instead, the TSN input is enhancing the strength of SN-MN signaling in response to the siphon being touched. In other words, a tail shock sensitizes the gill withdrawal reflex.

Perfect. The *Aplysia* can retract the gill in response to the siphon being perturbed, has a backup system for that just in case, has a means to reverse the process back to where things started, and can make the circuit more jumpy and vigilant if bad things are happening to other parts of the *Aplysia*.

Why do we know so much about the inner life of an *Aplysia*? Because of the work of one of the gods of neuroscience, Eric Kandel of Columbia University. Here is a figure from his 2000 Nobel Prize lecture:[1]

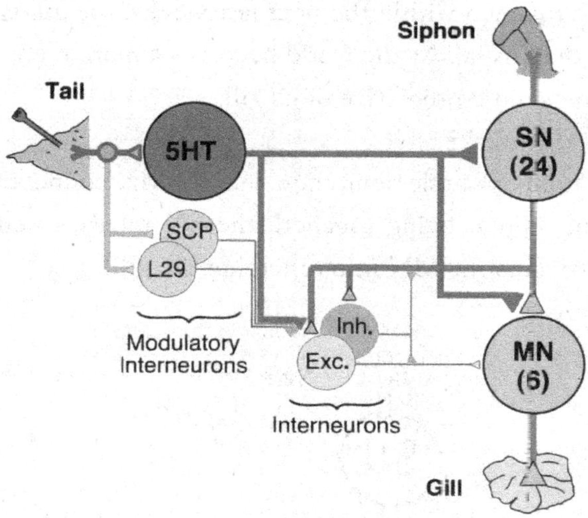

Some minor details: 5HT is the chemical abbreviation for the neurotransmitter (serotonin) used by the TSN. SCP and L29 fine-tune the system; we've ignored them, for simplicity. There are 24 SNs in a siphon, converging on to 6 MNs.

This is cool beyond description, just the clarity of this wiring system that this slug evolved. Unfortunately, though, it is also irrelevant to our interests; it has more in common with how your microwave works than with what's going on in us when we erroneously believe that we are acting out of free will. For that, we need to look at something much more interesting that happens in an *Aplysia*—this circuit will change in response to experience. It can be trained. It learns.

THE LEARNED *APLYSIA*

As we've seen, here are two basic rules. First, if an *Aplysia*'s siphon is touched, the gill retracts for a bit; second, if the siphon is touched within a minute of the tail being shocked, the gill is retracted for twice as long. But there's more. How about if the tail has been shocked four times? If the siphon is touched within four hours of that happening, the gill is retracted three times longer than usual. Shock the tail a cluster of times, and if the siphon is then touched within the next few weeks, the gill is retracted ten times longer than usual. As the world becomes a more menacing place, an *Aplysia* becomes more protective of its gill.

How does that work?

We know from our basic neuro how the SN-MN connection works—as a result of the siphon being touched, the SN releases neurotransmitter (which then triggers the MN into retracting the gill):

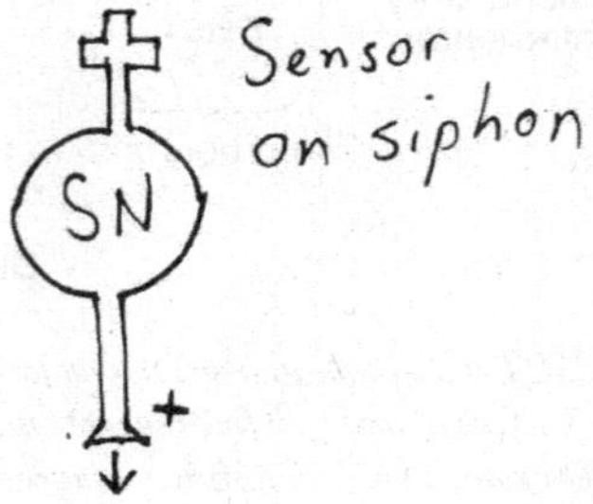

Now we need to see what happens inside the SN when the tail is getting shocked. The SM and MN are drawn very differently now, with little packets of neurotransmitter lined up at the bottom of the SN (the little circles), and with the MN and its neurotransmitter receptors (little horizontal lines) on the lower side of the synapse. The tail sensory neuron has been activated by one shock, causing it to release its neurotransmitter, which binds to a receptor on the SN. As a result of a single shock, some sort of "TSN activity–dependent stuff" (which we'll call Stuff) is released inside the SN:

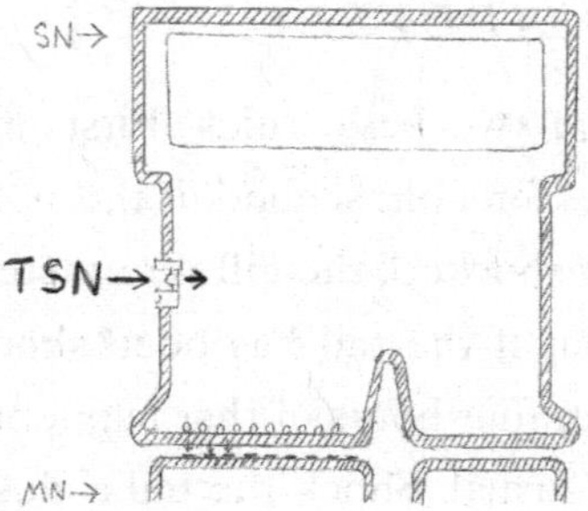

That Stuff within the SN glides to the bottom, where it beefs up the amount of neurotransmitter stored there (step #1). As a result, if the siphon is touched, enough additional neurotransmitter is released by the SN to cause the gill to retract for twice as long as usual. Within a minute or so of the single shock, the extra neurotransmitter stored in the SN is degraded, and things go back to normal:

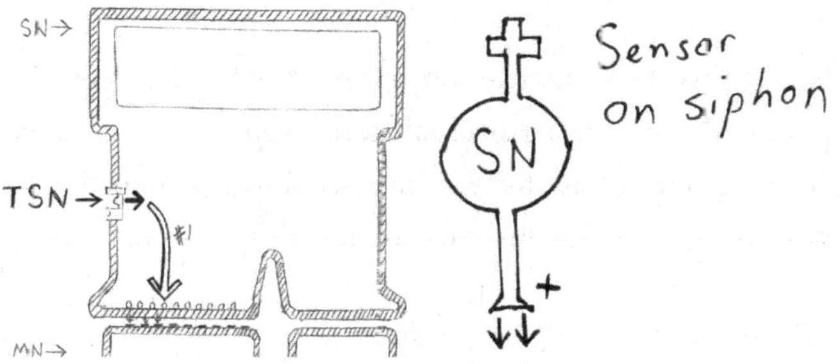

What if the tail is shocked four times in rapid succession? As a result, a whole lot more Stuff is liberated inside the SN than with one shock. Not only does this trigger the events of step #1, obviously, but also the surplus Stuff is enough to trigger step #2—that additional Stuff activates a gene on the DNA that produces a protein that stabilizes the neurotransmitter so that it is resistant to degradation. As a result, the neurotransmitter sticks around longer, and if the siphon is touched, enough additional neurotransmitter is released by the SN to cause the gill to retract for three times as long as usual. By four hours after that quartet of shocks, the degradation-inhibiting protein is itself degraded; as a result, the extra neurotransmitter is degraded, and things go back to normal (see the top figure on the next page).

Now, what if the tail is shocked with an intense, sustained cluster of shocks on a few successive days? Humongous amounts of Stuff are released, enough to activate not only steps #1 and #2 but #3 as well. For that final step, Stuff activates a whole string of genes* whose resulting proteins,

*Just as a reminder, all the DNA is in a single, continuous stretch, rather than broken into separate parts; the DNA was drawn this way for clarity; also, I don't know why the DNA gets smaller toward the right in my drawing, but it's not like that in real life.

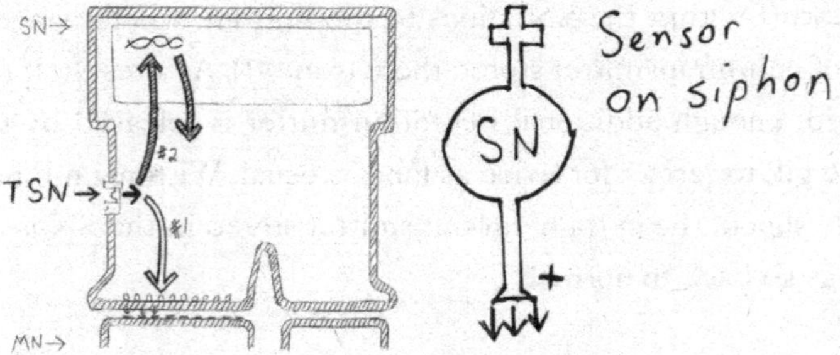

collectively, lead to the construction of an additional synapse. Now, if the siphon is touched, enough additional neurotransmitter is released by the SN to cause the gill to retract for *ten* times as long as usual. Weeks to months later, the new synapse is deconstructed, and things go back to normal:*

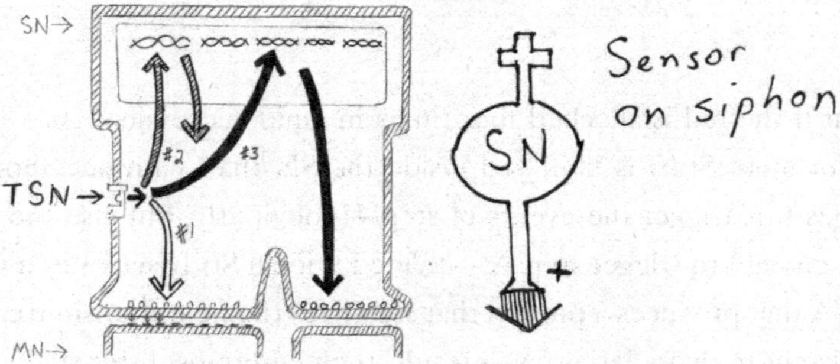

*Two subtleties. First, after all that effort to construct that second synapse, why not just keep it around, assuming that it will be useful at some point in the future for dealing with another cluster of high-intensity shocks? Because maintaining a synapse is expensive—repairing wear-and-tear damage to proteins there, replacing them with new models, paying rent, the electric bill, etc. And here there's been an econometric evolutionary trade-off for *Aplysia*—if there are going to frequently be shockful circumstances where the *Aplysia* will need to retract its gill ten times longer than normal, might as well retain that second synapse; in contrast, if it's a rare event, it's more economical to degrade the second synapse, and just make another one of it somewhere in the distant future when needed. This is a common issue in physiological systems, having to choose between keeping an emergency system on all the time versus making it inducible as a function of the frequency of emergencies. For example, should a plant expend energy making a costly toxin in its leaves to poison an herbivore munching on it? Depends—is it some sheep coming to graze every day or a cicada coming once every seventeen years?

An even subtler issue: Suppose the tail has been shocked once, and a smidgen of Stuff is liberated inside the SN. How does that small number of Stuff molecules "know" to activate step #1 rather than #2 or #3? Why that hierarchy? The way it is solved is a common theme in biological

Thus, we have a hierarchy. For a single shock, you add more copies of some molecule that already exists; for four shocks, you generate something novel to interact with that molecule that already exists; for a massive cluster of shocks, you start a whole construction project. All very logical. And this is precisely what Kandel showed as well (taken from that same Nobel Prize talk):

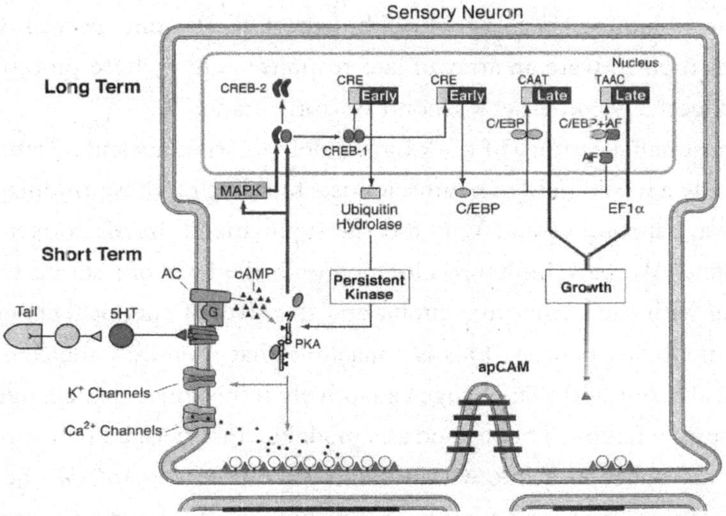

What exactly did he show goes on in an *Aplysia* SN when this is happening? Just skim this paragraph, and don't memorize a word of it. In fact, probably don't even read it—just know how to find it again later. The details: (A) What's actually happening in step #1? The neurotransmitter 5HT triggers the release of cAMP, which activates previously inactivated PKA, which works on the K^+ channel to trigger an influx of Ca^{2+} through Ca^{2+} channels, which results in the release of a greater amount of neurotransmitter. (B) Step #2: Enough cAMP has poured in not just to activate step

systems: The molecules that are triggered by Stuff in the step #1 pathway are much more sensitive to Stuff than the relevant molecules in the step #2 pathway, which in turn are more sensitive than those in step #3. Thus, it's like a layered fountain: it takes X amount of Stuff to activate #1; more-than-X to spill over and also start activating #2; lots-more-than-X to spill over into #3 as well.

#1 but also to spill over and cause MAPK to cleave CREB-2 from CREB-1, freeing the latter to dimerize into pairs of CREB-1, which interacts with the CRE promoter, which turns on an early-phase gene that leads to the synthesis of the enzyme ubiquitin hydrolase, which stabilizes PKA, allowing it to have its effects longer. (C) Step #3: The influx of cAMP is large enough that not only are steps #1 and #2 activated, but #3 is as well; this leads to liberation and dimerization of enough CREB-1 to not only activate the ubiquitin hydrolase gene, but the C/EBP gene as well; C/EBP proteins then activate an array of late-response genes whose protein products collectively construct a second synaptic branch.*

Almost half a century of work by Kandel, his students and collaborators, eventually a whole field of neuroscientists building on those findings, all to answer a just-so question: Why did the traumatized *Aplysia* retract its gill for so long? We have built a machine on both the level of neurons communicating with each other in a circuit and the level of chemical changes inside a single key neuron. This is a machine that is entirely mechanistic in biological terms and that changes adaptively in response to a changing environment; it has even been used as a model by roboticists. I dare anyone to invoke the concept of free will in making sense of this *Aplysia*'s behavior. No *Aplysia*, encountering another one, would say, "It's been a tough season, thanks for asking, lots and lots of shocks, no idea why. I had to build new synapses on every neuron in my siphon. I guess my gill is safe now, but *I* sure don't feel safe. This has been hell on my partner." We're watching a machine that did not choose to change its behavior; its behavior was changed by circumstances via logical, highly evolved pathways.[2]

And why is this the most gorgeous piece of neurobiological insight ever? Because pretty much the same thing goes on in us when we have become the sort of person who would pull a trigger, or run into a burning building to save a child, or steal an extra cookie, or advocate hard incom-

*Just to inundate you more, here's what the abbreviations are for: 5HT = serotonin; cAMP = cyclic adenosine monophosphate; PKA = protein kinase A; CREB = cAMP response element–binding protein; MAPK = mitogen-activated protein kinase; C/EBP = CCAAT-enhancer-binding protein. On and on.

patibilism in a book destined to be read only by two people who will hate it. The circuits and molecules of the *Aplysia* are all the building blocks we need to make sense of behavioral change in us.

Which, no doubt, seems absurd, totally implausible, leaping from *Aplysia* to us. Thus, we're going to get there with a few in-between examples (but in less agonizing detail than has gone into understanding *Aplysia* behavioral machinery). When we're done, the hard reality is that we are unimaginably more complex than an *Aplysia* but are biological machines with the same building blocks and the same mechanisms of change.

Aplysia californica. *As should be obvious, the one on the left is happy, in an unreflective kind of way. The one on the right is a wonderful* Aplysia *stuffie that could be your child's comfort object all the way until their freshman year of college.*

DETECTING A COINCIDENCE

Our next neuronal machine blinks its eye. Go up to it, spritz a little puff of air at its eyelid, and the eyelid blinks automatically, as a protective reflex. We already know the simple circuitry needed to pull this off. There's a sensory neuron that has an action potential in response to an air puff. This then triggers an action potential in a motor neuron, causing the eyelid to blink (see the following page).

Now let's add a totally useless additional piece of circuitry. We have a second sensory neuron. This one doesn't respond to the tactile stimulation of an air puff. Instead, it responds to an auditory stimulus, a tone. Neuron

3 projects to the blink motor neuron, where it isn't excitatory enough to cause an action potential in neuron 2. Play the tone, and nothing happens in neuron 2:

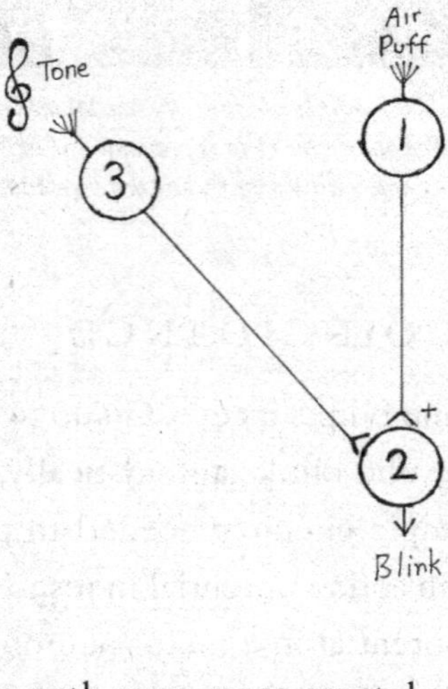

Let's make that side path even more ornately useless. Now the tone is played, activating neuron 3. As before, neuron 3 isn't able to cause an action potential in neuron 2; however, it does so in neuron 4. But as it turns out, the neuron 4 action potential has only about half the excitatory power needed to evoke an action potential in neuron 5. So stimulate neuron 3 with a tone, and the net result is that nothing happens in either neuron 2 or neuron 5; a tone still does nothing to blinking:

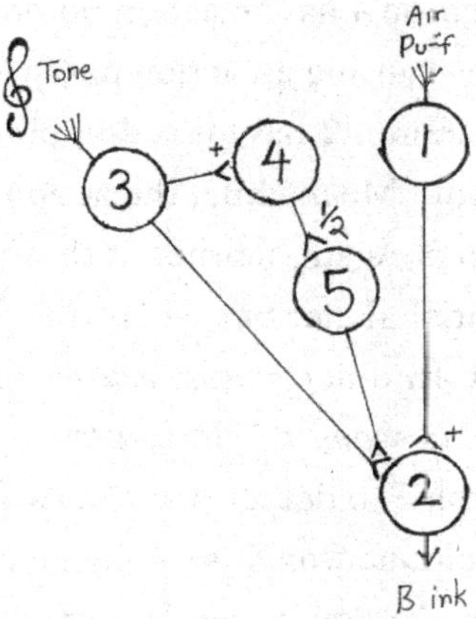

Let's add another useless projection to this circuit. Now neuron 1 sends a projection to neuron 5 (along with its usual projection to neuron 2). But when an air puff triggers an action potential in neuron 1, it gets only half-way to the excitation needed for neuron 5 to have an action potential. So: air puff, neuron 2 activates, nothing happens in neuron 5:

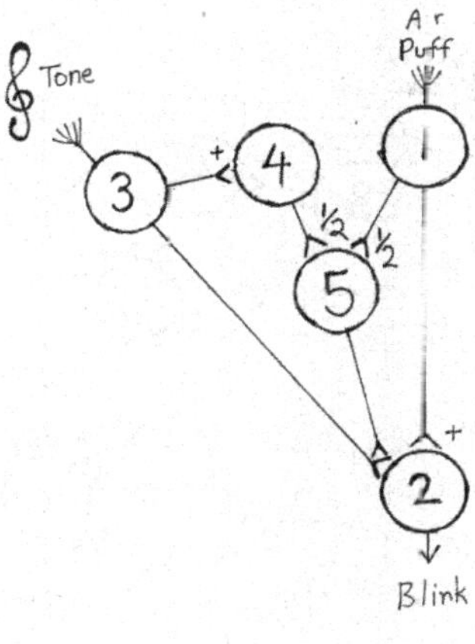

But now let's activate neuron 1 *and* neuron 3. Play a tone and release an air puff. Crucially, the tone comes one second before the air puff, and it takes a second for any action potential to reach the axon terminals. So:

At time zero: play the tone, neuron 3 has an action potential.

After one second: neuron 4 has an action potential (thanks to neuron 3), while the air puff is now causing an action potential in neuron 1.

After two seconds: neuron 2 has an action potential (thanks to neuron 1), triggering an eye blink. Meanwhile, the action potentials from neurons 4 and 1 arrive at neuron 5. Again, neither of those two inputs is enough to trigger an action potential alone, but when they are combined, neuron 5 has an action potential. In other words, *neuron 5 has an action potential if and only if the tone is played and is followed by an air puff one second later.* The circuit allows neuron 5 to detect that the two stimuli coincided. Or to use the jargon of the field, neuron 5 is a coincidence detector.

After three seconds: neuron 5 has its action potential, causing it to stimulate the axon terminals of neuron 3. Which, as it turns out, accomplishes nothing—it isn't strong enough to, say, cause those axon terminals to dump much neurotransmitter.

But play the tone followed by the air puff a second time. A tenth time, a hundredth time. Each time neuron 5 stimulates the axon terminals of neuron 3, it slowly causes neuron 3 to build up more neurotransmitter there, release more of it each time, until . . . finally . . . when neuron 3 is

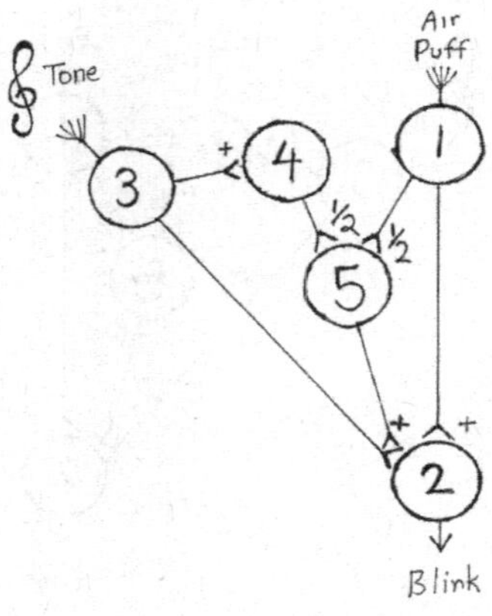

stimulated by the tone, it triggers an action potential in neuron 2. And the machine blinks *before* the air puff happens, blinks in anticipation of it (see figure on previous page).

It's called eyeblink conditioning, and it works this way in mammals—in lab rats, rabbits, in *humans.* It's useful, adaptive—it's great to be conditioned to close your eyelids protectively before, rather than after, a noxious stimulus occurs. We know the underlying circuit in a different, famous setting, one that gives the phenomenon its name: Pavlovian conditioning. Ol' Doc Pavlov lets his dog smell dinner; dog salivates. This is the circuit of neurons 1 and 2. Neuron 1 smells the food, neuron 2 stimulates salivary gland, dog drools. Thus we have an unconditioned stimulus (the smell), which automatically evokes the unconditioned response of salivating. Now ring the bell just before the food arrives; pair the two over and over and, thanks to neurons 1, 3, 4, and 5, you establish a conditioned stimulus and a conditioned response—ring the bell and the dog salivates in anticipation of the smell of the food.

The key spot where change happens is in the place where neuron 5 terminates on neuron 3. How does the former repeatedly stimulating the axon terminals of latter result in the latter increasing the amount of neurotransmitter released, eventually gaining the power to elicit an eyeblink on its own? Go back to page 277's description of the inner workings of the SN in an *Aplysia.* How does eyeblink conditioning work? By neurotransmitter from neuron 5 releasing cAMP inside neuron 3, which frees PKA from its brake, which activates MAPK and CREB, which activates certain genes, culminating in, among other changes, the formation of new synapses.* This is not "Neuron 5 causes the intracellular release of chemicals that kind of function like stuff in *Aplysia.*" *It's the same chemical messengers.*[3]

Think about this. Humans, being conditioned to blink their eyes, and marine sea slugs, conditioned to withdraw their gills, haven't shared a common ancestor for more than half a billion years. And here we are, with their neurons and ours using the same intracellular machinery for

*And the more new synapses, the stronger the conditioning.

changing in response to experience. You and an *Aplysia* could trade your cAMPs, PKAs, MAPKs, and so on, and things would work just fine in both of you.* And you'd both be using serotonin to kick-start this. These *Aplysia*/human similarities should demolish anyone's skepticism about evolution.[4]

More important for our purposes, these findings show that (just as with the *Aplysia* gill-withdrawal reflex) we can build ourselves deterministic circuits with deterministic neurons that explain an adaptive *change* in human behavior in response to experience.† All without having to invoke the notion of our "choosing" to start blinking our eyelids when we hear a tone.[5]

Booyah, we've crushed any philosopher whose lifework is premised on the notion that we have free will because we can be eyeblink conditioned. Yeah, yeah, this is not a very fancy outpost of human behavior. Nevertheless, it's fancier than you might think.

To appreciate that, what happens to lab rats if, when they were pups, they were intermittently separated from their mothers for a while? Rats that experienced such "maternal separation" early in life are, as adults, a mess. They are more anxious, show more of a glucocorticoid response to mild stress, don't learn as well, are easier to addict to alcohol or cocaine. It

*And implicit in this is that we and *Aplysia* share the genes that code for cAMP, PKA, MAPK, and so on. In fact, we share at least half our genes. To give a sense of just how pervasive this overlap is, we share roughly 70 percent of our genes with sponges—and they don't even have neurons.

†Just to be clear, the circuit is more complex than in the figure, and this has forced me to look up all sorts of obscure places in the brain in a neuroanatomy textbook that I open once a decade. Neuron 1, which signals the air puff, is actually a sequence of three classes of neurons—first neurons in the trigeminal nerve, which stimulate neurons in the trigeminal nucleus, which stimulate neurons in the inferior olivary nucleus. Neuron 2, which turns the air-puff signal into an eyeblink, is also actually a sequence of three classes of neurons—the first being neurons in the interpositus nucleus within the cerebellum, which activate neurons in the red nucleus, which activate facial nerve neurons in the facial nucleus, which cause the eyeblink. Neuron 3 is also a series of neurons in real life, starting with the neurons of the auditory nerve, which stimulate neurons in the vestibulocochlear nucleus, which stimulate neurons in the pontine nucleus. Logically, projections from the inferior olivary nucleus (carrying air-puff information) and the pontine nuclei (carrying tone information) converge on the interpositus nucleus. Neurons 4 and 5 are a circuit in the cerebellum involving granule cells, Golgi cells, basket cells, stellate cells, and Purkinje cells. There, I've done my neuroanatomical duty and have already forgotten what I wrote three sentences ago.

is a model for how one type of early-life adversity in humans produces dysfunctional adults, and people know tons about how each of those changes comes about in the brain.[6]

So get this—take a rat pup and maternally separate him, and as an adult, it will be harder to do eyeblink conditioning on him. In other words, along with all the other deleterious consequences of maternal separation, you have animals that don't acquire this adaptive response as readily. It is caused by an epigenetic change in the brain, such that forever after, there are elevated levels of receptors for glucocorticoid stress hormones in the equivalents of neuron 2. Block the effects of glucocorticoids in that adult rat, and eyeblink conditioning becomes normal.* Conclusion: early-life adversity impairs this circuit by making a key neuron in the circuit more sensitive to stress.[†,7]

Take one lone heroic rat that, for some reason, can save the world from disaster by developing a conditioned eyeblink response. And he screws up, doesn't do it, lets the world down. Afterward, everyone is pissed at the rat, blaming him for not conditioning. To which he can say, "It's not my fault—I didn't get conditioned because, one second before, my interpositus nucleus wasn't as responsive to the conditioned stimulus; because a few hours before, my stress hormone levels were elevated, which guaranteed that the interpositus would be particularly resistant to conditioning; because back in my childhood, my mother was taken from me, and this changed gene regulation in the interpositus, permanently increasing levels of a hormone receptor there; because back millions of years ago, my species evolved to be highly dependent on maternal care after birth, and the genes needed to make lifelong changes in circuitry if the mother is absent." A change in behavior due to specific changes that can be identified in a circuit, arising from circumstances a second, an hour, a lifetime, an

*How glucocorticoids disrupt the function of neurons like those in the interpositus is understood as well but is more detail than we need.

†As far as I know, no one has seen if adult humans who underwent a lot of childhood adversity have impaired eyeblink conditioning, but it seems perfectly plausible. Which would obviously be the least of their long list of life-altering problems.

evolutionary epoch earlier over which the organism had no control. No rodential moral responsibility involved, no grounds for everyone blaming the rat.

But still, this is just about blinking your eyes. On to the sorts of scenarios that this whole book is about.

WHEN THEY BECOME THEMS

Not many of the world's problems arise from the fact that a neutral stimulus can be conditioned to evoke an eyeblink reflex. But a lot of them sure arise from the same going on in the amygdala.

Take a lab rat or a human volunteer and give them a shock. The amygdala activates; you can show this in the rat by recording the activity of neurons in the amygdala with electrodes, while in humans you show the same with brain imaging. To prepare us for the subtleties to come, right off the bat, the link between shock and amygdaloid activation is modulated in all sorts of interesting ways. For example, in both the rat and the human, the amygdala activates more if the shock occurs unpredictably, rather than if you know when the shock is coming.

Once the amygdala activates, it triggers a variety of responses. The sympathetic nervous system is activated, the heart beats faster, blood pressure rises. Glucocorticoids are secreted. Your typical rat or human freezes in place. Nontrivially, if that rat has a smaller and weaker rat next to them, the rat that has been shocked becomes more likely to bite the other—which lessens their own stress response.

So this is a version of an SN-MN circuit, by now familiar. Now, before each shock, play a tone as a conditioned stimulus. Do it a bunch of times and you know what happens—the tone itself will eventually have gained the power to activate the amygdala, and we have a conditioned fear response. Beautiful work by Joseph LeDoux of New York University has revealed the circuitry to explain this. Look at it closely and, what do you know, it's the same basic wiring as for conditioning an eyeblink or a gill withdrawal. If timed right, information about the unconditioned stimulus (the shock), mediated by the somatosensory thalamus and cortex, and about the conditioned one (the tone), mediated by the auditory branch, just as with conditioned eye-blinking, simultaneously converge on the amygdala. Local neurons there act as coincidence detectors, repeated stimulation of the auditory branch induces all sorts of changes in the amygdala involving cAMP, PKA, CREB, all the usual, and a tone now elicits the same terror that a shock does.[8]

We saw that something as simple as eyeblink conditioning reflects a nervous system that has been sculpted by all that came before it (e.g., early maternal experience). The acquisition, consolidation, and extinction* of the conditioned fear of something neutral like a tone reflects the organism's history even more. Extinction will occur faster if, in the seconds before, there are high levels in the amygdala of endocannabinoids (whose receptor also binds THC, the most active component of cannabis)—this makes it easier to stop being afraid of something. The amygdala becomes less likely to store away a conditioned fear response as a stable memory if, in the previous hours, the individual has taken an SSRI antidepressant like Prozac (which makes people ruminate less about negative thoughts). The amygdala will be less active and harder to condition if, in the days before, it was exposed to high circulating levels of oxytocin, which helps explain how oxytocin can promote trust. In contrast, if the organism has been exposed to high levels of stress hormones in the previous month, it becomes easier to generate a conditioned fear response (thanks to the hormones increasing activity of the gene that produces the mammalian version of C/EBP, which appears in the figure on page 277). And pushing way back in our "one second before, one minute before" arc, if an organism was exposed to lots of Mom's alcohol back during fetal life, it has a harder time remembering a conditioned fear. And of course, what versions of the genes related to those

*We've unpacked the features of fear conditioning: acquisition of the response (acquiring the conditioned response in the first place); consolidation of the response (remembering it long afterward); extinction of the response (gradually losing the response after being exposed to the tone a bunch of times where it *isn't* followed by a shock).

in that figure are present, and whether the individual's species evolved those genes in the first place, will influence how readily conditioning occurs. How easily an organism learns to be afraid of something as simple as a tone is the end product of all these influences on the workings of this circuit, all factors over which the individual had no control.[9]

All this for a tone.

Consider something else that activates your amygdala. In this case, hearing the word *rapist.* You're not genetically programmed to activate your amygdala in response to it, not the way it would automatically activate if, say, you were dangled upside down by a thread up in the air while covered with spiders and snakes. Instead, the amygdala came to respond to the word through learning—you learn what the two syllables mean, what the act is; you've learned about its impact in general, how being raped, as it's been said, is like living through your own murder; you know someone who was or, unbearably, you yourself were. In any case, you now have an amygdala that activates automatically in response to the word, as surely as if you were given a shock.

Now let's take a neutral stimulus and rely upon the coincidence detectors in our amygdalas to generate a conditioned fear response. Something more complex than a bell that would make Pavlov's dogs salivate or a tone that would cause a lab rat to freeze:

> When Mexico sends its people, they're not sending their best. They're not sending you. They're not sending you. They're sending people that have lots of problems, and they're bringing those problems with us. They're bringing drugs. They're bringing crime. They're rapists.
>
> —Donald Trump, in the speech that famously opened his presidential campaign, June 16, 2015

Students of history and current events: Let's play a game called "Match the Conditioned and Unconditioned Stimulus." Get them all right and you win a prize, so have fun!

CONDITIONED STIMULUS AND THE PEOPLE WHO LABORED TO GENERATE THAT ASSOCIATION	UNCONDITIONED STIMULUS
1. Muslims, according to European nationalists	*a. Vermin, rodents*
2. Jews, according to the Nazis	*b. Thieves, pickpockets*
3. Indo-Pakistanis, according to half the Kenyans I know	*c. Opium addicts*
4. Irish immigrants, according to nineteenth-century WASPs	*d. A malignancy, a cancer*
5. Roma, according to centuries of Europeans	*e. Violent superpredators*
6. Mexicans, according to Donald Trump (this is a freebie thrown in)	*f. Rapists*
7. Young African American men, according to swaths of White America	*g. Shop owners who cheat you*
8. Chinese immigrants, according to nineteenth-century America	*h. Cockroaches*
9. Tutsi, according to the Hutu architects of the Rwandan genocide	*i. Drunken Papists*

Yes, yes, it's hard because there are overlaps but, come on, give it your best shot.*

The question now becomes how readily you come to associate *Mexicans*

*According to historical records, current events, and the thread of *See also*s, starting with the Wikipedia page "Ethnic and national stereotypes": 1d, 2a, 3g, 4i, 5b, 6f, 7e, 8c, 9h.

with *rapists* while undergoing Trumpian conditioning—how resistant or vulnerable are you to forming that automatic stereotype in your mind? As usual, it depends on what happened one second before hearing his statement, one minute before, and so on. Here are all sorts of circumstances that increase the odds of your being successfully conditioned by the man if you are your basic white-bread American: If you are exhausted, hungry, or drunk. If something frightening happened to you in the previous minute. If, as a male, your testosterone levels have been soaring over the last few days. If, in recent months, you've been chronically stressed by, say, unemployment. If, when you were in your twenties, your musical tastes led you to become an überfan of some musician who espoused that stereotype. If you lived in an ethnically homogeneous neighborhood as a teenager. If you were psychologically or physically abused as a child.* If your mother's values were those of a xenophobic rather than a pluralistic culture. If you were malnourished as a fetus. If you have particular variants of genes related to empathy, reactive aggression, anxiety, and responses to ambiguity. All things over which you had no control. All things that sculpted the amygdala you will have in this instant of being exposed to a stereotype, all the way down to how many molecules of cAMP each neuron releases, how tight the brakes are on PKA, and so on. Because there are millions of neurons involved, with gazillions of synapses, the process is subject to a lifetime of influences that are staggeringly more complex and nuanced than what goes into conditioning an eyeblink or changing how an *Aplysia* protects its gill. But it's all the same mechanistic building blocks that will determine whether your views will be changed by some demagogue's toxic attempt to form a conditioned association in you.[†,10]

*Interestingly, this turns out to be a significant predictor of growing up to believe that COVID-19 vaccines are part of a conspiracy to harm you.

†Just for clarification, there is actually little reason to think this was a circumstance where a lot of people were indeed conditioned to make this association solely as a result of that single statement. Instead, much of its success was in signaling the people who already thought this way that Trump was their kind of guy. So this is just a simple model system of the reality, which requires repetition.

Time to finally move to the sort of split in the road that this book is ultimately about, examining the biology of our moral behaviors being changed, rather than of our freely choosing to change our behaviors.

SPEEDING UP AND SLOWING DOWN

I'm driving down the freeway. I pass a car or truck here or there. Some pass me. I'm listening to music. And then a guy passes in a sensible electric car that I note has one of those COMMIT RANDOM ACTS OF KINDNESS bumper stickers. In the next few seconds, I probably have the microexpressive start of a smile, along with a number of thoughts. "Well that's nice." "I bet I would like the guy." "I wonder who he is." "I bet he has an organ donor sticker on his driver's license." And then I tease myself for having such a macabre thought. I think that he no doubt listens to NPR. Then I think how ironic it would be if he were on his way to rob a bank. And then something on the radio catches my attention, and I go back to listening, thinking about something else.

Then, about thirty seconds later, the car ahead of me to the right signals that it wants to merge into my lane. Being a jerk, I think, "Oh no you don't! I'm in a rush," and am just about to put my foot on the gas when I briefly flash on the bumper sticker. I stop from pressing the accelerator. And half a second later, I shift my foot to the brake, allowing that car to merge, briefly basking in a sense of my profound nobility.

What went on in those seconds after I saw the bumper sticker? It's deterministic *Aplysia* all the way down.

There's that classic image of us in a moral quandary: an angel on one shoulder, a devil on the other.* We have a motor output, the neuron(s) that

*I'm apparently easily distractible right now, since, while looking for a good angel/devil image, I wound up looking at two hundred such pictures to confirm a spur-of-the-moment hypothesis that a disproportionate percentage of the images have the devil on the left shoulder and the angel on the right. And that was the case 62 percent of the time in my sample. As a leftie, I'm slightly offended—I've come to terms with being gauche, but being satanic is another thing.

triggers our muscles to push down on the gas. And on a metaphorical level, there's neural circuitry whose net output is to stimulate that neuron, a "Do it" signal, while a different circuit prompts an inhibitory "Don't; slow down instead."

What is the "Do it" circuit about? The usual—the outcome of influences from one second ago to millions of years ago. You're hungry. There was just a mysterious throbbing pain on the left side of your butt, and you're briefly worried that you have left-side-of-your-butt cancer, and thus feel entitled to drive selfishly. You're going to an important meeting and can't be late. You've gone a few months without getting a decent night's sleep. In middle school way back when, the tough kids bullied you a lot, and from that you have a vague, unspoken belief that letting someone merge in front of you on the highway equates to your being an inadequate pushover. It's the time of day when your testosterone levels are elevated, thus strengthening the signaling of neurons in the "I'm a weakling if I let someone merge in front of me" circuit (regardless of your sex). You have this or that variant of this or that gene. You're male and a member of a species in which there's a moderate but significant correlation between male-male competition and male reproductive success. All of those push in the direction of "Do it."

Meanwhile, the "Don't; slow down instead" neuron has its inputs: You like to think of yourself as a kind person. You went to Quaker meetings for a while in college. Something in the news this morning made you feel slightly less jaundiced and helpless about the idea that the incrementalism of small good acts can make the world better. There's that Christian rock song that you really like, to your formless atheistic embarrassment. You were raised by parents who, each week on the Sabbath, gave you a dime to put in the charity box for an orphanage, and then, on behalf of the orphans, hugged you in a way that you can still feel sixty years later. Et cetera.

The two circuits sit there, prompting you toward opposing neurobiological outputs. At this moment, the "Don't; slow down instead" prompt has a little more oomph than usual. Why? Because the neurons activated by that bumper sticker, still rumbling in a reverberating loop that cycles for a minute or so in what we call short-term memory, have added a faint but decisive voice that tips the balance in favor of the "Don't" circuit.[11]

How did each of those circuits form to gain the collective neurotransmitter-ish power to influence our motor output? By a whole lot of neurons forming positive or negative associations with something or other. In other words, a whole lot of neurons where the likes of cAMP, PKA, or MAPK did this or that.

Let's consider a hypothetical neuronal circuit, one straight out of the appendix that introduces the basics of the nervous system. Suppose we have a network consisting of two layers of neurons. Layer 1 consists of neurons A, B, and C, while layer 2 consists of neurons 1–5. Note the wiring pattern, in that neuron A projects to neurons 1–3, neuron B to neurons 2–4, neuron C to neurons 3–5. Stated a different way, neuron 3 gets inputs from three other neurons; neurons 2 and 4 from two; neurons 1 and 5, a single input each:

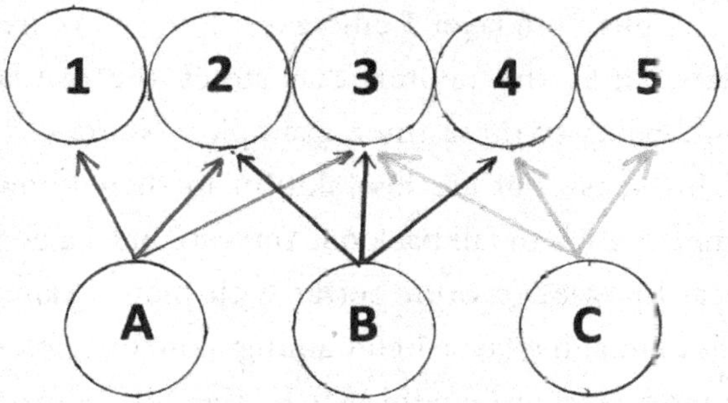

Now let's give layer 1 some unlikely specializations. Neuron A responds to pictures of Gandhi, neuron B to Martin Luther King Jr., neuron C to the Mirabal sisters. Neurons don't really come that way, but let's allow those three neurons to stand in for three complex networks of specialized recognition:

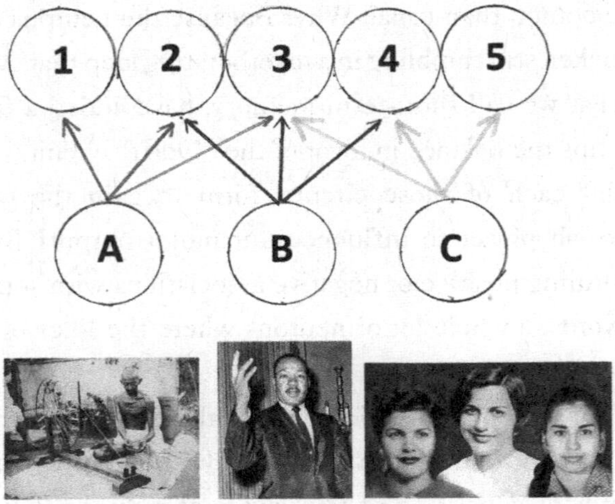

What is going on in layer 2? At one extreme are neurons 1 and 5; each is as specialized as any of the first-layer neurons, responding to Gandhi and the Mirabal sisters, respectively. How about neuron 3, at the other extreme? It is a generalist neuron, sitting at the intersection of knowledge among the three layer 1 neurons. What does it know about? Out of the overlap of projections from layer 1 emerges a category of people who died for their beliefs.* It is the neuron that stores the overlapping knowledge and commonality of those three examples. Neurons 2 and 4 are also generalist in this sense but are less skillful in their knowledge, having only two exemplars each to fall back on. You can make a generalist neuron with categorical knowledge even better with more examples—it would not be difficult to imagine layer 1 containing more examples and neuron 3 thus at the intersection of Gandhi, MLK, the Mirabal sisters, plus, say,

*The Mirabal sisters, Patria, Minerva, and Maria Teresa, were murdered in 1960 for their political opposition to the dictator of the Dominican Republic, Rafael Trujillo. An extra level of poignancy is added by the fact that there was a fourth sister, Dede, who was relatively apolitical and escaped death and who lived another fifty-four years without her sisters. Our household got obsessed with the Mirabals awhile back when one of our kids read a book about them.

Socrates, Harvey Milk, Saint Catherine of Siena,* Lincoln. Neuron 3 is that much more knowledgeable about this category of people dying for their beliefs.[12]

Uh-oh, you have a slightly irreverent thought, recognizing that this network of Gandhi, MLK, the Mirabals, Socrates, Harvey Milk, Saint Catherine, and Lincoln is just as accurately described as concerning people who have been the subjects of biopic movies. In other words, the string of examples in layer 1 could simultaneously be embedded (a) along with Sid Vicious, in the biopic movie category or (b) in the people-dying-for-their-beliefs category, now including a great-uncle killed in Normandy, whose

*Imagine a teenager, off at her freshman year of college. During that first semester, her friends begin to notice with concern that she isn't eating much—she's always insisting that she feels full halfway through dinner, or that she feels a bit unwell and has no appetite. She'll even fast two, three days at a time; on more than one occasion, her roommate catches her forcing herself to throw up after a meal. When told by friends that she is becoming too thin, needs to eat more, she insists instead that she has a huge appetite, eats like a glutton, feels like that is a personal shortcoming to be overcome—that's why she fasts. She's constantly talking about food, writing about it in letters home. While she has many female friendships, she seems to recoil from men—she says she plans to be a virgin her whole life, says that fasting is actually helpful to her in that it takes her mind off any sexual feelings. She's long since stopped menstruating, and her reproductive axis has shut down from starvation.

We know exactly what that is—anorexia nervosa, a life-threatening disease that is often interpreted in the context of our Westernized lifestyle as lying at the intersection of our overabundance of food and lives filled with interest in food consumption (*Iron Chef*, anyone?) on the one hand, and on the other, the corrosive, nonstop sexualizing of women in the media, which drives so many women and girls into body image problems.

Makes sense. But consider Catherine of Siena, born in 1347 in Italy. As an adolescent, and to her parents' consternation, she started limiting her food intake, always insisting that she was full or feeling infirm. She started having frequent, multiday fasts. Joining the Dominican Order of the church, she took a vow of celibacy; now married to Christ, she reported a vision in which she wore Christ's wedding ring . . . made from his foreskin. She would force herself to throw up when she felt she had eaten too much, explained her fasting as a display of her devotion and as a means to curb and punish herself for her "gluttony" and "lust." Her writings are full of imagery of eating—drinking the blood of Christ, eating his body, nursing from his nipples. Eventually, she got to a point where (wait for it . . .) she committed herself to eating only the scabs of lepers and drinking their pus, and wrote, "Never in my life have I tasted any food or drink sweeter or more exquisite [than pus]." She starved to death at thirty-three and was canonized in the next century, and her mummified head is on display in a basilica in Siena. An irresistible history. I even teach about her in one of my classes; the details about pus and scabs are always a crowd-pleaser.

memory still evokes tears in his adoring little sister, your grandmother, age ninety-five.[13]

Thus, the same layer 1 neuron can be part of multiple networks. Forgo Sid Vicious and add Jesus to the layer and, according to lots of earth's humans, we still have the category of people who died for their beliefs (as well as being subjects of biopic films). Meanwhile, Gandhi and Jesus plus Johnny Weissmuller could project as a trio on to a separate layer 2 array coding for guys in loincloths:

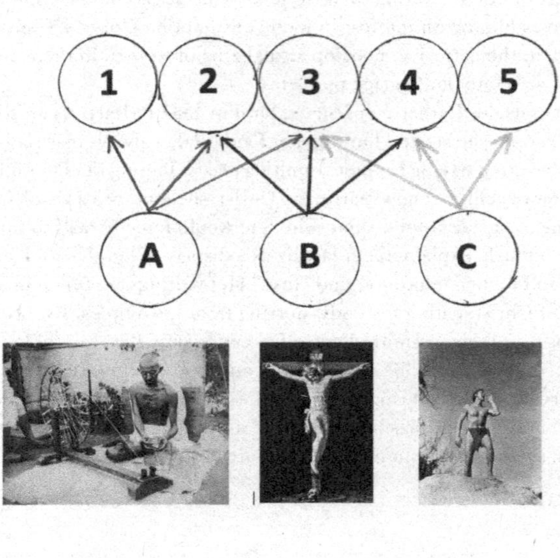

Let's take things one step further. To make things easier, let's ignore neurons 1, 2, 4, and 5 from the second layer, stripping things down to only the generalist neuron 3:

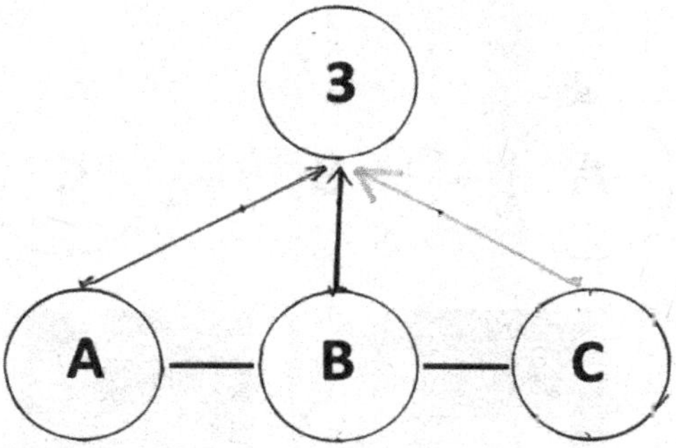

So we've got the Gandhi, MLK, and Mirabal neurons converging on the #3 "people who died for their beliefs" neuron. Adjacent to it is *another* network (again, ignoring neurons 1, 2, 4, and 5 for simplicity's sake). What does neuron A in this second network code for? The time when, despite being terrified by heights, you made yourself go off that diving board and felt great about yourself afterward. Neuron B in that second network? That semester where you were close to failing in geometry early on but then worked like mad and wound up with a good grade. Neuron C? All those times when you were a kid when your mother would tell you that you could grow up to be anything you wanted, if you put your mind to it. What is neuron 3 in this second network about? A category that can be roughly framed as "reasons why I feel a sense of optimism and agency about life." (See figure on the next page.)

Next to these is a third network. Its neuron 3 is about "peace has happened in some really unlikely places," and its layer 1 A/B/C neurons are the Good Friday Agreement in Northern Ireland, the Camp David Accords between Egypt and Israel, and the Christmas truce of World War I.

Thus, three adjacent networks, where neuron 3 in the first network is about "people have died for their beliefs," neuron 3 in the second is about

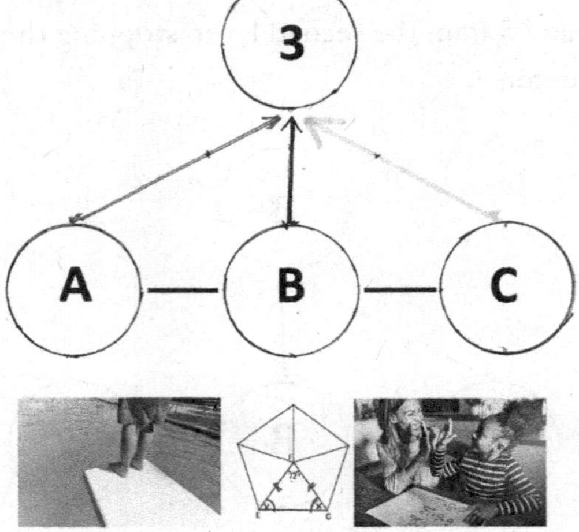

"why I feel a sense of optimism and agency in life," and neuron 3 in the third is about "peace has happened in some really unlikely places."

And as a final step, the three different neurons, in turn, form their own layer 1, projecting onto their own über-3 neuron:

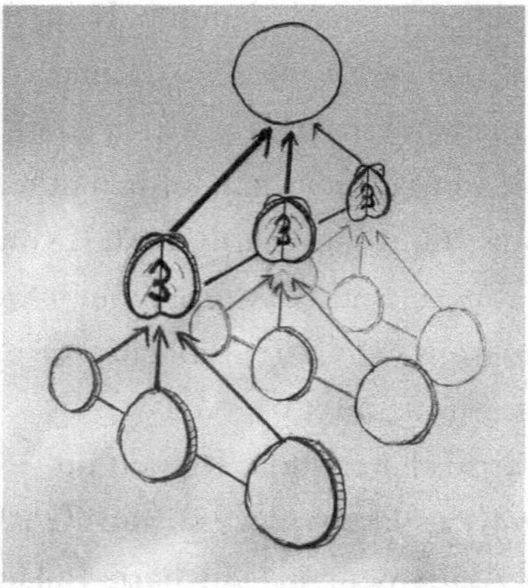

What's at the top of this three-layer network? Some emergent conclusion along the lines of "Things can get better; there are people who have heroically made things better; even *I* can make things better." There is hope.

Yes, yes, yes, this is crazily simplified. But it's still an approximation of how the brain works—exemplars converging on nodes out of which emerge the capacity to categorize and associate. Each node being a part of multiple networks—serving as a lower-layer element in one while simultaneously serving as a higher-layer element in another, a central player in one, peripheral in another. All built on wiring principles identical to those of an *Aplysia*.

And where the events around us alter the strengths of various synapses—another tyrant seizes control in a country that was crawling toward democracy, and a network like this last one is weakened by this counterexample. You slow down to let someone merge into your lane and it is strengthened. There are even loops where there's feedback, such that the positive affective content of the output of a hierarchical network motivates you to obtain more exemplars as inputs—"*Hotel Rwanda* was *so* inspiring that it made me start learning about truth and reconciliation commissions"—strengthening it further.

Change occurring, accomplished with the same molecules that make for a learned *Aplysia*, all without invoking the sort of willful agency and freedom that we intuitively attribute change to. You learn about how experience changes the nervous system of an *Aplysia*, and as a result, your nervous system changes. We don't choose to change, but it is abundantly possible for us to be changed, including for the better. Perhaps even by having read this chapter.

13

We Really Have Done This Before

The preceding chapters have left us with a clear path that every one of us needs to take, in roughly the following sequence of events:

Step #1. You're living a fine life. There are people whom you love and who love you; your days are filled with meaningful activities and sources of pleasure and happiness.

Step #2. Someone does something unimaginably awful, violent, destructive to a loved one. You are shattered, bereft of meaning in life. You barely function, will never feel pleasure or feel safe again. You will never feel love again because of the lesson that a loved one can be ripped from you like that.

Step #3. Some scientist sits you down and gives you a PowerPoint presentation on the biology of behavior, including violence; they go on and on in an irksome manner about how "we are nothing more or less than the sum of biology over which we had control, and its interactions with the environmental circumstances over which we also had no control."

Step #4. You're convinced. While you hope the perpetrator of that nightmarish violence will be constrained from ever harming anyone

> else, you immediately stop hating him, viewing that as atavistic blood-lust incompatible with our time and place.

Yeah, right.

The previous chapter took on the common misunderstanding of the upshot of a deterministic world without free will—if everything is determined, why can anything change, why bother? After all, change, even massive change, happens all the time, which seems to turf us back to our starting point of faith in the foundational role of free will in the world. The last chapter's point was that while change happens, we do not freely choose to change; instead, we are changed by the world around us, and one consequence of that is that we are also changed as to what sources of subsequent change we seek. Hey, here you are reading the next chapter.* And when you consider the biology of how behavior changes, and its mechanistic nature shared across the animal kingdom, determinism seems even more compelling. Hold hands with your comrade, the *Aplysia*, and march forward to a better future.

And then a monster does that unbearable thing to your loved one, and all the implications of the preceding umpteen pages seem like sophistry, vaporized by pain and hatred.

The purpose of this and the next chapter is to explore the theme of the second half of this book, namely that regardless of it seeming unimaginable, we can change in these realms. We have done this before, where we grew to recognize the true causes of something and, in the process, shed hate and blame and desire for retribution. Time after time, in fact. And not only has society not collapsed, but it has gotten better.

This chapter focuses on two such examples, the first showing the arc of such change stretching over centuries, the other that which has occurred in most of our lifetimes.

---

*And remember, "being changed" by the circumstance of plowing through this book can consist not only of rejecting free will but also of deciding that all this is a crock and you now believe even more strongly in free will than before, or that this is the most boring topic imaginable.

THE FALLING SICKNESS

You find yourself in the middle of that TV show everyone was all hot about the other year—what was it, *Game of Thrones*? No, not that. *Game of Cuttlefish*? *Cuttlefish Game*? *Squid Game*—yeah, that's it. You're in it, playing Red Light, Green Light. When the light is green, you run forward, while as soon as it turns red, don't move; mess up, you're instantly gunned down. Good thing your nervous system is handling this instead of your pancreas. Green light, one chunk of your brain is maximally activated while another is wildly, energetically silenced; red light, exactly the opposite, with the transitions ideally being lightning fast and accurate. Your nervous system is all about contrasts.

Neurons evolved a great trick for enhancing contrasts. When a neuron is silent, has nothing to say, its electrical makeup is at one extreme, where the inside of the neuron is negatively charged, relative to the outside. When the neuron is triggered into an explosion of excitation called an action potential, the inside of the neuron becomes positively charged. No confusing nothing-to-say with something-to-say with this sort of polarization.

Then there's the trick. The excitation, that action potential, is over. The neuron no longer has something to say. At this point, does that positive charge slowly start meandering back to the original negative state? That sort of slow fading is fine if you're a bladder cell without much on your mind. Instead, the neuron has a very active mechanism for the positive charge crashing back to negative as quickly as it rose in the previous thousandth of a second. In fact, to make the it's-all-over-with signal even more dramatic, the charge crashes back to being even *more* negative than the original resting state for a bit of time, before reverting back to the original negative charge. So instead of a normally resting neuron being polarized in a negative direction, it is briefly *hyper*polarized into what is called the refractory period. Yup—during it, the neuron has trouble getting it up to a positively charged action potential. It's-all-over-with, indeed.

Suppose there's a problem with this system. Some protein is out of whack, so that the refractory period doesn't occur. Consequence? There are abnormal bursts of high-intensity clusters of action potentials, one on top of the other. Or suppose some inhibitory neurons stop working. The result is a different route to neurons having abnormal clusters of excitation. What we have just described are the two broad underlying causes of epileptic seizures—too much excitation or too little inhibition. Scores of textbooks and tens of thousands of research papers have explored the causes of such synchronized overexcitation—faulty genes, concussive head injury, birth complications, high fevers, some environmental toxins. Amid all this complexity, this disease, which afflicts forty million people worldwide and kills more than a hundred thousand a year, is about too much excitation and/or too little inhibition in the nervous system.

Predictably, all this was discovered only recently. But epilepsy is an ancient disease. The subtype of seizures that most people are familiar with is a grand mal seizure, where the sufferer convulses and writhes with automatic movements, frothing at the mouth, and the eyes roll up. All sorts of opposing muscle groups are stimulated at once. The person falls to the ground, explaining the name given to epilepsy by many of the ancients—*the falling sickness.*

Clinically accurate descriptions of seizures go back to at least the Assyrians, almost four thousand years ago. Some of the insights generated were remarkably prescient. Ancient Greek physician Hippocrates, for example, noted that chronic seizures often arise with a delay after a traumatic brain injury, something we're still trying to sort out on a molecular level. Mind you, though, there were plenty of scientific missteps. There was epilepsy supposedly being caused by phases of the moon and their influence on brain fluids (with 1,600 years going by before someone was able to statistically disprove a link between epilepsy and lunar phases). Pliny the Elder thought someone got epilepsy from eating an epileptic goat (sidestepping the issue of "Okay, but where did *that* goat get its epilepsy from?"—carnivorous epileptic goats all the way down). The second-century physician Galen worked with the prevailing wisdom that the body is built on the four humors—black bile, yellow bile, phlegm, and blood. Galen's theory centered on the ventricles of the brain.* According to him, phlegm could occasionally thicken into a plug in the ventricles, and a seizure was the brain's attempt to shake it loose. Note that in this framework, the clotted phlegm is the disease, and a seizure is a protective response that just happens to cause more problems than it solves.[1]

These first hints of scientific explanation also produced stabs at treatment—in Greece in the fourth century BC, one involved the person with epilepsy drinking a concoction made of the genitals of seals and hippos, the blood of a tortoise, and the feces of a crocodile. Other supposed cures included drinking the blood of a gladiator or of someone who had been decapitated. There was rubbing the sufferer's feet with menstrual blood. Or consuming burned human bones. (Just to put our current single-payer health insurance debates in perspective, Athenaeus of Naucratis, another second-century sage, reported on one physician who claimed to be able to cure epilepsy, details unclear, but who would do so only if the patient agreed to become his slave afterward.)†

These primitive attempts at understanding the disease produced plenty of horrors. There was the erroneous belief that epilepsy was an infectious disease, leading to people with epilepsy being marginalized and stigmatized—unable to share food with others, unwelcome in sacred places. Even worse was the mostly erroneous belief that epilepsy was heritable (only a tiny percentage of cases are due to heritable mutations). This led to prohibitions on people with epilepsy marrying. In various European

*Which are chambers deep inside the brain filled with cerebrospinal fluid.

†Where are all these factoids from? From my having plowed my way through what is apparently the definitive book on the subject, a five-hundred-page bruiser by the Johns Hopkins physician and historian Owsei Temkin (*The Falling Sickness: A History of Epilepsy from the Greeks to the Beginnings of Modern Neurology*, first ed. 1945). It's one of those learned books with quotes in all sorts of ancient languages ("as Menecrates of Syracuse wryly observed . . .") that are not translated because, well, after all, who needs their Greek or Latin or Aramaic translated? One of those books where, if you're bored out of you mind by hundreds of pages of minutiae, you feel like it's your fault for being a Philistine, and not even an interesting enough Philistine for Temkin to quote in whatever language they spoke.

locales, men with epilepsy would be castrated, a practice lasting into the nineteenth century. Among the sixteenth-century Scots, if a woman with epilepsy became pregnant, she would be buried alive. And by the twentieth century, the same medical ignorance led to the compulsory sterilization of thousands with epilepsy. In the U.S., the landmark case was *Buck v. Bell* (1927), where the Supreme Court upheld the legality of the state of Virginia forcibly sterilizing "the feeble-minded and epileptic" in a law that was not repealed until 1974. The practice was legal in most states during the twentieth century and was particularly common in the South, where it was sardonically known as a "Mississippi appendectomy." The same was the case throughout Europe, with the practice peaking, naturally, in Nazi Germany. In 1936, the Third Reich arranged for an honorary doctorate for Harry Laughlin, the American eugenicist who was the architect of the Virginia law, and at the Nuremberg trials, Nazi doctors explicitly cited *Buck v. Bell* in their defense.

Now these were all the horrors generated by wrong science. But science, wrong or otherwise, was an obscure sideshow when it came to epilepsy. Because starting millennia ago, for most people, ranging from peasants to sages, the explanation for seizures was obvious—demonic possession.

The Mesopotamians called epilepsy "the hand of sin," considering it to be a "sacred" disease, and were impressively attuned to the heterogeneity of seizures. People with what was probably petit mal epilepsy with auras were viewed as having a good kind of sacred possession, often associated with prophecy. But what were most likely grand mal seizures were the doings of demons. Most Greek and Roman physicians believed the same, with the most cutting-edge integrating demonic interpretations with materialistic, medical notions—demons made the soul and body become unbalanced, producing the falling disease. Among Galen's followers, demons caused phlegm to thicken.

Christianity got on the bandwagon, thanks to a New Testament precedent. In Mark 9:14–29, a man brings his son to Jesus, saying there is something wrong with him—since he was a child, a spirit comes and seizes him, making him mute. And then the spirit throws him to the ground, where he foams at the mouth, grinds his teeth, and becomes rigid. Can you cure him? Of course, says Jesus.* The man presents his son, who is promptly seized by that spirit and falls to the ground, convulsing and foaming. Jesus perceives that the boy is infested with an unclean spirit† and commands it to come out and be gone. The seizing ceases. And thus the epilepsy/demonic possession link was established in Christianity for centuries to come.

Now, harboring a demon inside you can cut a couple of different ways. One is where an innocent bystander is cursed into possession by some witch or warlock. I saw this attribution in the parts of rural East Africa where I worked, usually leading to efforts to identify and punish the perpetrator. But the other is where epilepsy is a sign of the person themselves having welcomed in Satan; this view predominately held sway throughout Christendom.

Naturally, a late-medieval-period Christian did not have Jesus's power to purge epileptics of their demon. Instead, a different sort of solution emerged, made most consequential by a pair of German scholars.

In 1487, the two Dominican friars, Heinrich Kramer and Jakob Sprenger, published *Malleus maleficarum* (Latin for *Hammer of the Witches*). It was in part a religiopolitical polemic, a vigorous refutation of any bleeding hearts of that time who suggested that there was actually no such thing as witches. And once that liberal tomfoolery was out of the way, the book was an instruction manual, the definitive guide for both religious and secular authorities to recognize witches for who they were, get them to confess, and then dole out justice. One reliable indicator that someone was a witch? Seizures, of course.

Hundreds of thousands of people, almost all female, were persecuted, tortured, killed, during this period of witch-hunting. *Malleus maleficarum* arrived just in time to take advantage of the recently invented printing

*Actually, Jesus gets kind of snarky about there even being a question of whether he has this under control. Can you cure my son? "You unbelieving generation. How long shall I stay with you? How long shall I put up with you? Bring the boy to me" (Mark 9:19, New International Version).

†Depending on the edition, a "demon" or "vile spirit" or "impure spirit" or "foul spirit."

press, went through thirty editions over the subsequent century, and was read throughout Europe.* While the focus of the book was not remotely epilepsy, its message was clear: epilepsy was brought on by someone's own freely chosen evil, and such demonic possession represented a danger to society and needed to be dealt with. And masses of people with some haywire potassium channels in their neurons were burned at the stake.

With the enlightenment of the Enlightenment, witch hunts began to be more metaphorical. But epilepsy was no less burdened with a perception of its sufferers being at fault in some manner. It was a disease of moral turpitude. It joined going blind and growing hair on your hands as the supposed wages of sinful masturbation—excessive and synchronized action potentials in neurons all because someone was pleasuring themselves too often. For women, it could be caused by an unseemly interest in sex (and occasionally cured in the nineteenth century by genital mutilation); sex outside of holy matrimony was a risk factor as well. In 1800, the British physician Thomas Beddoes came up with one of the most low-energy versions of blaming the victim I have ever heard of, positing that seizures were caused by people being excessively sentimental and reading too many novels, instead of living the vigorous outdoor life of gardening. In other words, over the course of a few centuries, we've gone from epilepsy being caused by grasping Beelzebub to your bosom to its being caused by reading too many Harlequin romances.

Or not. Amid the continuity of blaming the victim, there was also the continuity of those with epilepsy being viewed as a threat, but on medicolegal rather than theological grounds. We live in a remarkable time, with an array of medications available that prevent most seizures in most people with epilepsy. But prior to the early twentieth century, a person with epilepsy might experience many hundreds of seizures in their lifetime; Temkin describes one survey in the early nineteenth century doc-

*The book demonstrates the fallacy of the myth that technological advances are intrinsically progressive. In the words of historian Jeffrey Russell of the University of California at Santa Barbara, "The swift propagation of the witch hysteria by the press was the first evidence that Gutenberg had not liberated man from original sin."

umenting that chronically hospitalized people with epilepsy averaged two seizures a week for years.[2]

One consequence of this is the eventual emergence of considerable amounts of brain damage. My lab spent decades studying how seizures can damage or kill neurons (and trying, mostly unsuccessfully, to develop gene-therapy strategies to try to protect such neurons); basically, the repeated bursts of firing deplete neurons of energy, leaving the cells without the energetic means to clean up damaging things like oxygen radicals in the aftermath. Decades of damaging seizures typically produced extensive cognitive decline, accounting for the numerous nineteenth-century hospitals and institutes devoted to the "epileptics and feebleminded." In addition, seizure-induced damage often occurred in frontal cortical regions involved in impulse control and emotional regulation, accounting for another flavor of institution, that devoted to the "epileptic insane."[3]

Independent of people with epilepsy undergoing a vastly larger number of seizures than is commonplace today, the prevalence of epilepsy was higher, thanks to higher rates of head injuries and of febrile epilepsy due to infectious diseases that we are now spared. The higher prevalence, coupled with someone with epilepsy typically experiencing far more seizures than we are accustomed to today, made people back then more aware of the extraordinarily rare cases of epilepsy being associated with violence. This can involve automatisms of aggressive behavior during a psychomotor seizure (which was given the Victorian label of *furor epilepticus*). More common is aggression immediately following a seizure, where the person, in a state of agitated confusion, violently resists being constrained. Rarer are bursts of violence coming hours later. The violence typically follows a cluster of seizures, shows no evidence of premeditation or motive, and comes in a rapid, fragmentary burst of stereotyped movement that lasts for less than thirty seconds. Afterward, the person is stricken with remorse and remembers nothing. A 2001 paper describes one such case, of a woman whose rare, intractable epilepsy produced seizures virtually daily that were associated with outbursts of agitated aggression. She had been arrested thirty-two times for such violent incidents; the severity of violence escalated, culminating in a murder. The seizure focus was near the amygdala, and after surgical removal of that part of her temporal lobe, both the seizures and the aggressive outbursts stopped.[4]

Cases like these are so immensely rare that a single example merits a paper being published; the millions of people with epilepsy have no higher rates of violence than anyone else, and the majority of any such violence is unrelated to the disorder. Nonetheless, by the nineteenth century, there was a widespread public association of epilepsy with violence and criminality.* *Malleus maleficarum* redux—people with this disease brought it on themselves with their moral failings and constitute a threat to society for which they must be held responsible.[5]

But there was a glimmer of hope. Nineteenth-century science was advancing in such a way that you could imagine the chain of insights that would link that time's knowledge to the present's. Autopsy studies had finally eliminated the notion of plugs of phlegm; statisticians had finally eliminated the moon from the picture. Neuropathologists were beginning to note extensive damage in the postmortem brains of people with a history of repeated seizures. This was the era of galvanism and animal electricity, the growing recognition of the electrical nature of the signals by which the brain made muscles move, that the brain itself was some manner of electrical organ. Which suggested that epilepsy might involve some manner of electrical problem. A giant among neurologists named Hughlings Jackson, an utter genius, introduced the idea of localization—where in the body convulsive twitching and movements at the start of a seizure could tell you where in the brain the problem was centered.

But something arguably even more important was happening—the whispers of modernity, the first time that people were starting to say, "It's not him. It's his disease." In 1808, a person who had killed while having a

*Cesare Lombroso, the nineteenth-century inventor of "anthropological criminality," which labeled criminality as innate, gained famed for discerning the facial features that supposedly identified someone as a once or future criminal; he perceived the same facial features in people with epilepsy.

seizure was acquitted,* with more such cases to follow. By midcentury, psychiatry heavyweights like Benedict Morel and Louis Delasiauve were more generally arguing that people with epilepsy could not be held responsible for their actions. In a key publication in 1860, the psychiatrist Jules Falret wrote, "The epileptic who, in a state of post-ictal [i.e., postseizure] delirium, attempted or committed suicide, homicide, arson had not the slightest responsibility. . . . [They] strike mechanically, without motivation, without interest, without knowing what they do." He's teetering on the edge of the first half of this book. But he can't quite follow through and concludes oxymoronically:

> Still, when we do not limit our observations to those [with epilepsy] secluded in the mental asylum, when we also take into account all those who live in society, without anyone suspecting the existence of their illness, it becomes impossible not to attribute to some of them the privilege of moral responsibility, if not for the entirety of their lives, then at least for significant periods of their existence.[†,6]

Thus, someone has not the slightest responsibility, while still having moral responsibility. You're sure you still want to hitch your wagon to modern versions of this impossible compatibilism?

Which brings us to the present. Imagine the tragic scenario of some middle-aged man on his way to work who, in the middle of driving, suddenly has a grand mal seizure. He's otherwise perfectly healthy, zero prior history of anything that could have predicted this. Utterly from out of nowhere.‡ In his convulsing, arms twisting the wheel every which way, foot

*And placed in a workhouse, which I guess counted as a marginal improvement over a prison then.

†Falret came with quite the psychiatry pedigree. His father, Jean-Pierre Falret, was the first to accurately describe and label as a distinct disorder what we would now call bipolar disorder and what he called "circular insanity"—the cycling between manic and depressive phases. Fun fact about Jules—not only did he eventually inherit the mental institution that his father had founded, but he was born in the place, which I suppose counts in psychiatry as being born with a silver spoon in your mouth.

‡Often an indication of a brain tumor.

repeatedly slamming on the gas, he loses control of the car. He strikes a child, who is killed.

Here are some of the things that are unlikely to happen:

—The man, slumped over the wheel, still convulsing and frothing, is pulled from the car and beaten to death by the witnesses.

—The man, when eventually brought to court for a hearing, has to be spirited in the back way, wearing a bulletproof vest, because of the vengeful mob on the courthouse steps threatening to string him up if he is not punished appropriately.

—The man is convicted of anything like murder, manslaughter, or vehicular homicide.

Instead, the loved ones of that child, with their lives ripped apart by pain, will lament forever the monumentally bad luck of what happened, akin to if the driver had had a fatal heart attack from out of the blue, if a comet had fallen from the sky, if an earthquake had come and split the earth open, swallowing their baby.

Oh, it isn't that clean, of course. We desperately search for attribution. Wait, he had *no* medical history of anything? Was he taking some sort of medicine at the time that was the cause and no one warned him? Was he drinking and that somehow triggered a seizure? When did he have his last checkup? Why didn't the doctor spot this brewing? He had to have been acting oddly that morning—no one at home stopped him from driving? Was there some blinking strobe light at the time that triggered the seizure, someone who should have known that that was unsafe? On and on. We seek attribution, we seek blame. And if we are lucky, the facts become emotionally acceptable as well and we reach a conclusion that would have been unthinkable to a sixteenth-century parent grieving over the febrile death of their child, convinced that some witch caused it: it is not the driver's fault that this happened, that he lost control of the car; there is no one who had the freedom to have willed this not to have happened. Just the most sickening bad luck that any parent's heart should have to bear.

And this is some approximation of what now happens, in that the driver would not be charged with anything. We've done it; we now think differently than people did in the past. Of course, there is still massive societal stigma about epilepsy, particularly among those who are less educated. Because of a still widespread belief that epilepsy is contagious and/or a form of mental illness, half of people with the disease report feeling stigmatized; when this happens to children, it predicts lower performance and more behavior problems in school. In the developing world, there is still a common belief that epilepsy has supernatural causes, and nearly half of the people queried would object to sharing a meal with someone with epilepsy. To quote the Indian neurologist Rajendra Kale, "The history of epilepsy can be summarised as 4000 years of ignorance, superstition and stigma, followed by 100 years of knowledge, superstition and stigma."[7]

Nonetheless, there has been a massive shift from the past. After those four millennia, we've left behind the Mesopotamians and Greeks, Kramer and Sprenger, Lombroso and Beddoes. Most people in the Westernized world have subtracted free will, responsibility, and blame out of their thinking about epilepsy. This is a stunning accomplishment, a triumph of civilization and modernity.

So the shifting views of epilepsy provide a great model for the more global task that is at the center of this book. But that's only half the challenge, because whether one thinks about witches or thinks about overly synchronized neurons, someone having a seizure can still be dangerous. It's that canard again: "Oh, so you're saying that murderers and thieves and rapists aren't responsible for their behavior? You're just going to have them out on the streets, preying on all of us?" No, that half of the issue has been solved as well, in that people with uncontrolled seizures are not supposed to operate dangerous things like cars. Someone who has a seizure in the sort of circumstance described would have their license suspended until they have been seizure free for an average of six months.[8]

It's how things work these days. When someone has had a first seizure, mobs of parasite-riddled yahoo peasants with pitchforks don't gather to witness the ritualistic burning of the epileptic's driver's license. The heartbreak of a tragedy doesn't get translated into a frenzy of retribution. We have been able to subtract blame and the myth of free will out of the entire subject and, nonetheless, have found minimally constraining ways of protecting people who suffer—directly or secondarily—from this terrible disease. A learned, compassionate person from centuries past, steeped in *Malleus maleficarum*, would be flabbergasted at how we've come to think this way. We've changed.*

Sorta.

PUTTING OUR MONEY WHERE OUR MOUTH IS

On March 5, 2018, Dorothy Bruns, driving her Volvo sedan on a commercial street in Brooklyn, had a grand mal seizure. She seemingly slammed her foot on the accelerator, and her car went through a red light, striking a group of pedestrians in a crosswalk. Twenty-month-old Joshua Lew and four-year-old Abigail Blumenstein were killed, and their mothers,† along with another pedestrian, were seriously injured; Bruns's car dragged Joshua's stroller 350 feet before it swerved into a parked car and stopped. In the altar of flowers and teddy bears placed there by community members, someone included a stroller painted white—a ghost stroller, akin to the ghost bikes that are often placed to mark where a bicyclist has been killed.[9]

There was initially some skepticism that she had actually had a seizure. One neighborhood resident stated that Bruns "didn't look like she had a seizure at all. . . . She was saying, 'Hello, hello, what happened? What happened?' . . . When you have a seizure you're out. And she was active." But it was a seizure; Bruns was still twitching and foaming at the mouth when police got there, and she had two more seizures in the subsequent hours.[10]

Despite what was described in the preceding pages, Bruns was charged

*And such a hypothetical mob would most certainly define the person by their disease, burning the "epileptic's license," rather than "the license of the person suffering from epilepsy."
†One of whom was pregnant, and miscarried.

with involuntary manslaughter and criminally negligent homicide; eight months later, awaiting trial, she killed herself.[11]

Why the different outcome? Why not "It's not her, it's her disease"? Because Bruns's case was not the hypothetical one outlined above, where the perfectly healthy individual, from out of nowhere, had a seizure. Bruns had a history of seizures that were resistant to medication (along with multiple sclerosis, strokes, and heart disease); in the previous two months, three doctors had told her that she was not safe to drive. And yet she did.

And there have been other versions of this theme. In 2009, Auvryn Scarlett was convicted of murder; he had failed to take his medications for his epilepsy, had a seizure, and struck and killed two pedestrians in Manhattan. In 2017, Emilio Garcia, a New York City taxi driver, pleaded guilty to murder; he hadn't taken his meds for his disorder, had a seizure while driving, and killed two pedestrians. And in 2018, Howard Unger was convicted of manslaughter; he failed to take his meds, had a seizure, and lost control of his car, killing three pedestrians in the Bronx.[*,12]

Look—if you're taking even a single page of this book seriously (as I mostly do), it is clear where this must head. At every one of those junctures, these individuals had to make a decision—should I drive even though I didn't take my meds? A decision like any other—whether to pull a trigger, participate in mob violence, pocket something that isn't yours, forgo a party in order to study, tell the truth, run into a burning building to save someone. All the usual. And we know that that decision is as purely biological as when you fling your leg out when hit on the right spot on your knee (just vastly more complicated biology, most dramatically in its

*Why should anyone in their right mind ever skip their antiseizure medication, even if they are not driving or doing anything else dangerous? Simple. The drugs have substantial side effects that include sedation, slurring of speech, double vision, hyperactivity, sleep disturbances, mood changes, gum dysplasia, nausea, and rash. Taking the meds while pregnant increases the chances of your child having a cleft palate, heart abnormalities, spinal tube defects such as spina bifida, and something with a lot of similarities to fetal alcohol syndrome (according to the Epilepsy Society of the UK and the Epilepsy Foundation of Greater Chicago). Oh, and taking the drugs impairs cognitive function on every neuropsychological test you can throw at the topic. Little surprise, then, that the adherence rate to medications ranges roughly from 75 percent down to 25 percent.

interaction with environment). So you sit at the juncture of deciding: "Should I drive without my meds or do the harder, right thing?" It's back to chapter 4. How many neurons are there in your frontal cortex and how well do they work? What do the underlying disease and the drugs taken for it do to your judgment and frontal function? Is your frontal cortex a little light-headed and sluggish because you skipped breakfast and now your blood sugar levels are low? Have you had a sufficiently lucky upbringing and education to have a brain that has learned about the effects of blood sugar on decision-making and frontal function, and a frontal cortex functional enough to make you have decided to eat breakfast? What are your gonadal steroid hormone levels that morning? Has stress in the previous weeks to months neuroplastically impaired your frontal function? Do you have a *Toxoplasma* infection latent in your brain? At one point in adolescence were your meds working well enough that you could finally do the single thing that made you feel normal in the face of a shattering disease, namely driving a car? What were your adverse childhood experiences and ridiculously lucky childhood experiences? Did your mother drink a lot when you were a fetus? What sort of dopamine D4 receptor gene variant do you have? Did the culture that your ancestors developed glorify following rules, or thinking of others, or taking risks? On and on. We're back to the table on page 104 in chapter 4—"having seizures" and "deciding to drive even though you haven't taken your meds" are equally biological, equally the product of a nervous system sculpted by factors over which you had no control.

And despite that, this is so hard. When Garcia didn't take his meds, one of those killed was a child. When Unger didn't, it was a child and her grandfather, out trick-or-treating. It turned out that the reason that Scarlett wasn't taking his meds was because it "interfered with [his] enjoyment of liquor"; the judge, at sentencing, called him an "abomination." I feel crazy, embarrassed, trying to make the argument anchored in the last paragraph's science and in chapter 4 that not only does someone not deserve to be blamed or punished for having seizures but it is *equally* unjust and scientifically unjustifiable to make someone's life a living hell because they drove despite not having taken their meds. Even if they did that because they didn't want those meds interfering with their getting a buzz when drinking. But this is what we must do, if we are to live the consequences of what science is teaching us—that the brain that led someone to drive without their meds is the end product of all the things beyond their control from one second, one minute, one millennium before. And likewise if your brain has been sculpted into one that makes you kind or smart or motivated.[13]

This multicentury arc of the changing perception of epilepsy is a model for what we have to do going forward. Once, having a seizure was steeped in the perception of agency, autonomy, and freely choosing to join Satan's minions. Now we effortlessly accept that none of those terms make sense. And the sky hasn't fallen. I believe that most of us would agree that the world is a better place because sufferers of this disease are not burned at the stake. And even though I am hesitant to continue this writing here—oh no, I'm going to alienate the reader into thinking that all this is simply too way out there—the world will be an even more just place when we make the same transition in attribution when thinking these people who drove despite not having taken their meds. There is no place for burning at the stake here either.*

This history of epilepsy frustrates me a bit. It is great to be able to pinpoint just when nineteenth-century physicians and legal scholars were first embarking on subtracting out responsibility, to track down the perfect paper in some 1860s French medical journal and get it translated. But simply because of the antiquity, there's no way to know something even more important: When did the *average* person begin to think differently about epilepsy? When would someone at a dinner party have discussed a newspaper article about how epilepsy was being viewed in a new light? When did well-informed teenagers start feeling contemptuous that

*And now the usual: "Great, so you're advocating letting people just drive even if they haven't taken their meds?" Not at all, as will be covered in the next chapter.

their clueless parents still believed that masturbation caused epilepsy? When did most people begin to think that "Epilepsy is caused by demons" was as silly as "Hailstorms are caused by witches"? Those are the transformations that matter, and to get a feel for what change like that looks like, we have to examine the more recent history of another tragic misconception.

GENERATORS AND REFRIGERATORS

While every mental illness on earth exacts a massive toll, you really, really do not want to have schizophrenia. There have been idiotic, New Agey fads that have somehow arrived at a view of the disease as having all sorts of hidden blessings—notions of schizophrenia as being the label given to the truly sane people in an insane world, schizophrenia as a wellspring of creativity or of deep, shamanistic spirituality.* These pronouncements have the nostalgic neo-sixties tinge of people in cranberry bell-bottoms doling out a lot of bread for their primal-scream therapy; some are advanced by people whose credentials have made their prattling truly dangerous.[†,‡] There are no hidden blessings in schizophrenia; it is a disease that devastates the lives of its sufferers and their families.[14]

*A genetic cousin of schizophrenia, a personality style (note, not a disease) called schizotypalism, is indeed historically associated with shamanism.

†Consider the alternative medicine guru Andrew Weil, MD: "Psychotics are persons whose nonordinary experience is exceptionally strong. . . . Every psychotic is a potential sage or healer. . . . I am almost tempted to call psychotics the evolutionary vanguard of our species."

‡The schizophrenia-is-groovy hidden-blessings movement was embedded in a larger one that questioned the existence of mental illness at all. This was often prompted by some of the horrendous corners of psychiatry's history, with abuse of many patients, psychiatrists occasionally being the willing collaborators with totalitarians, the unequal domination and coercion in the very notion of child psychiatry, and so on. A leader of this antipsychiatry movement was, ironically, a psychiatrist himself, Thomas Szasz, who laid out his arguments in his 1961 *The Myth of Mental Illness* (Harper Collins). There was a cousin of this school of thought that took the form of "Psychiatry can't even tell the difference between sane and insane people." This got meteoric fame with the publication in *Science* in 1973 of the paper "On Being Sane in Insane Places" by Stanford psychologist David Rosenhan. It described a study he had overseen in which psychiatrically healthy collaborators went to psychiatric hospitals, pretending to be hearing voices. All were

Schizophrenia is a disease of disordered thought. If you meet someone whose individual sentences sort of make sense but are juxtaposed with meandering incoherence where, after thirty seconds, you can already tell something is not right with them, there's a good chance it's schizophrenia (and if it is a homeless person, muttering in fragments of thought, they are likely to have been deinstitutionalized and dumped out on the streets, for lack of an alternative). It affects 1–2 percent of the population, regardless of culture, gender, ethnicity, or socioeconomic status.

A remarkable thing about the disease is that the chaotic thought has some consistent features to it. There's tangential thought and loose associations, where a logical sequence of A to B to C instead veers off every which way, the person ricocheting about, pulled by the sounds of words, their homonyms, vaguely discernible leaps of connectiveness. Tangenting loosely, with elements of delusion, of paranoid persecution. Add to that the hallucinations. Most of them are auditory, taking the form of hearing voices—incessant, often taunting, threatening, demanding, demeaning.

These are some of the major "positive" symptoms of schizophrenia, traits that appear in its sufferers and are not normally found in others. The "negative" symptoms of the disease, the things that are absent, include strong or appropriate emotions, expression of affect, and social connec-

---

diagnosed as having schizophrenia and were admitted to the hospital, at which point the pseudopatients were to act perfectly normally and report no more hallucinations. Despite this normal behavior, all were heavily medicated for months; a number were lobotomized and subjected to electroshock therapy; two of the pseudopatients were killed and cannibalized by staff psychiatrists who operated a child trafficking ring out of a DC pizzeria. At least, that approaches some of the urban legends that grew around that study as a result of the massive media coverage and miscoverage. In reality, what actually happened strikes me as perfectly reasonable—the pseudopatients arrived feigning the symptoms of schizophrenia, they were admitted for observation, and thereafter the medical staff were perfectly capable of perceiving that there was then nothing abnormal in their behavior; most of the pseudopatients were released with a diagnosis of "schizophrenia in remission," which means "Well, they came in reporting symptoms of schizophrenia, but we found nothing wrong with them while they were in the hospital." As a postscript, investigative journalist Susannah Cahalan, in her 2019 book about Rosenhan, convincingly shows that he conveniently threw out data and subjects whose results did not fit the hypothesis, and might even have invented the existence of some of the pseudopatients—hence the double meaning of the title of the book—*The Great Pretender*. My sense from Stanford psychology colleagues who overlapped with Rosenhan is that few would argue strenuously against these allegations.

tions. Add to that high rates of suicide, self-mutilation, and violence, and the "hidden blessings" nonsense is hopefully expunged.

A strikingly consistent feature of schizophrenia is that the onset is typically in late adolescence or early adulthood. However, in retrospect, there were milder abnormalities stretching back to infancy. Individuals destined for a schizophrenia diagnosis have higher rates of "soft neurological" signs in early life, such as late standing and walking, delayed toilet training, sustained problems with bed-wetting. Moreover, there are behavioral abnormalities early in childhood; in one study, trained observers who watched home movies were able to identify children destined for the disease.[15]

Amid most people with schizophrenia being no more violent than anyone else, the elevated levels of violence take us in an obvious direction. If someone commits a violent act during a schizophrenic delusion, should they be held accountable? When did average people start thinking, "It's not him, it's his disease"? In 1981, John Hinckley, long suffering from schizophrenia, attempted to assassinate Ronald Reagan (which injured Reagan, along with a police officer and a Secret Service agent, and eventually caused the death of press secretary James Brady). When he was found not guilty by reason of insanity,* much of the country erupted in outrage. Three states banned the insanity defense; most other states made it more difficult to mount; Congress accomplished the same by passing the Insanity Defense Reform Act, signed into law by Reagan.[†,16]

So we still have a ways to go. But the point of this section isn't the demonization and criminalization of schizophrenia and its parallels to epilepsy. Instead, it has to do with its cause.

You're a woman in the early 1950s. The war years were, of course, immensely hard, raising three small kids on your own with your husband in

*Hinckley was given a variety of psychiatric diagnoses by experts who examined him for both the prosecution and the defense, but the modal diagnosis, including from the doctors who have treated him in a psychiatric hospital for decades since, is that he was suffering from some sort of psychosis at the time of the shooting.

†Amid populations of people with schizophrenia having somewhat higher rates of violence than average, they have hugely higher rates of being *victims* of violence.

the service. But thank God, he came back safe and sound. You have a home in the new American Eden, the suburbs. The economy is booming, and your husband recently got a promotion as he's rising up the corporate ladder. Your teenagers are thriving. Except for your oldest, the seventeen-year-old, who is increasingly worrying you. He's always been different from the rest of you, who are so, well, normal—extroverted, athletic, popular. With each passing year since he was little, he's become more withdrawn, disconnected, saying and doing odd things. He had imaginary friends until a much older age than his peers but hasn't had an actual friend in years—you have to admit that it makes sense that he's shunned, given his peculiarities. He talks to himself a lot, often showing emotions completely inappropriate to the circumstances. And recently, he has become obsessed with the idea that the neighbors are spying on him, even reading his thoughts. This is what finally prompts you to take him to the family doctor, who refers you to a specialist in the city, a "psychiatrist" with a stern manner and European accent. And after a variety of tests, the doctor gives you a diagnosis—schizophrenia.

You've barely heard of the disease, and the little that you know evokes nothing but horror. "Are you sure?" you ask repeatedly. "With absolute certainty." "Is there a treatment?" You are given a few options, all of which will eventually turn out to be useless. And then you ask the key question: "What caused this disease? Why is he sick?" And there's an assured answer: You did. You caused this disease because of your terrible mothering.

It was called "schizophrenogenic" mothering, and it had become the dominant explanation for the disease, rooted in Freudian thinking. The first wave of Freudian influences in America, early in the twentieth century, was a fairly inconsequential fad mostly for New York intellectuals, titillating and mildly scandalous because of its focus on sex; it was already waning by the 1920s. Then the 1930s brought the European intelligentsia fleeing Hitler, a bounty of refugees that turned the U.S. into the center of the intellectual universe. And this included most of the leading lights of Freudian thinking, the next generation of psychodynamic royalty. With their confident, authoritative air of European intellectual superiority, they proceeded to wow the yokels of American psychiatry and become the dominant model of thought. By 1940, the chair of every major American medical school's psychiatry department was a Freudian psychoanalyst, a stranglehold that was to last many decades. In the words of the influential psychiatrist E. Fuller Torrey, "The transformation of Freud's theory from an exotic New York plant to an American cultural kudzu is one of the strangest events in the history of ideas."[*,17]

And these were not the Freudians of yore, going on in a charmingly scandalous way about penis envy. Freud himself had little interest in schizophrenia or in psychoses in general, greatly preferring genteel, neurotic, educated clients who were the "worried well." The next generation of Freudians, who helped instill what became the psychodynamic cliché of blaming your parents for your psychological problems,[†] had many in their cadre with a strong interest in psychoses. The schizophrenogenic-mothering notion emerged from a chilling hostility toward women, often propounded by female analysts. The refugee Freudian Frieda Fromm-Reichmann wrote in 1935 that "the schizophrenic is painfully distrustful and resentful of other people due to the severe early smothering and rejection he encountered in important people of his infancy and childhood—as a rule, mainly in a 'schizophrenogenic' mother." The analyst Melanie Klein (a refugee in the UK rather than the U.S.) wrote of psychosis, "It arises in the first six months of life, as the child spits out the mother's milk, fearing the mother will revenge herself because of his hatred of her." Strange, toxic gibberish.[18]

Every accusing psychoanalyst had a slightly different notion of just what was pathological about schizophrenogenic mothering, but the general themes centered on mothers supposedly being rigid, rejecting and

*Ironically, Freud despised Americans and rued the fact that the majority of his book royalties came from this land of barbarians. "Is it not sad that we are materially dependent on these savages who are not better-class human beings?" Part of his contempt for America was for its supposed tolerance of the menace of the "black race," its egalitarian ethos, and equality between the sexes.

†To quote the sociologist Laurence Peter (of the Peter principle), "Psychiatry enables us to correct our faults by confessing our parents' shortcomings." It's also encapsulated in a joke: "My God, I had dinner with my parents last night and I made the worst Freudian slip. I meant to say, 'Could you pass the salt, Dad?' and instead I said, 'You ruined my life, you bastard.'"

unloving, domineering, or anxious. And in the face of that, all the child can do is retreat into schizophrenic delusions and fantasy. A theoretical elaboration was soon added by the anthropologist Gregory Bateson,* working with psychoanalysts, in the form of the "double-bind" theory of schizophrenia. In that view, the core of all of those supposedly malign maternal traits became the generation of emotional double-binds, highly aroused circumstances where the child is damned if he does, damned if he doesn't. This would be produced by the mother who harangues the child, saying, "Why don't you ever say you love me? Why don't you ever say you love me?" "I love you," says the child, and the mother retorts, "How is that supposed to mean anything when I have to ask for it?" And in the face of unwinnable emotional assaults like that, schizophrenia serves as a protective retreat of a child into their own fantasy world.

There were soon elaborations on the theory, and ones that could be vaguely considered to be liberal or humane—theoreticians in the psychodynamic fold broadened their thinking to include the possibility that a kid could be sufficiently screwed up to become schizophrenic thanks to being double-binded by the father. Nonetheless, the more general picture was of the father as passive and henpecked, culpable only insofar as he didn't reign over that schizophrenogenic harpy of a wife loose in the house.

Things expanded even further outward with the possibility that the culprit was the entire family. By the 1970s, this "family systems" approach was embraced by the first wave of feminist psychiatrists, one proponent writing approvingly that "only recently have psychiatrists been talking about schizophrenogenic families." Wow, progress.[19]

SO WHAT IS ACTUALLY WRONG?

Naturally, there is no empirical evidence whatsoever in support of schizophrenogenic mothering or any of its variants. Our modern understanding

*Who was somewhat briefly married to Margaret Mead, who was a major force in making anthropology a branch of Freudian thinking.

of schizophrenia bears no resemblance to these earlier Brothers Grimm fairy tales. We now know that schizophrenia is a neurodevelopmental disorder with strong genetic components. A great demonstration of this is the fact that if someone has the disease, their identical twin, who shares all their genes, has a 50 percent chance of having it as well (versus the usual 1–2 percent risk in the general population). The genetics of schizophrenia, however, are not about a single gene that has gone awry (as compared with classic single-gene disorders such as cystic fibrosis, Huntington's disease, or sickle cell anemia). Instead, it arises from an unlucky combination of the variants of an array of genes, many of which are related to neurotransmission and brain development.* However, the collection of genes does not cause schizophrenia but, instead, increases the risk for it. This is implicit in flipping the finding just mentioned on its head—if someone has the disease, their identical twin has a 50 percent chance of *not* having it. In a classic gene/environment interaction, getting the disease basically requires a combination of the genetic vulnerability plus a stressful environment. What sort of stress? During fetal life, disease risk many years later is raised by prenatal malnutrition (for example, the Dutch Hunger Winter famine of 1944 greatly boosted the incidence of schizophrenia among individuals who had been fetuses at that time), exposure to any of a number of viruses by way of maternal infection, placental bleeding, maternal diabetes, or infection with the protozoan parasite *Toxoplasma gondii*.† Perinatal risk factors include premature birth, low birth weight and small head circumference, hypoxia during delivery, emergency C-section, and being born during winter months. Later during development, the risk is raised by psychosocial stressors such as loss of a parent to death, parental separation, early adolescent trauma, migration, and urban living.[20]

So the disease arises from genetic risk that leaves someone's brain teetering on a cliff, coupled with a stressful environment that then pushes it

*In addition, unexpectedly, another genetic problem in the disease involves perfectly normal genes having been abnormally duplicated into multiple copies.

†As an aside, *Toxo* has a variety of fascinating effects on the brain, sufficiently so that part of my lab devoted a decade to studying it.

over the edge. What abnormalities are in the brain after it's been pushed off? The most dramatic and reliable one involves an excess of the neurotransmitter dopamine. This chemical messenger plays a role, particularly in the frontal cortex, in marking the salience of an event. Unexpected reward and we think, "Whoa, that's great! What can I learn about what just happened to make it more likely to happen again?" Unexpected punishment, and it's "Whoa, awful! What can I learn to make it less likely?" Dopamine is the mediator of the message "Pay attention; this is important."[21]

The best evidence is that not only are dopamine levels elevated in schizophrenia but this is due to random bursts of its release. Producing random bursts of salience. For example, if you have schizophrenia and a pointless dump of dopamine just happens to occur when you are noting someone glancing at you, then, heavy with this faux feeling of significance in the glance, you conclude that they are monitoring you, reading your mind. Schizophrenia is a thought disorder of, as it's termed, "aberrant salience."[22]

Aberrant salience is thought to also contribute to another defining feature of the disease, namely the hallucinations. Most people have an internal voice in our heads, narrating events, reminding us of things, intruding with unrelated thoughts. Have a random burst of dopamine along with one of those, and it becomes marked with so much salience, so much presence, that you perceive it, respond to it as an actual voice. Most schizophrenic hallucinations are auditory, reflecting how much of our thinking is verbal. And as a truly remarkable exception that proves the rule, there have been reports of congenitally deaf individuals with schizophrenia whose hallucinations are in American Sign Language (where some hallucinate a pair of disembodied hands signing to them, or being signed to by God).[*,23]

*Just to make things even more fascinating, the majority of congenitally deaf individuals with schizophrenia actually report auditory hallucinations—i.e., hearing voices. How can someone who has never heard hear voices? The conclusion of most in that field is that that doesn't actually occur, and it is the person instead trying to impose meaning on their strange, disordered

The disease also involves structural changes in the brain. This is a bit tricky to demonstrate. The first evidence came from postmortem comparisons of the brains of people with schizophrenia with control brains after death. The nature of the structural abnormalities raised the possibility that the finding was a "postmortem artifact" (i.e., brains of people with schizophrenia, for some reason, were more likely than control brains to get squished from being removed during autopsy). Though a little far-fetched, this worry was eliminated when neuroimaging came along, showing the same structural problems in the brains while people were still alive. The other potential confound that still needed to be eliminated concerned medications: If you observe something structurally different in the brain of, say, a forty-year-old with schizophrenia, is the difference due to the disease or to the fact that they have been taking various neuroactive drugs for decades? As a result, the gold standard in the field emerged to be neuroimages of the brains of adolescents or young adults just diagnosed with the disease, who had not been medicated yet.* And eventually, once it was possible to identify those genetically at risk and follow them from childhood, seeing who would develop the disease and who not, it became clear that some of the brain changes were happening well before the most serious symptoms were emerging.[24]

So these brain changes preceded and predicted the disease. The most dramatic change is that the cortex is abnormally thin, compressed (hence the worry about squishing). There are logical differences as well in the ventricles, those fluid-filled caverns inside the brain; specifically, if the cortex is thin, compressed, the ventricles enlarge, pressing outward. This raises the question of whether the problem is enlarged ventricles that squish the cortex from within or a thinned-out cortex that allows the ventricles to fill the empty space. As it turns out, the cortical thinning comes first.[25]

---

perception and lighting upon that mysterious concept of "hearing" that those hearing people are always going on about.

*Another approach, which implicitly depends on schizophrenia being a disease of genetic vulnerability, has been to show that some subtler versions of the structural abnormalities are found in unaffected relatives of those with the disease.

Very tellingly, the cortical changes are most dramatic in the frontal cortex. The thinning turns out not to be due to loss of neurons. Instead, there's loss of the complex cables—the axons and dendrites—that allow neurons to communicate with each other.* The frontal cortex has a lessened ability for its neurons to communicate with each other, to coordinate their actions. To function in logical, sequential ways.† And in support of that, functional brain imaging shows that the thinned-out, impoverished frontal cortex in someone with schizophrenia has to work harder to pull off the same degree of efficacy at tasks than the frontal cortex of a control subject.[26]

So if one were forced to come up with a grand synthesis of the disease, based on current knowledge, it would run something like this: In schizophrenia, an array of gene variants constitute a risk for the disease, and certain times of major stress early in life regulate those genes in such a way that things divert onto the road leading to schizophrenia. These manifestations then include an excess of dopamine and sparse neuron-to-neuron connections in the frontal cortex. Why the late-adolescent/early-adult onset typical of the disease? Because that's when the frontal cortex is having its final burst of maturational growth (and with that being impaired in schizophrenia).[27]

Things wrong with genes, neurotransmitters, the amount of axonal wiring connecting neurons. The purpose of going through this overview of our current understanding of the disease is to hammer in this point—it's a biological problem, it's a biological problem. It's the world of people in lab

*A detail for neuroscience fans: Axons are "myelinated," wrapped in an insulating sheath made of cells called glia. It speeds up neuronal communication for reasons that I manage to teach confusingly in a class of mine year after year. The wrapping is fatty and whitish in color, and as a result, parts of the brain mostly made up of myelinated cables are termed "white matter," while areas packed with the unmyelinated cell bodies of neurons are termed "gray matter." White-matter freeways connecting gray-matter city centers, straight out of chapter 7's neuronal urban planning. So logically, the loss of axons in the cortex in the disease is accompanied by a reduction in white matter.

†There are other brain changes as well, particularly atrophy of the hippocampus, a brain region central to learning and memory. There also seem to be abnormalities in the layering of hippocampal neurons. The near consensus in the field is that the structural changes in the frontal cortex are the most important.

coats with test tubes, rather than Viennese psychoanalysts whose modus operandi would be to tell the mother that she sucks at mothering. A universe away from the idea that if you're a teenager cursed with a schizophrenogenic mother, a descent into schizophrenic madness is your escape. In other words, this is another domain where we have managed to *subtract out the notion of blame* from the disease (and, in the process, become vastly more effective at treating the disease than when mothers were being given scarlet letters).

As I said, learning about the transition of epilepsy from being what happens when you enlist with Satan to being a neurological disorder is frustrating, because there's next to no information about how the average person started thinking about the disease differently in the eighteenth and nineteenth centuries. But we know about how the transition most likely occurred in the case of schizophrenia.

A PICTURE IS WORTH A THOUSAND WORDS—ON TELEVISION

The change in the view of schizophrenia should have happened in the 1950s, when the first drugs that helped lessen the symptoms of schizophrenia came online. When dopamine is released by a neuron intent on sending a "dopaminergic" message to the next neuron in line, it works only if that next neuron has receptors that bind and respond to dopamine. Basic neurotransmitter signaling. And the first effective drugs were ones that blocked dopamine receptors. These were termed "neuroleptics" or "antipsychotics," the most famous being Thorazine (aka chlorpromazine) and Haldol. What happens when you block dopamine receptors? The first neuron in line can release dopamine till the cows come home and still no dopaminergic signal is going to get through. And if people with the disease start acting less schizophrenic at that point, you have to logically conclude that the problem was too much dopamine on the scene in the first place.* The case was strengthened even more by the demonstration of the flip side—take a drug that drastically increases dopamine signaling, and people develop many schizophrenia-like symptoms; this is an amphetamine psychosis. Findings like these jump-started the dopamine hypothesis, still the most credible explanation for what is going wrong in the disease. It also caused a drastic reduction in the numbers of people with schizophrenia warehoused for life in psychiatric institutions tucked away at a genteel distance from everyone else. It was the end of asylums.[28]

This should have stopped the schizophrenogenic voodoo right in its tracks. High blood pressure can be lessened with a drug that blocks a receptor for a different type of neurotransmitter, and you conclude that a core problem was too much of that neurotransmitter. But schizophrenic symptoms can be lessened with a drug that blocks dopamine receptors, and you still conclude that the core problem is toxic mothering. Remarkably, that's what psychiatry's psychoanalytic ruling class concluded. After fighting the introduction of the medications tooth and nail in America and eventually losing, they came up with an accommodation: neuroleptics

*There's a problem lurking here that is subtle and cool, in an abstract sort of way (but definitely not in real life). So, in schizophrenia, there appears to be an excess of dopamine in parts of the brain related to logical thought, and a key treatment is to throw in a drug that blocks dopamine signaling. Meanwhile, Parkinson's disease is a neurological disorder in which sufferers have trouble initiating movement, where the core problem is a *loss* of dopamine in a completely different part of the brain, and a key treatment is to give people a drug (most often L-DOPA) that will *boost* dopamine signaling. You don't inject any of these drugs directly into the relevant brain region. Instead, you take the drug systemically (e.g., by mouth or by injection), which means it gets into the bloodstream and has its effect all over the brain. Give someone with schizophrenia a dopamine receptor blocker, and you decrease the abnormally high levels of dopaminergic signaling in the "schizophrenic" part of the brain back to normal; but at the same time, you decrease the *normal* levels elsewhere to below normal. Give L-DOPA to someone with Parkinson's, and you *raise* dopamine signaling in the "Parkinsonian" part of the brain to normal but *boost* signaling to above-normal levels elsewhere in the brain. So if you treat someone with Parkinson's using high and/or prolonged doses of L-DOPA, do you increase their risk of a psychosis? Yes. If you treat someone with schizophrenia using high and/or prolonged doses of a dopamine receptor blockers, do you increase their risk of a Parkinsonian movement disorder? Yes—it's called "tardive dyskinesia," and its symptoms are referred to in a slangy way as the "*Thorazine* shuffle." (The Southern rock band Gov't Mule even has a song about it called "Thorazine Shuffle," whose final lyrics are "Ain't no need to worry today, Thorazine shuffle make everything OK." Not quite, but it's a good, Allman Brothers–esque song, and it's nice to see popular music less antiquated than "Lucy in the Sky with Diamonds" teaching about neurochemistry.)

weren't doing anything to the core problems of schizophrenia; they just sedated patients enough so that it is easier to psychodynamically make progress with them about the scars from how they were mothered.

The psychoanalytic scumbags even developed a sneering, pejorative term for families (i.e., mothers) of schizophrenic patients who tried to dodge responsibility by believing that it was a brain disease: *dissociative-organic types*. The influential 1958 book *Social Class and Mental Illness: A Community Study* (John Wiley), by the Viennese psychiatrist Frederick Redlich, who chaired Yale's psychiatry department for seventeen years, and the Yale sociologist August Hollingshead, explained it all. Dissociative-organic types were typically lower-class, less educated people, for whom "It's a biochemical disorder" was akin to still believing in the evil eye, an easy, erroneous explanation for those not intelligent enough to understand Freud.* Schizophrenia was still caused by lousy parenting, and nothing was to change in the mainstream for decades.[29]

The breakthrough, in the late 1970s, came at the intersections of public advocacy, neuroimaging, the influence of the media, money, and schizophrenia in the family's closet of powerful people.

In some ways, it started with a murder. In the early seventies, a young man suffering from schizophrenia killed two people in Olympia, Washington, while in a delusional state. A local woman named Eleanor Owen, the mother, sister, and aunt of people with schizophrenia, did something that was a catalyst. She resisted the usual response of someone touched by the disease, which was to retreat into the shame and guilt that was always there but particularly searing when the rare violence committed by someone with schizophrenia confirmed the stereotype. Owen contacted seven other local people she knew who had a close family member with the disease, and they contacted the family of the killer to offer support and comfort.

*Many psychoanalysts approved of mothers being tarred with schizophrenogenic mothering, not just because they thought it was correct but also because the guilt made mothers more prompt in paying the shrink in a timely manner. Some did advocate, however, that these guilt-riddled parents be treated with some humaneness, but most seemed to view this as sentimentality.

Owen and cohort felt empowered by the act, and rather than shame and guilt, the main emotion they felt was rage. The antipsychotic revolution had emptied the psychiatric hospitals of chronic-care schizophrenic patients who were not much more normal and healthy in their behavior. The laudable plan was to construct community mental health clinics throughout the country that would care for these individuals and help them reintegrate into their communities. Except the funding was way slower in coming than what was needed to keep pace with the numbers of people being deinstitutionalized. By the Reagan years, funding had basically completely stopped. Most of the people deinstitutionalized, if they were lucky, wound up being dumped back on their families; otherwise, the streets. Thus the rage was at the irony of this: We're such toxic family members that we caused the disease in the first place, and now we're being entrusted with their care because various agencies couldn't figure out what else to do with them? Moreover, as a group, it was easier for them to air the real source of their rage—their increasing conviction that the idea of a schizophrenogenic mother or family was sheer nonsense.

I had the opportunity to talk with Owen a few years ago, a two-hour conversation with this ninety-nine-year-old who remembered it all well. "On a primal level, I knew it was not my fault. I was operating on sheer emotional rage."* Her group soon formed the Washington Advocates for the Mentally Ill, basically a support group tiptoeing into the realm of advocacy.

Meanwhile, a similar group, called the Parents of Adult Schizophrenics, had formed in San Mateo, California; it scored an early victory in winning the right for family members of individuals with schizophrenia to be on every county mental health board in the state. In Madison, Wisconsin,

*Eleanor DeVito Owen was extraordinary. Over the course of her lifetime, she was a journalist, playwright, professor, costume designer, successful actress, and immensely successful mental health advocate. And our conversation was delayed for a stretch while she traveled across the country alone to visit her nonagenarian kid sister. She died in early 2022, a few weeks after the publication of her memoir, *The Gone Room*, on her 101st birthday. In our conversation, she was vibrant, passionate about the political past and present, and self-effacing about her role in righting one of the travesties of psychiatry's past. If my belief system were a very different one, I would say that I was blessed by having gotten to briefly be in her orbit.

another group had formed, founded by Harriet Shetler and Beverly Young. They all eventually got word of each other, and by around 1979, the National Alliance on Mental Illness (NAMI) had formed. One of their first actual hires was Laurie Flynn, who became director from 1984 to 2000. A homemaker with some experience with community volunteering, she had a daughter who had starred in her high school musical and been on track to be valedictorian when a variant of schizophrenia destroyed her. She and Owen were soon joined by Ron Honberg, a lawyer and social worker who wound up running NAMI's policy work for thirty years, despite having no family member touched by schizophrenia. The pull for him was a sense of justice: "Someone's kid gets diagnosed with cancer, that's one thing. Someone's kid gets diagnosed with schizophrenia, neighbors did not come over with casseroles."*

They had some successes, getting a few state legislatures to push in the direction of more medical insurance coverage of schizophrenia. Owen was the bulldog. "I have no idea how I managed to threaten them [legislators]," she recalled later. "I was a monster. It was from the pain." Flynn described the members as "furious, in their nice Midwestern way."

And then a catalyst happened when NAMI connected with the perfect hybrid of an individual, a first-degree family member of a schizophrenic person who also happened to be one of the world's experts in the emerging field of biological psychiatry. E. Fuller Torrey, mentioned earlier, had decided to become a psychiatrist when his younger sister was diagnosed with schizophrenia. Schizophrenogenic theorizing struck him as deeply wrong for the same reason it did the early NAMI members, with a number of them in effect saying, "Wait, my mother mothered nine of us kids, but she only schizophrenogenically mothered one of us?" It turned him into a scathing critic of the psychoanalytic school of psychiatry. With degrees from Princeton, McGill, and Stanford, he could have settled into a comfortable, lucrative private practice. Instead, he spent some years as a

*I had the pleasure and privilege to have long conversations with Flynn and Honberg as well. Now in their later years, as they reflected back on the uphill battle that they had waged, one gets the sense of what lives well lived look like.

physician in Ethiopia, then the South Bronx, and then an Inuit community in Alaska. He eventually became a psychiatrist in the National Institute of Mental Health, and at St. Elizabeths, the oldest federal psychiatric hospital in the U.S. In the process, he became a fierce critic of the psychodynamic stranglehold, authoring the superb books *The Death of Psychiatry* and *Freudian Fraud* (along with a highly regarded biography of Ezra Pound, a longtime patient at St. Elizabeths, and . . . eighteen other books). His outspokenness cost him at least one position, and he eventually quit the federal psychiatry establishment as well as the psychodynamically dominated American Psychiatric Association, and founded his own mental health research institute with a focus on the biological causes of schizophrenia. It was inevitable that he and NAMI would connect.

Torrey was a godsend to them. "Fuller spoke for us when no one in the medical community would," said Flynn—because he was one of them. He became NAMI's medical spokesperson, lectured and taught NAMI groups all over the country (including getting many of its members to drop their embrace of various unproven alternative-medicine treatments for the disease, such as megavitamin therapy). He wrote the bestselling primer *Surviving Schizophrenia: A Manual for Families, Consumers, and Providers* (HarperPerennial, 1995), which has gone through five editions. Torrey donated more than a hundred thousand dollars in royalties from the book to NAMI and persuaded a philanthropist to hire a DC lobbyist for NAMI instead of funding Torrey's own research.*

And then another piece of the puzzle fell into place, one that I suspect is enormously important for the battles to come in removing blame from our thinking about the worst and most troubled human behavior. It's what Harvard biologist Brian Farrell would label a case of "applied celebrity"—famous and/or powerful people touched by schizophrenia in their own families who became involved. Two were Senators Paul Wellstone (D-Minnesota) and Pete Domenici (R-New Mexico; Flynn recalls thinking, "Oh good, a Republican"). Both became supporters in Congress,

*Yeah, in case you can't tell, I admire Torrey immensely and consider him an inspiration; he's also a very kind, decent man.

pushing for more medical insurance coverage of schizophrenia care and advocating in other ways (Honberg recalls the day he rented a truck, filled it with more than half a million paper petitions calling for more federal funding for the biological roots of mental illness, and deposited them on the steps of the Capitol, standing alongside Domenici).*

And then lightning really struck. On December 9, 1988, Torrey appeared on *The Phil Donahue Show*. Donahue was then the king of daytime talk shows and quietly had a family member with the disease. Guests included Lionel Aldridge, the famed Green Bay Packer, who had descended into misdiagnosed schizophrenia and homelessness after his Super Bowl days. He was now successfully medicated, as were a number of other guests on the show who, along with similar audience members with comments and testimonies, appeared, well, fairly normal. And then there was Torrey, emphasizing how schizophrenia was a biological disease. It has "nothing to do with what your mother did to you. Just like multiple sclerosis. Like diabetes." Not because of an unloved childhood. He showed the brain scans of a pair of twins, one with the disease and one without. The enlarged ventricles jumped out in a powerful demonstration of a picture being worth at least a thousand words. At the end, Torrey gave a shout-out to NAMI.

In the days afterward, NAMI received "a dozen bags of mail a day" from family members of people with schizophrenia. Membership soared to more than 150,000, donations poured in, and NAMI became a powerful lobbying force, pushing for public education about the nature of the disease, advocating for medical schools to change their curriculum about schizophrenia and to shift psychiatry departments away from psychoanalysis and toward biological psychiatry,† funding the next generation of

*One can be jaundiced and/or grateful when a politician with a track record of little sympathy for underdogs selectively develops some for a particular topic that they are personally touched by. Just to take that jaundice to the next level, many a scientist says, in effect "Oh, please, please, let the loved one of some Republican senator come down with the awful disease I study so there'll finally be enough funding for us to figure out how to cure it."

†When I was being recruited to Stanford in the mid-1980s, people bragged about the quality of biological psychiatry in the Bay Area—Stanford had already purged the psychoanalysts from leadership positions in its psychiatry department, and UCSF was in the process of doing the same. It was definitely a draw.

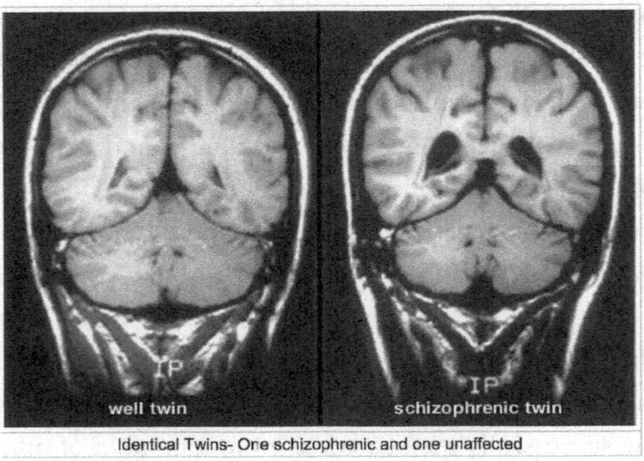

The photograph displayed by Torrey

young researchers in the field. Torrey and Flynn appeared repeatedly on *Donahue*, on *Oprah*, and in an influential PBS documentary. Celebrities came forward with stories about the mental illness struggles that they or family members had endured. *A Beautiful Mind* won a Best Picture Oscar for its depiction of John Nash, the Nobel laureate economist who struggled his entire adult life with schizophrenia.

And along the way, the myth of schizophrenogenic mothers, fathers, and families died. No credible psychiatrist would counsel someone anymore that their toxicity caused their loved one's schizophrenia or take a schizophrenic patient on a journey of free-associative psychoanalysis to uncover the sins of the mother. No medical schools teach it. Close to no one in the public believes it. We're still maddeningly unsuccessful in understanding the nuts and bolts of the disease and in devising new and more effective treatments. Our streets teem with homeless, deinstitutionalized schizophrenia sufferers, and families are still devastated by the disease, but at least no family member is being taught that it is all their damn fault. We've subtracted out the blame.[30]

The picture isn't perfect, of course. A few gray eminences of psychoanalysis recanted their views in technical journals, and some even did studies showing that psychoanalytic approaches did nothing to help with the disease. But to the bitterness of the NAMI members I spoke to, no leader in that field ever came to them to apologize (bringing to mind the quip of physicist Max Planck that "science progresses one funeral at a time"). The bitterness still resonates forty-three years later from a brilliant piece of sociopolitical theater by Torrey, published in 1977 in *Psychology Today*. In "A Fantasy Trial about a Real Issue," he imagined a trial of the psychoanalytic establishment for the harm done to mothers of people with schizophrenia. "No trial since Nuremberg has stirred so much public interest," he facetiously reported about the supposed mass trial held in a stadium in DC. He noted the charges: "The accused did willfully and with forethought but no scientific evidence blame the parents of patients with schizophrenia . . . for their children's condition thereby causing great anguish guilt pain and suffering." Defendants included Fromm-Reichmann, Klein, Bateson, and Theodore Lidz, who claimed that parents of schizophrenics are "narcissistic" and "egocentric." All were convicted and sentenced to spend ten years reading their own writings. He finished with an acidic flourish, "Relatives wept openly. Nobody had expected that harsh a sentence." Eleanor Owen had a movingly different take on it. Despite the fury that drove the advocacy that ultimately helped move mountains, despite the shame and guilt heaped on people like her by ideologues preaching a judgmental pseudoreligion free of facts, she still says, "But there were no villains."*[31]

*The vanquishing of the idea of schizophrenogenic mothering might appear to have a substantial problem. As it was formulated, schizophrenogenic mothers (or fathers or family members) drove their loved one into late-adolescent schizophrenia through the toxic ways that they interacted with them. But then the discovery of things like elevated dopamine levels, impoverished frontal cortical circuitry, and ventricular enlargement screamed that this is a biological disease. In other words, experience (such as the adversity of that style of mothering) can't be the cause of the disease if the disease involves structural and chemical changes in the brain. But experience does exactly that to the brain; just go back to some of the examples from chapters 3 and 4—childhood poverty thins the frontal cortex; chronic stress shrinks the hippocampus and enlarges the amygdala. So why can't it be the case that schizophrenogenic mothering causes schizophrenia *by way of* elevating dopamine levels, atrophying the cortex, and so on? That would seem like a sophisticated, contemporary view of biology and environment interacting. Uh-oh, have we just

SNAPSHOTS MIDMETAMORPHOSIS

There have been other success stories as well. Autism has undergone a remarkably similar shift. Once loosely termed "childhood schizophrenia," it was formalized into the diagnosis of "early infantile autism" by psychiatrist Leo Kanner. After considering the possibility of biological, specifically genetic, roots to the disease, he settled into the thinking of the time, which was, of course, once again blaming the mother. In this case, the presumed maternal toxicity was a coldness and inability to love; Kanner's sound bite that haunted generations of parents was "refrigerator mothers." There then followed the usual story: Decades of shame and guilt. Increasing scientific insight showing that there is zero evidence for the "refrigerator mothering" concept. First hints of advocacy and pushing back against the accusation. Increasing public awareness of the prevalence of the disease, making the refrigerator accusation tougher to maintain, with some applied celebrity thrown in. And the role of blame in autism has disappeared, as we now know it to be an alarmingly common neurodevelopmental disorder. Moreover, many with milder versions of autism (what used to be called Asperger's syndrome and is now labeled something like "high-functioning autism spectrum disorder" [ASD]) object to being pathologized with the concept of "disorder." Instead, they argue that ASD should instead be viewed as merely an extreme in the normal variation in human sociality, and that it brings many cognitive traits that compare favorably with those of "neurotypicals" (i.e., everyone else).*

A remarkably similar story, with three interesting differences.

The first involved Kanner. He was as much of a dead-White-male

---

reinvigorated schizophrenogenic mothering? Not at all. There's no science to show that the mothering style could produce those brain changes. Experts couldn't even reach consensus as to what the style consisted of. No one could demonstrate that supposedly schizophrenogenic mothers mothered dramatically differently when it came to their nonschizophrenic children. Neurological and neuropsychological markers of the disease are apparent as early in life as they can be studied. And oh, there are those genes involved. Schizophrenogenic mothering is dead ideology.

*The Joan of Arc of climate change, Greta Thunberg, is one such individual; she credits her Asperger's syndrome with sparing her from social distractions, allowing her to focus on saving the planet.

authority as you could find—professor at Johns Hopkins School of Medicine, the first certified child psychiatrist in the country, the author of the first textbook on the subject. And he appeared to have been a really good person. As another of the intellectuals who was able to escape Europe, he helped save the lives of many others, sponsoring their entry into the U.S., supporting them materially. He had a deep vein of social activism concerning psychiatric public health and community psychiatry outreach programs. Remarkably, he changed his view as more knowledge accrued. And in 1969 he did something extraordinary—he appeared at the annual meeting of the parent advocacy group Autism Society of America, and apologized: "Herewith, I acquit you people as parents."

Next, while Owen felt that there were no villains in the schizophrenogenic-mothering saga, that of refrigerator mothering indeed had one, in my opinion. Bruno Bettelheim had survived the concentration camps and made it to America, an Austrian intellectual of the psychoanalytic stripe who became the supposed definitive expert on the causes and treatment of autism (he also wrote influential books on the psychodynamic roots of fairy tales in *The Uses of Enchantment* and on child-rearing practices on Israeli kibbutzim in *The Children of the Dream*). He founded the Orthogenic School for autistic children, associated with the University of Chicago, and became the recognized pioneer in their successful treatment. He was lauded and revered. And he embraced refrigerator mothering with a venom that would have made Fromm-Reichmann or Klein blanch (Torrey included Bettelheim as a defendant in his fantasy show trial). In his widely read book about autism, *The Empty Fortress* (Free Press, 1967), his stated belief was "that THE precipitating factor in infantile autism is the parent's wish that his child should not exist." In words that take one's breath away, he wrote, "Whether in the death camps of Nazi Germany or while lying in a possibly luxurious crib, but there subjected to the unconscious death wishes of what overtly may be a conscientious mother—in either situation a living soul has death for a master."[32]

He was also emptier than the supposed fortress of autism. He faked his European credentials and training history. He plagiarized writing. His school actually had very few kids with autism and he fabricated his supposed successes. He was a tyrannical bully to his staff (I have heard people who had been in his training orbit refer to him sarcastically as "Betto Brutalheim") and, as is well documented, he repeatedly physically abused the children. And of course he apologized for nothing. It was only after his death that a spate of articles, books, and testimonials of scores of survivors of his wisdom came forward.[*,33]

The final difference from the schizophrenia story is why I consider "the vanquishing of blame" regarding autism to still be midmetamorphosis. This is the anti-vaxxer movement, which insists, in the face of every possible scientific refutation, that autism can be caused by vaccinations gone awry. Amid these often well-educated and privileged medieval witch-hunters being responsible for decreased vaccination rates, a resurgence of measles, and the deaths of children, I note what is often a secondary theme. There is, of course, the primary conspiracy theory of some sort of medico-pharmaceutical willingness to shower autistic hell down on the innocent for the sake of vaccine profits. But there is also often some additional, familiar finger-pointing: if your child has autism, it's your own damn fault because you didn't listen to us about vaccines.

We are in the midst of other transitions as well. In 1943, General George Patton famously slapped a soldier in the hospital for what we would now call post-traumatic stress disorder (PTSD) but which Patton interpreted as cowardice; Patton ordered his court-martial, which was fortunately

*Bettelheim had another domain of fraudulent, self-aggrandizing blaming that evokes particular revulsion in me, in that he was a classic anti-Semitic Semite, blaming his fellow Jews for the Holocaust. Addressing a group of Jewish students, he asked, "Anti-Semitism, whose fault is it?" and then shouted, "Yours! . . . Because you don't assimilate, it is your fault." He was one of the architects of the sick accusation that Jews were complicit in their genocide by being passive "sheep being led to the ovens" (ever hear of, say, the Warsaw Uprising, "Dr." Brutalheim?). He invented a history for himself as having been sent to the camps because of his heroic underground resistance actions, whereas he was actually led away as meekly or otherwise as those he charged. I have to try to go through the same thinking process that this whole book is about to arrive at any feelings about Bettelheim other than that he was a sick, sadistic fuck. (The quote comes from R. Pollack, *The Creation of Dr. B: A Biography of Bruno Bettelheim,* London, UK: Touchstone [1998], page 228.)

overruled by Ike. Even well after Vietnam, PTSD was officially viewed as psychosomatic malingering by most of governmental powers that be, and afflicted veterans were often denied health benefits to treat it. And then the usual—genetic links, identification of early developmental neurological issues and types of childhood adversity that increase the risk of succumbing to it, neuroimages showing brain abnormalities. Things are slowly changing.

In the early 1990s, about a third of the soldiers deployed in the first Gulf War complained of being "never quite right again," with a constellation of symptoms—exhaustion, chronic unexplained pain, cognitive impairments. "Gulf War syndrome" was generally viewed as being some sort of psychological disorder, i.e., not for real, a marker of psychologically weak, self-indulgent veterans. And then science trickled in. Soldiers had been administered a heavy-duty class of drugs related to pesticides as protection against the nerve gas that Saddam Hussein was expected to use. While these drugs could readily explain the neurological features of Gulf War syndrome, this was discounted—careful research in the run-up to the war had identified what doses could be given safely, would not damage brain function. But then it turned out that the drugs became more damaging to the brain during stress, something that was not considered beforehand. One of the mechanisms implicated was that stress—in this case, body heat generated by carrying eighty pounds of gear in 120-degree desert weather, coupled with basic combat terror—could open up the blood-brain barrier, increasing the amount of drug getting into the brain. It was not until 2008 that the Department of Veterans Affairs officially declared Gulf War syndrome to be a disease, not some psychological malingering.[34]

So many fronts of advances: Kids who are having trouble learning to read and keep reversing letters aren't lazy and unmotivated; instead, there are cortical malformations in their brains that cause dyslexia. Issues of free will and choice are irrelevant when it comes to any scientifically informed read of someone's sexual orientation. Someone insists that, despite evidence from their genes, gonads, hormones, anatomy, and secondary sexual characteristics that they are the sex they were assigned at birth, that is not who they are, has never been from as far back as they can remember—and the neurobiology agrees with them.[35]

And even further-reaching, sneaking into everyday life so subtly that we cannot readily see the change in mindset implied: Someone doesn't help you carry something heavy, and rather than being irritated, you recall their serious back problems. The person singing soprano in your choir keeps missing the notes, and you resort to knowledge of prenatal endocrinology for an explanation—oh, they're a baritone. Oddly, you have an unfortunate research assistant who searches for the one green sock in a mound of a hundred thousand red socks, at your request; they fail, and instead of holding this against them, you think, ah, that's right, they have red-green color blindness. And in a recent blink of a historical eye, the majority of Americans changed their minds and decided that, given the insufficiency of love in the world, the love between two same-sex adults should be permitted to be consecrated with marriage.

The long explorations in this chapter all show the same thing: we can subtract responsibility out of our view of aspects of behavior. And this makes the world a better place.

CONCLUSION

We can do lots more of the same.

14

The Joy of Punishment

JUSTICE SERVED I

In her 1987 classic *A Distant Mirror*, historian Barbara Tuchman famously described Europe in the fourteenth century as "calamitous" (and in ways that parallel the present). Mirror or not, by anyone's standards, the century sucked. One source of misery was the start of the Hundred Years' War between France and England in 1337, leaving destruction in its path. Christianity was roiled by the papal schism, which produced multiple competing popes. But above all, the calamity was the Black Plague, sweeping through Europe beginning 1347; over the next few years, nearly half the population died in bubonic agony. So severe was the loss that it took London, for example, two centuries to regain its preplague population.[1]

Things were pretty awful even earlier in the century. Take 1321—the average peasant was illiterate, parasite riddled, and struggling for existence. Their life expectancy was about a quarter of a century; a third of infants died before their first birthday. Poverty was made worse by enforced tithing of income to the church; 10–15 percent of people in England were starving to death in a famine. Moreover, everyone was still recovering from the events of the previous year, in which the Shepherds' Crusade rampaged through France rather than fulfilling its stated goal of rampaging among Muslims in Spain. At least no one thought that some out-group was poisoning the wells.[2]

In the summer of 1321, people throughout France decided that some out-group—lepers*—was poisoning the wells. The conspiracy theory soon spread to Germany and was accepted by everyone from peasants to royalty. Under torture, lepers soon confessed that, yes, they had formed a guild sworn to poison wells, using potions made from the likes of snakes, toads, lizards, bats, and human excrement.

Why were the lepers supposedly poisoning the wells? In one *Night of the Living Dead* version, people believed that the poisons caused leprosy—i.e., were a recruitment measure. In another interpretation, some empathically speculated that lepers were so embittered by the lack of empathy with which they were treated that this was their revenge. But some prescient individuals, centuries ahead of their time in appreciating the rot of capitalism, sensed a profit motive. Soon, under more "enhanced interrogation," the answer emerged—tortured lepers passed the buck, claiming between their shrieks of pain that they were being paid to poison the wells by their sidekicks, the Jews. Perfect—everyone believed that Jews couldn't get leprosy, allowing them to safely conspire with the lepers.†

But then the Jews passed the buck further. Despite their bloated wealth from venal usury and the selling of kidnapped Christian children for blood sacrifice, employing that many lepers cost them a bundle. Soon Jews being broken on the wheel proclaimed that they were just middlemen—they were being funded by the Muslims! Specifically the king of Granada and the sultan of Egypt, scheming to overthrow Christendom. Inconveniently, the mobs couldn't get their hands on those two. Settling for second best,

*In my writing and lectures, I try to refer to, for example, lepers, schizophrenics, or epileptics instead as, "people with" leprosy, schizophrenia, or epilepsy. It is a reminder both that there are actual humans involved in these maladies and that such people are not merely their disease. I'm dropping that convention in this section, reflecting the nature of this historical event—for the promulgators of this savagery, their actions did not concern "people with leprosy." They concerned "the lepers."

†Supposedly because Jews, unlike Christians, didn't have sex during menstruation, one of the supposed causes of leprosy.

mobs immolated lepers and Jews in town after town in France and Germany, killing thousands.

Having addressed what became known historically as "the Lepers' Plot," people returned to their daily struggle for existence; justice had been done.*[3]

THOSE BLEEDING-HEART LIBERALS

Reform isn't everyone's cup of tea. Maybe you're sitting pretty in the Vatican and there's this uncouth German monk going on about his ninety-five theses. Or if your taste runs in the "Things have to get worse before they get better" direction of the proletariat losing its chains, reform just undercuts revolution. Reform especially doesn't seem like the way to go when it accepts as a given a system that is utterly, brutally, indefensibly nonsensical. You can see where we're heading.

Yes, yes, there is so much to reform about the criminal justice system. Prisons are criminogenic, a training ground for revolving-door recidivism. Implicit bias makes a mockery of the notion of objective judges and juries. The system offers all the justice money can buy. All of this needs to be reformed, and the people in the trenches trying to do it—the Innocence

*Mind you, no actual wells were ever poisoned.

Project, candidates for district attorney intent on change from within, lawyers helping underdogs pro bono—are amazing. I've now had the chance to work on around a dozen murder cases with public defenders, and they're inspiring—underpaid, overworked, passing up the riches of the corporate world, losing most of their cases defending broken people who were usually already lost by the time they were second-trimester fetuses.

Yet if there's no free will, there is no reform that can give retributive punishment even a whiff of moral good.

Here is what criminal justice reform can look like:* In sixteenth-century Europe, a variety of tests were used to identify witches, all truly awful. One of the more benign ones was to read the suspect the biblical account of the crucifixion of Our Lord. If they weren't moved to tears, they were a witch. In 1563–68, Dutch physician Johann Weyer tried to reform the witch-justice system, publishing a book, *De praestigiis daemonum et incantationibus ac venificiis* ("On the Illusions of the Demons and on Spells and Poisons"). In it, Weyer calculated that Satan had an army of 7,405,926 devils and demons, organized in 1,111 divisions of 6,666 each. So Weyer had bought into the system big time. The book made three suggestions for reforming it. First, obviously nonwitches might confess to anything, including being witches, just because they were being flayed. The second, which caused Weyer to be viewed as one of the forefathers of psychiatry, was that someone might appear to be a witch but actually just be mentally unbalanced. The third referred to that tears test. By all means use it, urged Weyer, but keep in mind that lacrimal glands often atrophy in old age, so that the tearless old woman hearing the crucifixion story is organically impaired from crying, rather than a witch.[†,4]

That's what it looks like when you try to reform a system based on sheer gibberish. Ditto if reformist phrenologists excluded any potential subjects from their studies who had gotten a bump on the head from ice hockey, or if reformist alchemy journals required authors to list their

*This is an example that I covered at greater length in my book *Behave: The Biology of Humans at Our Best and Worst.*

†By the way, Weyer's book was condemned by both Catholics and Protestants.

funding sources. Or when reformers try to bring more equality to a criminal justice system; this is trying to make the actual meting out of justice more aligned with its Platonic ideal, when that very ideal is without scientific or moral justification. Just to start off things off in an understated kind of way . . .

JUSTICE SERVED II

Of the long line of King Louises in France, Louis XV was certainly underwhelming. He was ineffectual with his few policies and was scorned as a corrupt sybarite who had brought economic and military ruin to France; the celebration of his death by the citizenry in 1774 foretold the French Revolution fifteen years later. In 1757, an assassin stabbed him with what was essentially a penknife, which, after penetrating layers of clothing (it was outdoors in midwinter), caused a superficial wound; to help out the grievously injured monarch, the archbishop of Paris commanded forty hours of prayers for his speedy recovery.*

History is unclear as to the motives of his would-be assassin, Robert-François Damiens, a household servant dismissed from a series of jobs for stealing from his employers. One interpretation is that he was deranged, psychiatrically unwell. Another concerned a religious controversy at the time where Damiens was on the losing side, which was suppressed by Louis, and decided to take revenge. The king particularly feared that Damiens was part of a larger conspiracy, although Damiens didn't give up any names while being tortured. Motives aside, the only pertinent thing was that he had attempted to kill the king; Damiens was convicted, destined to be the last person drawn and quartered in France.

The execution, which took place in a public square in Paris on March 28, 1757, was well documented. Damiens's feet were first crushed with a

*Louis, apparently chastened by his brush with mortality, vowed to pay more attention to the affairs of state and to cavort less with mistresses; the latter resolution apparently lasted a few weeks.

torture device called the "boot." The offending hand with which he had held the knife was then scorched with burning pincers; a mixture of molten lead, boiling oil, burning resin, wax, and sulfur was then poured on his wounds. He was then castrated and the burning mixture applied there as well.

These actions, along with Damiens's wailing and begging for death, provoked cheers from the massive crowd that filled the square, as well as from the apartments above (which had been rented out to the wealthy as box seats at exorbitant prices*).

But these tortures were merely the warm-up act for the main event, which was the "quartering"—each of a victim's limbs would be tied to a horse, and the four horses would be led off in opposing directions, tearing off the person's limbs. Damiens apparently had tougher-than-expected connective tissue; his limbs remained intact, despite repeated attempts with the horses. Eventually, the overseeing executioner severed the tendons and ligaments in Damiens's four limbs, and the horses were finally successful. Damiens, reduced to a torso and still breathing, was flung onto

LE SUPPLICE DE DAMIENS

*Who included Giacomo Casanova—you know, *the* Casanova—who had rented an apartment with fellow partying friends (and who described a sexual act with one of the women there while she was leaning out the window to get a good view of the goings-on).

a fire, along with his severed limbs. When he was reduced to ash after four hours, the crowd dispersed, justice having been served.[5]

RECONCILIATION AND RESTORATIVE JUSTICE AS BAND-AIDS

Suppose trials were abolished, replaced by mere investigation to figure out who actually carried out some act, and with what state of mind. No prisons, no prisoners. No responsibility in a moral sense, no blame or retribution.

This scenario inevitably provokes the response "So you're saying that violent criminals should just run wild with no responsibility for their actions?" No. A car that, through no fault of its own, has brakes that don't work should be kept off the road. A person with active COVID-19, through no fault of their own, should be blocked from attending a crowded concert. A leopard that would shred you, through no fault of its own, should be barred from your home.

So then what should be done with criminals? There have been a few approaches that, while swell, still accept the premise of free will, but at least show that really smart, serious people are thinking about radical alternatives to our current responses to people who damage. One possibility is the "truth and reconciliation commission" model, first mandated in postapartheid South Africa and since used in numerous countries recovering from civil war or a violent dictatorship. With South Africa as the archetype, architects and henchmen of apartheid could appear before the commission, rather than go to jail. About 10 percent of applicants were granted the opportunity, where they were required to confess to every detail of their politically motivated human rights violations—whom they had killed, tortured, and disappeared—even the ones whom no one knew about, who hadn't been pinned on them. They would vow to never do it again (e.g., to not join the White militias that formed a threat to the peaceful transition to a free South Africa); family members of the victims who were in attendance essentially vowed not to take revenge. The killer would then be released rather than imprisoned or executed. Mind you, there was no requirement for remorse—no photo ops where some apartheid murderer, anguished with contrition, is hugged and forgiven by a widow he created. Instead, the approach was pragmatic (to the frustration of many family members), helping the country rebuild itself.* Most important, it provided a parallel to the police strategy of getting the goods on some entry-level organized-crime schnook and offering him immunity in exchange for implicating his higher-up, who would then be similarly squeezed, and so on, all the way up to implicating the shadowy crime boss. In this case, immunity was being offered to the soldiers of apartheid in order to implicate the crime boss at the top, namely the very foundation of the apartheid government. Unlike with the Holocaust or the Armenian genocide, there could never be repulsive apartheid deniers, insisting that the violence was exaggerated as propaganda or the work of unsanctioned lone wolves.[6]

While moving and surprisingly successful in preventing subsequent violence, the relevance of such commissions to our concerns is limited. Something similar might arise during the sentencing phase of a convicted crime, when the perpetrator takes responsibility for his crime and expresses remorse to his victims, often resulting in a lesser sentence. But this whole approach is just reform, where a criminal is simply punished less by a system that makes no sense. Basically, someone claims that their criminal actions to have been freely willed and that their current freely willed actions are those of a changed person. Not what we're dealing with here.

Another model with some similarities and the same ultimate irrelevance arises from the "restorative justice" movement, which concerns the relationship between criminal and victim, rather than between criminal and state. As with truth and reconciliation commissions, the criminal is expected to take responsibility for all the details of his actions. The

*As one measure of Nelson Mandela's status as a moral giant, he insisted that the commission also investigate human rights violations by African National Congress fighters (i.e., his "side").

emphasis then is on mutual understanding. For the perpetrator, it is to recognize the pain and suffering that he has caused—to understand, to feel, to the point of remorse. And for the victim, the goal is to understand the circumstances, often awful and completely alien, that made the offender the damaging person that he is. And from that point, the aim becomes for both parties (often with a mediator) to figure out what they can do to eliminate some of each other's pain, and to find ways to lessen the likelihood of this happening again.

Restorative justice seems to work, decreasing recidivism rates. That said, there's the likelihood of self-selection bias—a criminal who *chooses* to face their victim this way is almost certainly not your average prisoner, and is already heading in a good direction.

Restorative justice also seems to impact victims in salutary ways. Those who go through the process report less fear and hatred of the perpetrator, less anxiety about safety, better functioning, more enjoyment of everyday activities. Nice but, again, there's the probability of self-selection bias.[7]

But restorative justice has nothing to do with our focus. This is because it accepts the need for retribution as a given, with the prisoner, now understanding the pain they inflicted, more accepting of the legitimacy of being punished by an irrational system.

The approach that actually makes sense to me the most is the idea of "quarantine." It is intellectually clear as day and completely compatible with there being no free will. It also immediately sticks in the craw of lots of people.

As outlined by the hard incompatibilist philosopher Derk Pereboom of Cornell University, it's straight out of the medical quarantine model's four tenets: (A) It is possible for someone to have a medical malady that makes them infectious, contagious, dangerous, or damaging to those around them. (B) It is not their fault. (C) To protect everyone else from them, as something akin to an act of collective self-defense, it is okay to harm them by constraining their freedom. (D) We should constrain the person the absolute minimal amount needed to protect everyone, and not an inch more.

It's leper colonies, involuntary hospitalization in some cases of psychiatric illness, the late-fourteenth-century European requirement that ships sailing in from Asia sit in the harbor for forty days (hence the *quar* in *quarantine*) to avoid bringing another round of bubonic plague.

This medical quarantine model is a given in everyday life. If your kindergartener has a cough or fever, you're expected to keep them home from school until they're better. If you're a pilot, you can't fly if you're taking a medication that makes you drowsy. If your elderly parent is sliding into dementia, they can't drive anymore.

Sometimes quarantine is imposed out of ignorance—it turns out that not all forms of leprosy are particularly contagious, obviating the need for many of those quaint leper colonies. Sometimes it is imposed because of the profoundly unknowable—when the Apollo 11 astronauts returned from the first moon landing, they spent twenty-one days in quarantine, just in case of who knows what. Sometimes it is laden with abuse and bias—a striking example concerns Mary Mallon, the "Typhoid Mary" of history. As the first identified case of an asymptomatic spreader of typhoid, responsible for sickening more than a hundred people, Mallon was arrested in 1907 and forcibly isolated on a quarantine island in the East River in New York.[*,8]

From day one, medical quarantine has generated controversy, a battle between the rights of the individual and the greater good. We certainly saw just how incendiary this could be during early COVID-19, with those jackass don't-tell-me-what-to-do coronavirus parties, where superspreaders killed droves of people by practicing unsafe exhalation.

---

*Why "bias"? Mallon, an Irish immigrant at a time when her people occupied the lowest rung of New York's ethnic hierarchy, probably would not have been treated that way if her last name had been, say, Forbes or Sedgwick; as evidence, during the remainder of her lifetime, more than four hundred other asymptomatic spreaders were identified, with none forcibly quarantined in the same way. Actually, the bias had an additional motivation—Mallon's transgressions included not only being Pestilential While Irish, not only sickening her fellow tenement dwellers, but also sickening the wealthy families whom she served as a cook. She was released from the island in 1910 and returned to working as a cook under an assumed name, again spreading the disease; apprehended in 1915, she unwillingly lived out her days on the island for roughly twenty-five years. That business about using a fake name kind of besmirches the picture of her as blameless victim; on the other hand, her only other possible work was as a laundress, where her wage would have been half the starvation-level wage she received as a cook.

The extension of this to criminology in Pereboom's thinking is obvious: (A) Some people are dangerous because of problems with the likes of impulse control, propensity for violence, or incapacity for empathy. (B) If you truly accept that there is no free will, it's not their fault—it's the result of their genes, fetal life, hormone levels, the usual. (C) Nonetheless, the public needs to be protected from them until they can be rehabilitated, if possible, justifying the constraint of their freedom. (D) But their "quarantine" should be done in a way that constrains the least—do what's needed to make them safe, and in all other ways, they're free to be. The retributive justice system is built on backward-looking proportionality, where the more damage is caused, the more severe the punishment. A quarantine model of criminality shows forward-looking proportionality, where the more danger is posed in the future, the more constraints are needed.[9]

Pereboom's quarantine model has been extended by philosopher Gregg Caruso of the State University of New York, another leading incompatibilist. Public health scientists don't just figure out that, say, the brains of migrant farmworkers' kids are damaged by pesticide residues. They also have a moral imperative to work to prevent that from happening in the first place (say, by testifying in lawsuits against pesticide manufacturers). Caruso extends this thinking to criminology—yes, the person is dangerous because of causes that they couldn't control, and we don't know how to rehabilitate them, so let's minimally constrain them to keep everyone safe.* But let's also address the root causes, typically putting us in the realm of social justice. Just as public health workers think about the social determinants of health, a public health–oriented quarantine model that replaces the criminal justice system requires attention to the social determinants of criminal behavior. In effect, it implies that while a criminal can be dangerous, the poverty, bias, systemic disadvantaging, and so on that produce criminals are more dangerous.[10]

Naturally, quarantine models have been strongly criticized, in at least three major ways.

*Caruso frames this as "incapacitating" the person with the "least infringement."

The issue of indefinite detention. With prison, there's an upper limit to the length of incarceration (except in the case of life without parole), but a quarantine model could keep you constrained as long as was Mary Mallon. Thus, it resembles its disfigured troll cousin, sending a criminal who is not guilty by reason of insanity to a psychiatric hospital, where the average stay is often longer than if they had been jailed instead. Unfortunately, it makes sense that if the person continues to be dangerous, constraints have to continue for as long as necessary—but in the context of least infringement, where "constraint" might consist of having to register with the police whenever you move, or wearing a tracking bracelet. And notice that in this cheery, perfect world I'm imagining, if things have gotten to this point, people wouldn't be recoiling from this constrained person as a loathsome, blameworthy criminal anymore, but merely as someone whose problems in some domain require that they not be allowed to do this or that. Yeah, I know, we have a *long* way to go.*

The issue of preemptive constraint. If you can predict whether a quarantined ("Please—we don't call them criminals anymore") criminal is likely to offend again, you should have been able to see it coming even before they damaged someone in the first place. This raises the specter of creepy precrime apprehension (as well as the need to keep an eye on the biases of the folks predicting the future criminality). Definitely something we don't want to go near, even if Tom Cruise is willing to star in the movie adaptation. But yet. We do "precrime apprehension" all the time in public health. The rule for parents of schoolchildren is "if your child isn't feeling well, keep them home," not "if your child wasn't feeling well, infected everyone else in their class, and still feels crummy, then keep them home." Precough constraint. Ideally, you keep an individual increasingly impaired

*This raises an issue that really gets me into the weeds: If we've gotten to a point of recognizing that it is not right for anyone to be blamed or punished for something negative that they do, is it okay to not want to be around someone yearning for social contact because circumstance made them irritating, boring, irksome, chew with their mouth open, derail conversations with inane puns, whistle tunelessly in a crazy-making way, etc.? Are we teetering on the edge of the equivalent of convincing your child that *everyone* in their kindergarten class should be invited to their birthday party, including even the kid they don't like?

by dementia from driving anymore before they hit someone, rather than after. An equivalent of precrime apprehension is a standard in public health. What, then, should the same look like in our post–criminal justice world? Something decidedly undystopian, once we recall Caruso's emphasis that "least infringement" has to be coupled with a paramount focus on the social determinants of criminality—another version of "And where did that intent come from?" Identify the next high school shooter and, yes, make it impossible for him to buy an automatic weapon, or a big sharp knife, or an unregistered black-market shillelagh. But also do something about how he's getting bullied at school and home and is close to sinking from unaddressed psychological problems. Sure, spot the guy whose costly and growing addiction to a street drug is about to lead him to mug people, get him into rehab to writhe and shiver and puke in a safe setting, but also do something about how he was taught no skills and has no job options. I know, after trying to be a mixture of Emma Goldman in a bad mood and John Lennon singing "Imagine," I'm sounding like a mildly progressive candidate for the town council, complete with an endorsement from Mister Rogers. All I can say is any version of preemptive constraint would have to be in the context of a world in which people truly accept that terrible people are produced by terrible circumstances (one minute before, one hour before . . .). We have a *really* long way to go.

The issue of all that potential fun. A seemingly strong objection comes from Israeli philosopher Saul Smilansky, who argues that no matter how minimally you constrain someone's behavior to make them safe, they're still being constrained for something that is not their fault. Given this, the only morally acceptable stance must be to compensate the constrained person appropriately. In this view, if you're a convicted pedophile and thus, as is often the case, are constrained from coming within a certain distance of schools or parks, at least you should get discounts on drinks at strip clubs; if you're so violent that you have to be placed on a small island, at least make it a five-star resort with private golf lessons. If constraint, no matter how minimal, involves an adverse element that is undeserved punishment, quarantine advocates must provide, in Smilansky's words, compensatory "funishment."* And in his view, this will generate more crime—if you get away with it, you benefit; if you get caught, you're compensated; win-win. It would cause what he calls a "motivational catastrophe."[11]

Caruso's convincing response is based on solid empirical evidence from those fun funishers, the Scandinavians. Compared with the U.S., Norway, for example, has one eighth the murder rate, one eleventh the rate of incarceration, one quarter the rate of recidivism. Well, that must be due to a really draconian prison system. Quite the opposite; it is of the type that Smilansky anticipates with dread—in Norway's "open prison" system, criminals, even those under maximal security, have rooms rather than cells, computers and TVs in each, freedom of movement, kitchens to cook in communally, workshops for hobbies, music studios filled with instruments, art on the walls, trees on the campus-like grounds, a chance to ski in the winter and go to the beach in the summer. But what about the cost, which must be ruinous? It is true that the annual cost of housing a prisoner in Norway is about three times that in the U.S. (roughly ninety thousand dollars versus thirty thousand). Nonetheless, if you really analyze things, the overall per-capita cost of containing crime in Norway is far less than in the U.S.: fewer prisoners, who are educated enough in prison so that most eventually return to the outside world as wage earners rather than as likely recidivists; huge savings from smaller police forces; fewer families disrupted and driven into poverty by the incarceration of the primary source of income; heck, even the well-off save money, with less need for expensive home security systems with CCTV and panic buttons.† But what about Smilansky's motivational catastrophe, with folks lured into criminality in order to head off to a prison resort? The much lower recidivism rate

*As some sort of cosmic joke, spell-check keeps turning *funishment* into *punishment*. Also, when you Google *funishment*, you get sent not only to various philosophical debates but also to BDSM sites, plus some beer maker whose product is supposedly ideal for someone who is a glutton for funishment.

†A comparison between Norway and the U.S. is obviously complicated by apples versus oranges, because the government of someplace like Norway already perceives moral obligations to take care of its citizens to an extent that Americans currently can only dream of.

shows that no amount of art on the walls and well-equipped kitchens can outweigh the incalculable value of freedom. Apparently, we need not fear turpitude and mayhem caused by funishment.*[,12]

I really like quarantine models for reconciling there's-no-free-will with protecting society from dangerous individuals. It seems like a logical and morally acceptable approach to take. Nonetheless, it has a doozy of a problem, one often framed narrowly as the issue of "victim's rights." This is actually the tip of the iceberg of a gigantic problem that could sink any approach to subtracting free will out of dealing with dangerous individuals. This is the intense, complex, and often rewarding feelings we have about getting to punish someone.

JUSTICE SERVED III

Predictably, southern states lagged behind northern ones by a few decades, but, spurred by growing condemnation of the yahoo carnival atmosphere that typically ensued, public hangings were banned throughout the United States by the 1930s. Everywhere, that is, except Kentucky, where, in the town of Owensboro in 1936, there was what proved to be the final public hanging in American history.

The case was some mutation of "perfect." An elderly White woman, Lischia Edwards, had been robbed, raped, and murdered in her home. Soon came the arrest of Rainey Bethea, a twentysomething† African American with a record of house break-ins. The law had seemingly gotten their man. Bethea confessed, which obviously meant little when a Black man was interrogated by police in the Jim Crow South. But the perpetrator had

*An instructive lesson came from a couple honeymooning on a small resort island in the Maldives when the pandemic hit; because of the timing of different countries shutting down air travel, they were stranded there for months, the lone guests, along with the resort staff, also marooned there. A dozen otherwise bored waiters scrambled to fill their water glasses after each sip, their pillows were fluffed up by room staff hourly. Basically, it sounded like hell with a private cabana. "Everyone says they want to be stuck on a tropical island, until you're actually stuck. It only sounds good because you know you can leave," said one of the tanned captives.

†His year of birth is unclear.

stolen some of Edwards's jewelry, and after confessing, Bethea led police to where the jewelry was stashed. The trial lasted three hours; Bethea's lawyer neither cross-examined prosecution witnesses nor called any witnesses;* the jury deliberated for 4.5 minutes, and Bethea was condemned to be executed two months after committing the crime.

There's an extraordinary detail. Despite having both raped and murdered Edwards, he was charged only with rape. Why? Under state law, murderers were executed by the electric chair, within the prison. In contrast, a rapist could still be hanged publicly. In other words, the joy of getting to publicly hang a Black man for the rape of a White woman was irresistible.

The planned execution had a juicy detail that made the national news—Bethea would be hanged by a woman. In 1936, the long-standing sheriff of Owensboro, Everett Thompson, had died of pneumonia. In an act of "widow's succession," the county appointed Thompson's widow, Florence Shoemaker Thompson, to fill in. She had been sheriff for only two months when she presided over the hunt for Bethea, and now she was to preside over his hanging.

The press and public went wild. There was a national guessing game—would Thompson actually pull the lever, or would a professional executioner do it with Thompson officiating? Rumors spread, clairvoyants weighed in, people placed bets. The day before the hanging, Thompson announced that there'd be a professional executioner (something she had actually decided weeks earlier).†

During this period of frenzied speculation, Thompson became one of the most polarizing figures in the nation. To some, she was inspiring, a member of the delicate sex suited for embroidery and childcare but nonetheless willing to step into the breach as her civic duty. To others, she was an abomination, taking a man's job and neglecting her children; she

*Five African American lawyers attempted to appeal Bethea's conviction on grounds of incompetent legal representation. They were told that, sorry, the appeals court was closed for the summer; by the time fall rolled around, Bethea was long dead.

†The executioner arrived too drunk to spring the trap; a deputy sheriff stepped in.

received death threats. In an odd paleo-feminist spirit (it being merely sixteen years since women had won the right to vote), she was praised by some for showing that women were as capable as men in this occupational niche. Throughout, there was the powerful narrative of Thompson as some sort of retributive spirit animal for the slain Edwards—a Black man who had despoiled White Southern womanhood would be hanged by a White Southern woman. Newspapers fixated on her being a mother (KILLER TO BE HANGED BY MOTHER OF FOUR ran the headline in the Springfield, Massachusetts, *Republican*); the *Washington Post* called her "plump, middle-aged"; the *New York Times* referred to her as the "fair sheriff"; she was "matronly" in another newspaper account, while another noted that she was a good cook. In addition to the mountains of supportive letters and hate mail, Thompson received several marriage proposals.*

When the day arrived, every hotel room in Owensboro was taken by people from across the country. Bars stayed open all night in anticipation. The hanging venue had to be moved from the front of the town courthouse to a larger open square, as it was anticipated that the huge crowds would trample the recently planted flowers at the courthouse. People camped out the night before in hopes of a good view; fights broke out among attendees for prime spots (including between women holding babies); enterprising young men sold hot dogs and lemonade to the crowd. An Owensboro man who was a fugitive from the law was arrested when he returned to his hometown to see the hanging. Twenty thousand people jammed the square.

Bethea was led to the gallows. He paused at the bottom of the steps with

*Controversy on the national stage notwithstanding, Thompson was acclaimed in Owensboro. When she ran for reelection, she received all but 3 of the 9,814 votes cast.

an unlikely request—in his pocket was a new pair of socks, which he wanted to wear. After hurried consultations, the request was granted; shackled, he sat on the first step to make the clothing switch and was then led up the stairs shoeless, in new socks.

There were only scattered shouts from the crowd to hang him; most craned their necks in silence.

Bethea's head was hooded, and after the trap door failed to open on the first try, he was properly hanged. Some members of the crowd surged forward while Bethea was still breathing, to rip up the hood for bits of souvenir cloth. Despite this whiff of mob violence, most attendees peacefully dispersed, justice having been done.*[,13]

*Numerous reporters from the North covered the event, drawn by the chance to see Thompson spring the trap; robbed of that, they instead filed stories about Southern barbarity. Embarrassed, the Kentucky legislature soon banned public hangings.

The notoriety weighed on Owensboro for decades, and the town developed a bristly, self-serving revisionism in which the twenty thousand attendees fighting for good spots and souvenir cloth were entirely outsiders, and that the town itself had shunned the spectacle.

I can't resist noting that Owensboro's most celebrated native son is Johnny Depp. Make of that what you will.

PUNISHING CHEATERS

So now we've got the plan to abolish prisons and the idea of criminality and switch to quarantine approaches. All set. But likely to be unsuccessful because of those "intense, complex and often rewarding feelings we have about getting to punish someone." Which raises the key issue of how punishment evolved.

It's easy to get impressed with the extent of our own human sociality; 2.9 billion users of Facebook, Europe opening its doors to Ukrainian refugees,* Mbuti rainforest hunter-gatherers in Congo being up on the Kardashians. But we're not the only ones. Baboons live in groups of fifty to a hundred. A gazillion fish school together. A million wildebeest migrate as a herd in the Serengeti each year, leaving mountains of gnu dung. A mob of meerkats, pack of wolves, clan of hyenas. Social insects, slime molds, single-cell bacteria living in colonies.

A driving force on the evolution of sociality is the fact that it fosters cooperation, many hands making for light loads. African Cape hunting dogs pursue prey cooperatively, where some will cut a corner, run diagonally, to be ready if the prey changes direction. Ditto for chimps, where some drive potential prey, usually a monkey, in the direction where other chimps are ready and waiting. Female bats feed each other's babies; meerkats and vervet monkeys endanger themselves by revealing their location when they give predator alarm calls that benefit everyone. There are those social insects forgoing reproduction in obeisant loyalty to queen and colony. Single-cell bacteria cooperatively form multicellular structures that are needed for reproduction. Then there are the slime mold's constituent members, studying together for the maze-solving final. There's even the nascent field of sociovirology, concerning cooperation between viruses in better penetrating and replicating in a target cell. At the turn of the last century, scientists in the West were busy misinterpreting Darwin as

*Opening doors for people of color from one of the other hellholes on earth? Well, not so much. I suppose in that case, our striking human sociality is shown when nationalists cooperate in forming political parties charging that immigrants are destroying European culture.

showing that evolutionary success is built solely on competition, aggression, and domination. Meanwhile, Russian scientist (and historian, philosopher, ex-czarist prince, revolutionary, and gently fire-breathing anarchist) Peter Kropotkin published a book sixty years ahead of its time, *Mutual Aid: A Factor of Evolution*.[14]

The ubiquity of cooperation among social species raises a ubiquitous problem. Sure, it's great when everyone cooperates for the greater good, but it's even better when everyone else does that while you mooch off them. This is the problem of cheating. A lioness conveniently lags behind the others in a dangerous hunt; a bat doesn't feed the others' kids but freeloads on their cooperation; a baboon stabs his coalitional partner in the back. Two separate colonies of genetically identical social amoebas merge to form a multicellular structure called a fruiting body, which consists of a stalk, which gives stability, and a cap. Only the amoebas in the cap reproduce, and cooperation consists of each colony equally sharing the bummer of being nonreproducing stalk cells; instead, different strains will try to cheat, exploiting the other colony by preferentially hogging seats on the cap. Even mitochondria and stretches of DNA cheat in cooperative ventures.*[,15]

*Long biology digression: Mitochondria, seventh-grade biology's "powerhouses of the cell," are at the center of one of the coolest events in the history of life. Mitochondria were once tiny, independent cells, with their own genes, willing to attack larger cells for their own benefit; those larger cells would counterattack with proteins that would perforate mitochondria, or by engulfing mitochondria and harvesting their molecules. Then, in the "endosymbiotic" revolution some 1.5 billion years ago, swords were hammered into plowshares, and when a large cell engulfed a mitochondrion, rather than destroy it, it allowed the mitochondrion to live there, to their mutual benefit. Mitochondria had evolved the capacity to use oxygen to generate energy, a hugely efficient move; they shared the plentiful fruits of their oxygen-based metabolism with the enveloping cell, which in turn protected mitochondria from the wear and tear of the outside world. And in a move reminiscent of two medieval rulers making a peace treaty but, not trusting each other, sending their sons to be guests/spies/prisoners in the other kingdom, mitochondria and the host cells even traded some of their own original genes (although it was overwhelmingly mitochondria transferring genes to the host cell).

Where does cheating come in? When it's time for the cell to divide, you have to make new copies of everything, including the DNA in the nucleus, mitochondria, and so on. And some mitochondria will cheat, making way more new copies of themselves than they're "supposed to," dominating replicative resources for themselves. The cells' countermeasure? We'll get to that.

How about DNA cheating? The entire genome is a cooperative venture, individual genes and other DNA elements working collectively when replicating. It turns out that there are stretches of DNA called transposons that code for nothing useful and are usually derived from ancient

And sure as the day is long, the ubiquity of cheating drove the evolution of countermeasures to detect and punish it. Chimps that fail to support an ally in a fight are pummeled afterward. Wrens that don't feed the nestlings of the dominant breeding pair are attacked. Naked mole rat queens are aggressive toward workers that are slacking. In the mutualism where reefer fish are cleaned by wrasse fish that get to eat the parasites they harvest, some wrasses cheat to get an even better meal by taking a bite out of the reefer fish; they are then driven away and chastened, less likely to renege on their mutualistic contract afterward. Social bacteria won't form fruiting bodies with bacterial clonal lines that cheat. Green algae have developed means of not passing on egregiously selfish mitochondria when the cell divides. Cells evolve the means to silence all of the copies of a transposon whose self-serving replication has gotten out of hand—for example, a particular exploitative type of transposon invaded fruit flies in the 1970s, and it took forty years for the flies to evolve the means of punitively silencing it.[16]

Crucially, punishment *works* to maintain cooperation. In economic games involving a pair of players (e.g., the Ultimatum Game), one of the two is given the power to exploit the other. And putting the lie to the myth that we are nothing but *Homo economicus*, rational optimizers of self-interest, players in the driver's seat typically don't start off exploiting as much as they could. If the other player has the opportunity to punish the first player for being unduly exploitative, exploitation subsequently decreases further; in the absence of a mechanism for punishment, exploitation festers.[*,17]

viruses. And they are selfish, insofar as all they care about is making more copies of their useless selves, trying to monopolize the replication machinery. As a measure of the effectiveness of their cheating, about half of human DNA is derived from self-serving copies of useless transposons. And the cell's response to that selfishness? We'll get to that as well.

As a reminder, lionesses, fish, bats, bacteria, mitochondria, and transposons are not consciously plotting about how to cheat for their own benefit. This personifying language is just shorthand for things like "Over the course of time, transposons that evolved the capacity for preferential self-replication became more prevalent."

*What do unfettered exploitation, restrained exploitation, and punishment look like in the Ultimatum Game? There are two players. The first player gets $100 and then divides it between the

The right kind of punishment at the right time matters for enhancing cooperation. A monumentally influential example of this came from a game theory study in 1981 by political scientist Robert Axelrod and evolutionary biologist W. D. Hamilton, two titans in their fields. The experiment involved the Prisoner's Dilemma (PD), a game in which two players, unable to communicate, must each decide whether to cooperate with or cheat against the other player—if they both cooperate, each gains some brownie points; if they both cheat, they both lose. So obviously, one should always cooperate, right? But not so fast—if the other player cooperates while you stab them in the back by not reciprocating, they lose a bundle of points and you get the biggest reward of all; if it's you who is the overly trusting goat, the opposite ensues. Axelrod and Hamilton asked an array of game theorists to tell them what their PD strategy would be and ran a computerized round-robin tournament where each strategy was pitted against each of the other ones two hundred times. And amid some complex algorithms for contingent punishment, the strategy that won was the simplest—tit for tat. Start off by cooperating, and continue to do so unless the other player double-crosses you; at that point, do it back in the next round. If they continue cheating, continue punishing them back, but if they go back to cooperating, you resume the same in the next round. A strategy that has clear rules, that starts with cooperation, that is proportionately punitive against cheaters, and that can forgive. This study launched an industry of follow-ups exploring variants on tit for tat, their evolution, and real-world examples in various social species.[18]

The punishment scenarios in the Ultimatum Game or the PD are termed "second party" punishment—where a victim revenges themselves on a bully. An even more effective mechanism for suppressing cheating and fostering cooperation is "third party" punishment, when an outsider steps in and punishes the jerk. Think police. This is a much more so-

two of them however the first player pleases. Offering zero and keeping $100 is maximal exploitation. Fifty-fifty maximizes fairness. Most people start off somewhere around a restrained sixty-forty. Where does punishment come in? The only power that the second player has is to refuse the offer—in which case, neither gets anything.

phisticated domain of punishment; while infants show the rudiments of it, they take years to do so consistently, and it is unique to humans. This is an altruistic act, where you pay a cost (e.g., your effort) to punish someone for everyone's good. Reflecting that altruism, people who are prone to do this tend to be prosocial in other realms* and show disproportionate activation of a brain region involved in perspective-taking†—they excel at viewing the world from the standpoint of the victim. Moreover, treating subjects with oxytocin, a hormone that stimulates in-group prosociality, increases people's willingness to take on the burden of third-party punishment.[19]

Then there's fourth-party punishment, where a third-order witness is punished for not doing their job—think of honor codes where you get in trouble if you don't rat out the person you saw cheating, or of police arrested for taking bribes. And fifth-party punishment—punish the police review board for not punishing corrupt cops. Then sixth-, seventh- . . . at which point you're describing a network of people willing to punish to maintain cooperation.

Cool cross-cultural research shows that small, traditional cultures—say, hunter-gatherers or subsistence farmers—don't carry out third-party punishment (either in real life or when playing economic games). They fully understand when cheating occurs but just don't bother. Explanation: everyone knows each other and what they're up to, so you don't need fancy third-party enforcement to rein in antisocial behavior. Supporting this, the larger the society, the more formalized the third-party policing. In addition, fourth-party punishing of third-party cheating works best when there are only a small number of third-party enforcers—think of the chaos if, instead of police, things ran solely on citizens' arrests.[20]

Cross-cultural research casts light on the emergence of the ultimate form of third-party punishment, namely deities who monitor and judge

---

*The folks who are *really* prosocial are the ones who readily do third-party punishment without bothering to do self-serving second-party punishing.

†The temporal-parietal junction.

humans. As studied by psychologist Ara Norenzayan of the University of British Columbia, gods invented by cultures built on small social groups have no interest in human affairs. It's only when communities get large enough that there's the possibility for anonymous actions, or interactions between strangers, that we see invention of "moralizing" gods who know if you've been bad or good. Consonant with this, across an array of religions, the more that deities are viewed as punitive, the more people are prosocial to anonymous, distant coreligionists.[*,21]

Thus, punishment in its game-theory versions discourages cheating and facilitates cooperation. But there's a big problem—*punishment is costly*. Suppose you're playing the Ultimatum Game, and the other player makes you an offer of 99:1. If you reject their offer, you're giving up the opportunity to get one dollar, which, while not great, is better than nothing. Rejecting is irrational and costly . . . unless you're playing a second round with the person, where your rejecting a lowball offer will prompt the other player to come back with a better offer that produces a net gain for you. In a case like this, punishment isn't costly; instead, such self-serving punishment pays off in the future (assuming you have the privilege of being able to hold out for the future instead of accepting whatever crumbs are being offered).

Where game play is purely altruistic is if you give up one dollar, rejecting a 99:1 offer, in a single-round game, with the chastened other guy thus making a better offer . . . to the next person.

Third-party punishment is even more costly. It's the Ultimatum Game and you observe player A totally exploiting powerless player B. Outraged, you step in and spend, say, ten dollars of your own money to cost player A twenty dollars as punishment. Humbled, they're nicer to whomever they

*There's an additional kind of punishment that is really messed up. Termed "perverse" or "antisocial" punishment, this is when someone is punished for making *too* generous of an offer; it is motivated by how unpunished generosity will make the rest of us look bad, pressuring everyone else to start being generous. Cross-cultural studies show that you find this malignant kind of punishment only in cultures that you wouldn't want to live in—those with low social capital, with low levels of trust and cooperation.

play against thereafter, and if this doesn't include you down the line, your costly act is purely altruistic punishment.[*,22]

A way to lower the cost of punishing involves reputation, an incredibly reliable means of influencing behavior. In tests of game theory, cooperation is boosted if people know your history of play (i.e., open-book play that produces a shadow of the future); be known as a free rider, and others will start off not trusting you or refusing to play with you. This occurs among hunter-gatherers, who spend a huge amount of time gossiping about, among other things, who has cheated by, say, not sharing meat; get a reputation for that and you're ostracized, which can be life-threatening. In contrast, the costs of third-party punishment are reduced because your reputation is enhanced and people trust you more; if you're already viewed as socially dominant, being a third-party punisher makes you seem more formidable and likable.[23]

These are all distal solutions to the problem of the cost of punishment. As first introduced in chapter 2, there's "distal" (big-picture, long-perspective levels of explanation), in contrast to "proximal" (focusing on motivations and explanations in the moment). Why do animals mate, expending effort and calories, often risking their lives? Distal explanation: because it allows you to leave copies of your genes in the next generation. Proximal explanation: it feels good. Why punish cheaters when it's costly? The distal explanation is what we've been discussing—because reliably and collectively sharing the costs benefits everyone. But it is when we look for a proximal explanation that we see how it's going to be so damn hard to get people to proclaim the lack of free will and just quarantine the dangerous. Why punish cheaters when it's costly? Proximally, because we *like* to punish wrongdoers. It feels *great*.

*In traditional Fijian culture, being a third-party punisher of antisocial behavior isn't costly—it is understood that you can do things like steal possessions of the miscreant with impunity.

JUSTICE SERVED IV

It's a magnetic draw of our attention. We want to identify a perimeter; it's the concentration camp porn of sensing the outer limits of human depravity. It facilitates a feel-good experiment: "What if it were a loved one of mine?" followed by the relief of choosing to step back from the edge of the bottomless pit with the knowledge that it doesn't apply to us. Sometimes it's just primate voyeurism. It is our fascination with serial murderers, the filing away of numbers of victims and the grotesqueries of the killings.* Jeffrey Dahmer having sex with the corpses of his victims, cannibalizing them, proclaiming his love for them. John Wayne Gacy entertaining hospitalized kids, dressed as a clown. Charles Manson, the sixties cultural incarnation of Satan's son. The ones with the nicknames—Son of Sam, the Boston Strangler, the Zodiac Killer, the Night Stalker, the DC Sniper. The steampunk kitsch of Jack the Ripper.

Another serial murderer whose notoriety has persisted is Ted Bundy. To use a ghastly term, he was a run-of-the-mill serial murderer, killing roughly thirty women in the mid-1970s, far from a record holder. There was the usual sickening litany—rape, murder, necrophilia, cannibalism; he kept decapitated heads of victims in his apartment as mementos, shampooing their hair and putting makeup on them.

We're particularly fascinated with unlikely serial killers—responsible husband and father, Boy Scout leader, church elder—and Bundy is way up on that list. It's virtually required to describe him as "handsome and charismatic," which is exactly what he was in interviews. An honor student at the University of Washington, then a law student, active in politics (a delegate for Nelson Rockefeller at the 1968 Republican National Convention), a kind, empathic volunteer at a suicide hotline. He worked on someone's successful campaign to be governor of Washington State; the

*For example, a selection of the many books currently available on Amazon: *The Ultimate Serial Killer Trivia Book* (Jack Rosewood, 2022); *True Crime Activity Book for Adults* (making one wonder what the kids' edition looks like; Brian Berry, 2021); and of course, *Serial Killers Coloring Book with Facts and Their Last Words* (Katys Corner, 2022).

candidate expressed his gratitude with the breathtaking irony of appointing Bundy to the Seattle Crime Prevention Advisory Committee.

Somewhere around then he began killing. He targeted young women. Early on, Bundy simply broke into apartments and attacked people in their sleep. His approach evolved to luring someone to his car by requesting help carrying something, pulling this off with charm and a seemingly broken limb in a cast. Sometimes he buffed up the verisimilitude with crutches. He would then bludgeon the victim.

Eventually caught, Bundy was convicted for a number of murders (in one much-publicized instance, incriminated in part by the match between his teeth and the bite marks on the buttock of his victim) and given the death penalty. He escaped from prison twice but was eventually put to death in 1989.

Bundy fascinated criminologists and mental health professionals, who handed out a variety of diagnoses of psychopathy, reflecting his manipulativeness, narcissism, and remorselessness. He fascinated the public as well; books were written and movies made about him (two and one, respectively, while he was still alive). Numerous women wrote to him in prison, some devastated by both his death and the subsequent discovery that they were not actually his beloved one and only. Few remember the names of his victims.

Bundy was executed by electric chair. In 1881, a drunk worker had grabbed the wires of an electric dynamo in a power plant and been instantly killed. Hearing about this, a dentist named Alfred Southwick conceived of a machine for electrocuting people as a humane alternative to hanging. After practicing on stray dogs, he had perfected his invention. The chair part of the original "electric chair," bound for iconic status, was a dental chair modified by Southwick. It was the execution method of choice for most of the twentieth century.[24]

When things went right, the wave of electricity caused unconsciousness within seconds and fatal cardiac arrest within a minute or two. When things didn't go right, multiple rounds of electrocution might be required, or the prisoner would remain conscious and in extreme pain; in one case, the prisoner's face mask caught on fire. Bundy's execution, however, was routine.

The execution was much anticipated throughout the country, with celebratory barbecues held the evening before, many called "Bundy-cues," featuring "Bundy burgers" and "electric hot dogs." Particularly raucous partying occurred at a fraternity at Florida State University, a school attended by two of Bundy's victims. On the day of the execution, hundreds gathered across the street from the prison in Raiford, Florida, where he would be killed. The crowd, which included families with children, sang, chanted, "Burn, Bundy, burn," and set off fireworks. News of his death was greeted with cheers (the somber witnesses to the execution, as they exited the prison, were reportedly shocked by the revelry). Celebrations over, the crowd dispersed, justice served.[25]

DELICIOUS, WHETHER SERVED HOT OR COLD

Here's a thoroughly elegant study, carried out by German psychologist Tania Singer. Subjects were either six-year-old kids or chimps. One of the researchers comes into the room and either does something nice to the kid/chimp—offering some desirable food—or does something mean—teasing them by starting to give the food and then snatching it away. The researcher leaves and then enters an adjacent room, visible to the subject through an observation window. Someone sneaks up behind the researcher and—whoa!—seemingly starts hitting them over the head with a stick, with the researcher crying out in pain. After ten seconds, the assailant drags the researcher to an adjacent room and then resumes the hitting. The kid/chimp can go into their own adjacent room with another window, giving them the opportunity to watch. Do they move to do this? If the researcher being pummeled had been nice to them, only 18 percent moved to see the rest; if the researcher had been mean, 50 percent leaped at the opportunity. Both kids and chimps were particularly interested in seeing someone who was mean to them get punished.[26]

Importantly, getting into the adjacent viewing room was costly. Kids had been receiving tokens for some irrelevant task, which they could trade in for desirable stickers; they had to relinquish tokens to watch the continued punishment. For the chimps, the door to the next room was extremely heavy, requiring considerable work to watch the continued punishment. And when it was the mean person being punished, kids forked out the tokens and chimps moved mountains and heavy doors to see. In other words, the kids and the chimps were willing to incur costs—to *pay* in currency or effort—to continue basking in the pleasure of watching the antisocial person getting what they deserved.

While watching the continued punishment, kids typically had a facial expression long associated with Schadenfreude, the emotion of gloating over another's misfortune—an involuntary frown coinciding with the blows, coupled with a smile. If it was the antisocial teaser getting punished, that expression occurred about four times as often as if it was the kind, prosocial person being punished. And for the chimps, if it was the Good Samaritan getting punished, the chimps gave agitated vocalizations; if it was the mean human, not a peep from the chimps.

We pay for stuff that gives us pleasure—a terrifying slasher movie (if you're that kind of paradoxical individual), cocaine, bananas, a chance to read sexually arousing writing or look at sexually arousing pictures.* And here are both kids and chimps paying for the pleasure of watching the wicked get their just rewards.[27]

The study had another fascinating wrinkle, showing the sophistication of humans, even kids, relative to chimps. In this version of the experimental design, the kid/chimp watched the researcher being nice or mean to another human/chimp (that second chimp, termed the "stooge" chimp, was trained for the role, and presumably received coauthorship on the paper). Then, as before, the researcher got attacked and dragged to the other room. Kids would also pay to watch third-party punishment; chimps, who show no third-party punishment experimentally, had no interest in watching it occur.

A great study, showing how deeply seated, both developmentally and taxonomically, is our enjoyment of seeing righteous punishment served.

---

*Whether humans or rhesus monkeys; in a paper entitled "Monkeys Pay per View: Adaptive Valuation of Social Images by Rhesus Macaques," male rhesus monkeys were shown to be willing to "pay" the price of forgoing desirable juice in order to see, well, crotch shots of female monkeys. Meanwhile, female rhesus monkeys liked looking at pictures of high-ranking males (which, given the characteristic aggressiveness of male rhesus, is a bit like falling for the animal magnetism of Billy Bigelow) or crotch shots of both male and female rhesus. Okay, just to go further down the rabbit hole, when female rhesus are ovulating, they show a stronger preference for looking at the faces of male rhesus (but, in an oddly reassuring way, not the faces of male chimps or humans).

Good luck convincing people that blame and punishment are scientifically and morally bankrupt.

The same unsettling conclusion comes from neuroimaging studies. If someone makes you an unfair offer in the Ultimatum Game, your insula, anterior cingulate cortex, and amygdala activate, a picture of disgust, pain, and anger. The lowball offer puts you at a split in the road. If it's a single-round game, punish retributively or be purely logical and accept the offer that is better than nothing? The more activation of your insula and amygdala, and the more pissed off you report being by inequity in general, the more likely you are to reject the offer. This retributive irrationality is all about emotion—if people believe they are rejecting an unfair offer from a human rather than a computer, there is also activation of that emotional vmPFC; making a similar point, men with higher testosterone levels are more likely to reject such offers.[28]

The picture of altruistic third-party punishment is much the same, with the neuroimaging indices of anger and disgust activated. Along with that is what you'd also expect, namely activation of a brain region called the temporal-parietal junction (TPJ), that region involved in perspective taking. And the perspective taking isn't just about the victims—the more TPJ activation, the more likely you are to forgive transgressors or accept the role of mitigating factors (e.g., poverty) in explaining their behavior.[29]

So on a neurobiological level, second-party punishers are about disgust, anger, and pain, whereas third-party punishers have the same plus the perspective taking needed to view someone else's misfortune as akin to your own. But then there is the crucial additional finding in all these cases: retributive punishing in any of these guises also *activates the dopamine circuitry involved in reward* (the ventral tegmentum and nucleus accumbens). Activation by punishment of the brain region goosed by the likes of orgasm or cocaine. It feels good.[30]

Additional studies make the point even further. Symbolic punishment doesn't activate reward circuits as much as does the real thing (e.g., blasting someone with a loud noise). More punishment correlates with more activation of the nucleus accumbens, and lots of accumbens activation when you get to punish a cheater for free predicts a greater likelihood of paying to punish a cheater. The circuitry activates whether you are someone who is independently meting out punishment or a conformer joining the vengeful crowd.

Being altruistic can feel good—it decreases pain in cancer patients, blunts the activation of neural pain pathways in response to shock. It even literally gives you a warm glow (such that people estimate ambient temperatures as being higher after an altruistic act). Nice. But being able to righteously punish evildoers feels *really* good. But as will be seen in a bit, even that can be tamed.[31]

JUSTICE SERVED V

The United States began as an experiment in convincing a bunch of unlike-minded states to form, if not a perfect union, at least a functional one. This was an iffy proposition from the start; it took nearly a century for Americans to transition from statements like "The United States are doing X" to "The United States *is* doing X." And from the start, there has always been an opposition that views the very notion of a federated government as tyranny. That certainly describes the Confederacy. Likewise for those resisting federal mask mandates during the pandemic. Likewise on January 6, 2020, for those who believed that it was despotic for those DC pedophiles to insist that the person who loses an election doesn't get to be president.

The "patriot" antigovernment militia movement continues to grow, and provided the toxic ideology that motivated an American to declare war against the United States in 1995. Most proximally, he was outraged by the siege of White supremacist Randy Weaver and family in Ruby Ridge, Idaho, in 1992, and the siege of David Koresh's Branch Davidian cult in Waco, Texas, in 1993.* On the second anniversary of the Waco

*I'm not remotely going to try to summarize what happened at either, as they will be mired in controversy forever; both have taken on near-sacred significance to the antigovernment militia movement.

siege, he used a bomb made of five thousand pounds of ammonium nitrate to blow up the Alfred P. Murrah Federal Building in Oklahoma City.

Timothy McVeigh's act of terrorism was the most destructive in American history (until 9/11). He killed 168 people, injured 853. More than three hundred surrounding buildings were damaged and 400 people left homeless; the blast registered 6.0 on the Richter scale fifty-five miles away. And as the detail seared into everyone's memory, McVeigh's victims included 9 children in a day care center in the building.

Thanks to eyewitness descriptions, McVeigh was soon apprehended. His statements in the years after were conflicting: He claimed he didn't know there was a day care center in the building and that if he had, he would have shifted targets; he dismissed the dead children as "collateral damage." He described understanding the pain of victims' families; he said that he had no sympathy for them. He wondered if maybe he should have bypassed a bombing and instead used his army-acquired skill as a sniper to take out selected targets; he expressed regret at not killing more people. His 1997 trial was moved to Denver, because of the impossibility of a fair trial in Oklahoma; it was estimated that 360,000 Oklahomans knew someone who worked in the Murrah Building. He was found guilty of all charges and given the death penalty. He asserted his supposed dominance by describing his eventual execution as "state-assisted suicide."

He would be executed by lethal injection, which by then had become the technique of choice, viewed as more humane than the electric chair or gas chamber. The prisoner is strapped down, an IV line is put into the arm (with a backup line into the other), and a trio of drugs is infused that, sequentially, renders the person unconscious within seconds, paralyzes the person and thus stops their breathing, and stops their heart. The painless process kills the prisoner within minutes.

Naturally, it's not so simple. Trained medical professionals usually refuse to participate or are banned from doing so by their state professional board. As a result, the IV line is put in by a correctional officer, who often botches things, with multiple sticks required or the vein missed entirely so that drug is injected into muscle and then absorbed slowly.* The initial anesthetic, which rapidly induces unconsciousness, also wears off quickly, so the subsequent steps might be done to someone who is conscious and feeling pain but can't express that because they are paralyzed. Sometimes the second drug does not adequately stop breathing, minutes passing with the prisoner gasping for air. Moreover, many drug manufacturers, particularly in the European Union, refuse to sell or are banned from selling a medical drug that will be used for killing, and various states have had to improvise alternative drug cocktails, with varying degrees of success at inducing a painless death.

Despite those potential snafus, McVeigh's 2001 execution went off without a hitch. The night before, he met with a priest, watched some TV, and had his last meal. Incongruously, PETA, People for the Ethical Treatment of Animals, had written to the warden, stating that after the lives McVeigh had taken, animals should at least be spared and he be served a vegetarian meal. The warden, defending McVeigh's rights, told PETA to get lost and that he could eat whatever he wanted, so long as it didn't involve alcohol or cost more than twenty dollars; whether McVeigh heeded PETA's call is unknown, but his last meal consisted of mint chocolate chip ice cream.

Normally, the witness room has seats for relatives of the victim; more than three hundred applied to be there, along with survivors of the bombing. Room was made for ten, with the rest allowed to watch the execution

*The process begins with the seemingly bizarre step of cleaning the infusion site with alcohol. What, so the person won't get an infection after they're dead? Why not also try to sell them a new coffee maker, to be delivered within three to five working days? In actuality, the alcohol makes it easier to find a vein.

by a video link from the Terre Haute, Indiana, prison to Oklahoma City; a bug in the video system delayed the execution for ten minutes. The remaining witnesses were mostly reporters, and all gave the same account: McVeigh, from the gurney, made eye contact and nodded slightly to each witness; he lay on his back, stared at the ceiling, and died with his eyes open. While silent throughout, McVeigh had requested that copies be handed to witnesses of the 1875 poem "Invictus," by William Ernest Henley, a treacly, self-congratulatory paean to stoicism in which the author acclaims himself as unconquerable, unbowed, and with a fearless visage, ending with flourishy bragging about mastering his fate and captaining his soul. Screw you all, the mass murderer had said one last time.

In the press conference afterward, media witnesses varied in describing him as seeming arrogant, defeated, aged, or commanding the scene; one reporter appeared to believe that McVeigh had written the poem; they all struggled to flesh out the story, noting the number of times he took a breath at some juncture, the color of his shirt, the length of his hair; differing opinions were offered as to whether the curtain was green or bluish green.

Outside the prison, 1,400 reporters had been on-site for three days. The event was catered by a local meeting-and-event firm, their first execution. For $1,146.50, reporters were given a padded chair, a writing table with a cloth skirt guaranteed to be changed daily, chilled bottled water, phone service, and transportation around the prison grounds by golf cart. The hoi polloi reporters unwilling to pony up the money made do in tents without chairs, electricity, or phone lines. A *Washington Post* reporter, either sheepish or gloating, admitted in his coverage that his paper had sprung for three deluxe packages.

Four thousand feet away from the reporters were the prison grounds reserved for protesters, two separate areas for the anti–death penalty attendees, numbering approximately a hundred, and the handful of pro-death celebrants, driven to the spot on two different buses; no transportation here for ambivalent protesters. Prison authorities wanted to avoid the circus of vulgarians that had accompanied Bundy's death; protesters were allowed a protest sign, a candle with a windscreen, and a Bible. Other than some jeers from the pro–death penalty protesters, the crowd was quiet and dispersed peacefully. Justice had been served.[32]

We're at loggerheads. There's no such thing as free will, and blame and punishment are without any ethical justification. But we've evolved to find the right kind of punishment viscerally rewarding. This is hopeless.

Maybe not, though, as this chapter has shown an additional type of evolution. Frenzied mobs, intoxicated with conspiracy theories, slashing, stabbing, burning hundreds to right a supposed wrong. A huge mob spending four hours watching a man be slowly torn to pieces by horses, in order to right a wrong. Twenty thousand people watching someone be dropped through a trap door with a rope snapping his neck, another act of righting a wrong. Hundreds gathering to celebrate news that a wrong has been righted with an electric chair. A handful of people, outnumbered ten to one by death penalty opponents, gathering for news that a wrong has been righted by someone being quietly overdosed.

What accounts for these transitions? The replacement of violent mobs with mobs watching officials being violent is obvious, part of the centralization of power and legitimizing of the state, the first steps toward a criminal trial reified with a "The State of Whichever versus Jones." The transition from drawing and quartering to a quick public hanging? A standard explanation is that this reflected reformist pressure.* The shift from public execution to electrocution behind prison walls? This revolved around whom the killing was being performed for. Sociologist Annulla

*In *Discipline and Punish* (which begins with the execution of Damiens), Michel Foucault rejected this sanguine idea; instead, he framed this as part of the shift from the state asserting power by owning and breaking a person's body—execution—to asserting it by owning and breaking their spirit and soul long before that, thanks to years of moldering imprisonment and the ceaseless surveillance of the panopticon. Political theorist C. Fred Alford of the University of Maryland rejects this interpretation. He lost me, though, when he started discussing what he called the microphysics of power (I was pretty lost with Foucault as well, actually).

Linders of the University of Cincinnati has argued that this was another step in the state's quest for legitimacy—instead of taking its approval from a mob of observers often threatening to lynch the person if the state didn't do it for them, legitimacy was now coming from the approving presence of a handful of distinguished gentlemen quietly observing the event. In other words, acquiring this new source of legitimacy outweighed the moral rejuvenation of the mob, obtained by reminding it viscerally about who was the Man. Electrocution to lethal injection? With the U.S. in the ever-dwindling club of death penalty countries, with the likes of Saudi Arabia, Ethiopia, and Iran, it seemed prudent to switch from a method that could cause the person's face mask to explode in flames to something (ideally) resembling euthanizing an elderly dog.[33]

From our perspective, the transition can be framed much more informatively. At some point, authorities showed up and said, "Look, we know it's great fun for all of you to get to slaughter lepers and Jews, but the times they are a' changing, and from now on, we're the ones doing the killing, and you're just going to have to get your pleasure from watching the person be tortured for hours." And then the transition to "And you're just going to have to get your pleasure from watching us take a minute or two to kill someone by hanging." And then to "You can wait outside, and we'll tell you when it's done. We'll even let journalist witnesses tell you about the gory parts of electrocuting someone,* and that has to be sufficiently pleasurable." And then on to "Get your pleasure from knowing that we've killed the person, albeit relatively peacefully."

And with each transition, people got used to things.

Not always, not quickly, sometimes not ever, of course. Every crowd celebrating the news of some criminal's execution inevitably produces a quote to the effect of how the condemned is getting an easier death than he deserves, after what he put his victims through. And that must feel just searing in its injustice. There were probably people in the crowd who felt that Damiens was getting off easy for sticking a penknife into the king.

*Linders speculates that the decision was made to include the press as witnesses precisely for this reason.

So there are always people who feel like there was too little retribution. Importantly, retribution built on perceptions of free will does help some victims reach the unreachable state of "closure." One sticky way to respond to that is to question whether acts of retribution, reframed as compassion for the bereaved, should be a "right" of victims or their families. An easier response is to point out the well-documented but not widely known fact that closure for victims or their families is mostly a myth. Law professor Susan Bandes of DePaul University finds that for many, the execution and the accompanying media coverage are retraumatizing, impeding their recovery.* A surprising number reach the point of actively opposing the execution. Social workers Marilyn Armour of the University of Texas and Mark Umbreit of the University of Minnesota studied family members of homicide victims in their two states, the former leading the way in executing prisoners, the latter having banned it more than a century ago. They found that from the perspectives of health, psychological well-being, and daily function, the Minnesotans fared significantly better than the Texans.† Moreover, a recent, first-of-its-kind national survey of victims of violent crime reported, by a wide margin, a preference that criminal justice focus on rehabilitation rather than retribution, and that expenditures be increased for crime prevention rather than incarceration.[34]

Those victims and families who do favor retribution and beefing up of prisons may actually be looking for something very different and rarely stated. In justifying the death penalty, William Barr, attorney general for both George W. Bush and Donald Trump, wrote, "We owe it to the victims and their families to carry forward the sentence imposed by our justice system." What he is really saying is that a government is morally obligated to enact the strongest possible manifestation of its culture's values in that realm—whether drawing and quartering or quarantining.[35]

We can get a sense of this by taking the evolutionary arc from burning

*In the words of legal scholar Pete Alces, the challenge of the death penalty is that it can feel intensely like both too much and too little, often for the same person (personal communication).

†It's important, however, to point out the obvious factor that Texas and Minnesota differ from each other in lots of other dramatic ways, so these findings are merely correlative.

lepers to overdosing McVeigh one step further. In July 2011, Norwegian Anders Breivik carried out the largest terrorist attack in Norwegian history. Breivik, a lump of narcissism and mediocrity, had tried and failed at a string of personas, with his ideology completely malleable and his failures always someone else's fault; he had finally found his people among White supremacist troglodytes. Following the standard playbook, Breivik proclaimed that White, Christian European culture was being destroyed in his country by immigrants, multiculturalism, and the political progressives who supported it. He first set off a bomb near the office of the socialist-democratic prime minister, killing eight. He then drove twenty-five miles to a lake containing the small island of Utøya; on it was a summer camp for a youth organization associated with the Labour Party, an organization that over the decades had produced a string of left-leaning prime ministers and one Nobel Peace Prize winner. Breivik, dressed as a police officer, was ferried to the island and spent the next hour calmly gunning down sixty-nine teenagers.

At his trial, he gave long, meandering rants about how his Christian European volk were being destroyed, claimed to be a knight in a confabulated modern Knights Templar, and gave pseudo-Nazi salutes. He was found guilty of the mass murder and given the longest sentence possible in Norway—twenty-one years.

Breivik was then deposited in one of Norway's dens of funishment.* He has a three-room living space, computer, TV, PlayStation, treadmill, and kitchen (he was able to submit an entry to a prison gingerbread house competition). Amid some heated public debate, he was accepted by the University of Oslo to matriculate remotely as, unironically, a political science student.

Norway's response to the slaughter? Exactly what Barr unintentionally

*It is important to note that, despite the time-limited funishment, Breivik has spent much of his time in solitary, because of the danger of his interacting with other prisoners, and his twenty-one-year sentence can be extended if he is deemed to still be a danger to society. At one point, he sued the Norwegian government over the cruel nature of his isolation (he was ultimately unsuccessful). In searching for a solution, a psychiatrist working for the prison suggested that retired police officers visit Breivik to socialize, drink coffee, and play card games.

implied. One survivor appraised the trial, saying, "The ruling in the Breivik case shows that we acknowledge the humanity of extremists too." They continued, "If he [Breivik] is deemed not to be dangerous any more after 21 years, then he should be released. . . . That's how it should work. That's staying true to our principles, and the best evidence that he hasn't changed our society." Prime Minister Jens Stoltenberg, who knew a number of the victims and their families, stated, "Our answer is more democracy, more openness, and more humanity but never naivety." Norway's universities accept prisoners as (remote) students, and in explaining his decision to offer the same to Breivik, the rector of the university said that they were doing so "for our own sake, not for his." In the Norwegian version of Barr, survivors and families of the slaughtered were owed the knowledge that their nation had responded to their nightmare with the strongest possible expression of their values.

And what were the responses of the average Norwegian to the trial? The majority were satisfied with the outcome, felt it had preventative value and reaffirmed democratic values; perhaps as a measure of its efficacy, before the trial 8 percent wanted revenge, while after, only 4 percent did. And the response of Norwegians to Breivik himself? In his arraignment hearing, Breivik's claim to be a (literal) knight of the indigenous Norwegian people was met with a wave of derisive laughter from the gallery. Breivik had posted a photo of himself in his Knights Templar outfit,* and one newspaper reproduced it under the sardonic, contemptuous heading "That's How He Got His One-Man Army"; what he wore was described as a "costume" rather than a "uniform." A pathetic nobody playing dress-up who could now be forgotten.[36]

With Breivik, Norway joined the ranks of peoples who have had to figure out how not to hate those who have damaged them terribly. When it works, it is awe-inspiring. It is also fascinating, seeing the culture-specific paths to this state employed by different peoples who have had a lot of

*Breivik had bought the uniform and tchotchkes from military surplus dealers and sewed on the medals; it's unclear if he knew what they signified, but he had awarded himself medals for, among other things, valor in the U.S. Navy, Air Force, and Coast Guard.

practice. We saw that in Charleston, after the Emanuel African Methodist Episcopal Church massacre that left nine African American parishioners dead at the hands of a White supremacist whom they had welcomed in—in the days afterward, some of the survivors and their families publicly forgave him and prayed for his soul. "I will never be able to hold her again, but I forgive you," said the daughter of one of the victims. "You hurt me. You hurt a lot of people. But God forgives you. I forgive you." The sister-in-law of one of the victims faced the shooter and offered to visit him in prison in order to pray with him.* We saw a different cultural version of the same after another White supremacist opened fire and killed eleven people at the Tree of Life Synagogue in Pittsburgh. The shooter was injured in the process and taken to a hospital to be cared for by a mostly Jewish medical staff; when asked how they managed that, Dr. Jeff Cohen, president of the hospital, said predictable things about the Hippocratic oath but then gave a more revealing explanation—the shooter, he said, was a confused man who was easily exploited by online hate groups: "The gentleman didn't appear to be a member of the Mensa society." And in the aftermath of Breivik's attack, a survivor who had subsequently become deputy mayor of Oslo wrote to him, "It is my job to see that no one experiences the same social rejection that you did [experience]. Your fight against social rejection is the only fight we have in common, Anders."† How have you managed not to hate this person? No one cited the frontal cortex or stress hormones. Instead, they found more poetic, personal routes to the same.

*And of course, as with all these examples, none of these were monolithic group responses. "You are Satan. Instead of a heart, you have a cold, dark space," said the daughter of another victim, hoping that the shooter would "go straight to hell."

†Just imagine the implausibility of the same if circumstances were such that Osama bin Laden were living out his days in a supermax prison.

Why don't I hate him? Because he has a soul, whether it is soiled or not, and God forgives him. Because he is not smart enough to know that he's been used and manipulated. Because, starting from childhood, he became embittered from loneliness, with a desperate need to be accepted and belong, and I am willing to call him by his first name and acknowledge that to him.[37]

We all are perched on an edge, with head-shaking disbelief whether we look back or forward. My guess is that most Norwegians view American criminal justice as barbaric. Yet at the same time, most Norwegians view it as unattainable and undesirable to consider Breivik in a context of there being no free will. The early part of his trial was dominated by the issue of whether he was insane, and the judges showed the same mindset critiqued in chapter 4 when, having decided he was sane, they concluded that he thus had free will, could have chosen to do otherwise, and was responsible for his actions. One commentator, who had moved beyond the Norwegians, wrote, "If Breivik's actions on that fateful Friday were completely beyond any free will, then punishing him (as distinct from restraining him from further harm to the community) may be as immoral as our perception of Breivik's criminal acts themselves."

Meanwhile, Americans are perched on a different edge of disbelief. I'm going out on a limb, but I assume that most Americans would view a public execution, complete with twenty thousand gawkers, and mobs of them putting aside their hot dogs and lemonade afterward to fight over souvenirs, as savagery. Yet Americans were gobsmacked by Breivik's trial, beginning with astonishment at how it opened with the prosecutors shaking Breivik's hand. "Mocking Justice in Norway" was the title of a piece criticizing the national values that resulted in Breivik's kid-glove treatment. One (British) criminologist began his piece by writing, "Anders Breivik is a monster who deserves a slow and painful death." And on a different edge, no doubt some nineteenth-century professional hangman would be appalled by how justice is mocked by lethal injection, but would also think that drawing and quartering was a bit beyond the pale.[38]

The theme of the second half of this book is this: We've done it before. Over and over, in various domains, we've shown that we can subtract out a belief that actions are freely, willfully chosen, as we've become more knowledgeable, more reflective, more modern. And the roof has not caved in; society can function without our believing that people with epilepsy are in cahoots with Satan and that mothers of people with schizophrenia caused the disease by hating their child.

But it will be hugely difficult to continue this arc, so much so that I've spent a lot of the last five years procrastinating over this book because it seemed like a waste of time. And because I am endlessly reminded about how far I personally have to go. As I noted, I've worked with public defenders on various murder trials, teaching juries about the circumstances that produce brains that make horrible decisions. I was once asked if I would take on that role working on the case of a White supremacist who, a month after attempting to burn down a mosque, had invaded a synagogue and used an assault rifle to shoot four people, killing one. "Whoa," I thought. "WTF, I'm supposed to help out with this?" Members of my family died in Hitler's camps. When I was a kid, our synagogue was arsoned; my father, an architect, rebuilt it, and I had to spend hours holding one end of a tape measure for him amid the scorched, acrid ruins while he railed on in a near-altered state about the history of anti-Semitism. When my wife directed a production of *Cabaret*, with me assisting, I had to actively force myself to touch the swastika armbands when distributing costumes. Given all that, I'm supposed to help out with this trial? I said yes—if I believed any of this shit I've been spouting, I had to. And then I subtly proved to myself how far I still had to go. On these trials I've worked on, the lawyer has often asked me if I wanted to meet with the defendant, and I've instantly said no—I would have to admit during my testimony to having done that, and it would compromise my credibility as a teaching witness impartially discussing the brain. But this time, before I knew it, it was I who asked these attorneys if I could meet with the defendant. Was this because I wanted to figure out what epigenetic changes had occurred in his amygdala, what version of the MAO-beta gene he possessed? Because I wanted to understand his personal case of turtles all the way down? No. I wanted to see close up what the face of evil looked like.*

Perhaps when done with the writing, I should read this book.

It will be hard. But we've done it before.

*To my vast relief, the case never went to trial—there was a guilty plea for life without parole rather than the death penalty.

15

If You Die Poor

I was surfing the web, procrastinating to avoid doing some chore, and was looking at one of those sites where people ask a question and readers then weigh in. One asked, "After you've pooped, do you wipe front to back or back to front?" There was a long string of answers. Almost everyone said front to back, many doing so emphatically. Of those who said front to back, most cited their mother as the source of that advice. And there it was, someone in Oregon and someone continents away in Romania writing virtually the *same exact unlikely response*: "When I was a kid, my mother always told me that if I wiped back to front, I wouldn't have any friends."

I was thunderstruck. Were their mothers identical twins separated at birth? Had the Oracle of Delphi franchised so there were now also an Oracle of Portland and an Oracle of Bucharest? Why had both people given the same bizarre framing of advice about personal hygiene?

Someone named Bruce Stephan survived both the collapse of the San Francisco Bay Bridge during the 1989 Loma Prieta earthquake, and the attack on the World Trade Center on 9/11. Tsutomu Yamaguchi was in both Hiroshima and Nagasaki on the days they were bombed, yet lived another sixty-five years. On the other hand, Pete Best was dropped as the Beatles' drummer a few weeks before they had their first hit, and Ron

Wayne, who was one of the three cofounders of Apple Computer, didn't enjoy working with Steve and Woz (to show my Silicon Valley bro-ness), and quit after a few weeks. Meanwhile, there's Joe Grisamore, world record holder for having a mohawk that towers three feet above his head.

What does it mean that the universe converged on those two mothers giving advice to their children? Or that Stephan and Yamaguchi were lucky, Best and Wayne were arguably not, and Grisamore lives in Minnesota? What does it mean that the doctor who will someday tell you how many months you have left is currently standing in front of an open refrigerator eating cold pad thai noodles? And that Jennifer Lopez and Ben Affleck got back together, while Henry VIII and Catherine of Aragon never did? Most fundamentally, what does it mean that you can look at two five-year-olds and accurately predict which of the two will be elderly from diseases of despair by age fifty and which will be an eighty-year-old having a hip replacement in time for ski season?[1]

What the science in this book ultimately teaches is that there *is* no meaning. There's no answer to "Why?" beyond "This happened because of what came just before, which happened because of what came just before that." There is nothing but an empty, indifferent universe in which, occasionally, atoms come together temporarily to form things we each call Me.

A whole field of psychology explores terror management theory, trying to make sense of the hodgepodge of coping mechanisms we resort to when facing the inevitability and unpredictability of death. As we know, those responses cover the range of humans at our best and worst—becoming closer to your intimates, identifying more with your cultural values (whether humanitarian or fascist in nature), making the world a better place, deciding to live well as the best revenge. And by now, in our age of existential crisis, the terror we feel when shadowed by death has a kid sibling in our terror when shadowed by meaninglessness. Shadowed by our being biological machines wobbling on top of turtles that go all the way down. We are not captains of our ships; our ships never had captains.[2]

Fuck. That really blows.

Which I think helps explain a pattern. One compatibilist philosopher after another reassuringly proclaims their belief in material, deterministic modernity . . . yet somehow, there is still room for free will. As might be kinda clear by now, I think that this doesn't work (see chapters 1, 2, 3, 4, 5, 6 . . .). I suspect that most of them know this as well. When you read between the lines, or sometimes even the lines themselves in their writing, a lot of these compatibilists are actually saying that there has to be free will because it would be a total downer otherwise, doing contortions to make an emotional stance seem like an intellectual one. Humans "descended from the apes! Let us hope it is not true, but if it is, let us pray that it will not become generally known," said the wife of an Anglican bishop in 1860, when told about Darwin's novel theory of evolution.* One hundred fifty-six years later, Stephen Cave titled a much-discussed June 2016 article in *The Atlantic* "There's No Such Thing as Free Will . . . but We're Better Off Believing in It Anyway."†

He just might be right. Chapter 2 discussed a study in which a sense of "illusory will" could be induced in people. One subgroup of subjects, however, was resistant to this—individuals with clinical depression. Depression is often framed as a sufferer having a cognitively distorted sense of "learned helplessness," where the reality of some loss in the past becomes mistakenly perceived as an inevitable future. In this study, though, it was not that depressed individuals were cognitively distorted, underestimating their actual control. Instead they were accurate compared with everyone else's overestimates. Findings like these support the view that in some circumstances, depressed individuals are not distortive but are "sadder but wiser." As such, depression is the pathological loss of the capacity to rationalize away reality.

And thus, perhaps, "we're better off believing in it anyway." Truth doesn't always set you free; truth, mental health, and well-being have a

*The famous quote might actually be apocryphal; see quoteinvestigator.com/2011/02/09/darwinism-hope-pray/.

†A philosophical stance called illusionism, associated with philosopher Saul Smilansky, whose ideas were discussed in the previous chapter.

complex relationship, something explored in an extensive literature on the psychology of stress. Expose a test subject to a series of unpredictable shocks, and she will activate a stress response. If you warn her ten seconds before each shock that it is coming, the stress response is lessened, as truth girds predictability, giving time to prepare a coping response. Give a warning one second before each shock, and there's too little time for an effect. But give a warning one *minute* before and the stress response is worsened, as that minute stretches into feeling a year's worth of anticipatory dread. Thus, truthful predictive information can lessen, worsen, or have no effect on psychological stress, depending on the circumstances.[3]

Researchers have explored another facet of our complex relationship with truth. If someone's actions have produced a mildly adverse outcome, truthfully emphasizing the control he had—"Think how much worse things could have been, good thing you had control"—blunts his stress response. But if someone's actions have produced a disastrous outcome, untruthfully emphasizing the opposite—"No one could have stopped the car in time, the way that child darted out"—can be deeply humane.

The truth can even be life-threatening. Someone teetering on the edge of death in an ER, 90 percent of their body covered in third-degree burns, gathers their strength to ask in a whisper whether the rest of their family is okay. And most medical professionals would be mighty torn about telling the person the shattering truth. As some evolutionary biologists have pointed out, the only way humans have survived amid being able to understand truths about life is by having evolved a robust capacity for self-deception.* And this certainly includes a belief in free will.[4]

*Right around now, I'm concerned about my prattling on about "truth," rejecting so many other people's thinking about free will, worried that I'm going to come off as self-congratulatory. That I *am* self-congratulatory. Wow, all these super-smart people who run philosophical circles around me, and I'm one of the few who understand that you can't successfully wish for what you want to wish for or will yourself to have willpower. Wow, I'm awesome. The previous few paragraphs suggest an additional route to being self-congratulatory—wow, all these thinkers fleeing from unpalatable truths to the point of irrationality, and I'm the one with the bollocks to lick truth's smelly armpit.

This many pages into this book, I hope it's clear that I don't think it is valid for anyone to be self-congratulatory about anything. At some point in this writing process, I was struck with what

Despite that, I obviously think that we should face the music about our uncaptained ships. This, of course, has some substantial downsides.

WHAT YOU'D GIVE UP ALONG WITH FREE WILL

The most immediate area of distress is consistently the running-amok challenge, returning to chapter 11. For Gilberto Gomes, "[rejection of the idea of free will] leaves us with an incomprehensible picture of the human world, since there is no responsibility or moral obligation in it. If one could not have done otherwise, it cannot be the case that one ought to have done otherwise." Michael Gazzaniga recoils from rejecting free will and responsibility because "[people] have to be held accountable for their actions—their participation. Without that rule, nothing works" (and where the only thing that might constrain behavior is people not wanting to hang out with you if you run amok in particularly unwelcome ways). According to Daniel Dennett, if there were no belief in free will, "there would be no rights, no recourse to authority to protect against fraud, theft, rape, murder. In short, no morality. . . . Do you really want to return humanity to [the seventeenth-century English philosopher Thomas] Hobbes's state of nature where life is nasty, brutish and short?"[5]

---

seemed like the explanation for why I've been able to stick with an unshakable rejection of free will, despite the bummers of feelings it can evoke. A point made earlier in the chapter is personally very relevant. Since my teenage years, I've struggled with depression. Now and then, the meds work great and I'm completely free of it, and life seems like hiking above the tree line on a spectacular snow-capped mountain. This most reliably occurs when I'm actually doing that with my wife and children. Most of the time, though, the depression is just beneath the surface, kept at bay by a toxic combination of ambition and insecurity, manipulative shit, and a willingness to ignore who and what matter. And sometimes it incapacitates me, where I mistake every seated person as being in a wheelchair and every child I glance at as having Down syndrome.

And I think that the depressions explain a lot. Bummed out by the scientific evidence that there's no free will? Try looking at your children, your perfect, beautiful children, playing and laughing, and somehow this seems *so* sad that your chest constricts enough to make you whimper for an instant. After that, dealing with the fact that our microtubules don't set us free is a piece of cake.

Dennett bad-mouths neuroscientists along these lines by frequently telling his parable of "the nefarious neurosurgeon." The surgeon does some procedure on a patient. Afterward—because, hey, why not?—she lies to him, claiming that during surgery, she also implanted a chip in his brain that robs him of free will, that she and her fellow scientists now control him. Unburdened by a sense of being responsible for his actions anymore, unconstrained by norms of trust that make for the social contract, the man becomes a criminal. That's what neuroscientists do, concludes Dennett, in "nefariously" and "irresponsibly" lying to people about how they have no free will. Thus, along with the terrors of mortality and of meaninglessness, there's the terror that there's a nasty, brutish, short murderer standing behind you in line at Starbucks.

As we've seen, rejection of free will doesn't doom you to break bad, not if you've been educated about the roots of where our behavior comes from. Trouble is, that requires education. And even that doesn't guarantee a good, moral outcome. After all, most Americans have been educated to believe in free will and have reflected on how this produces responsibility for our actions. And most have also been taught to believe in a moralizing god, guaranteeing that your actions have consequences. And yet our rates of violence are unmatched in the West. We're doing plenty of running amok as it is. Maybe we should call this one a draw and, based on the sorts of findings reviewed in chapter 11, conclude at least that rejection of free will is unlikely to make things worse.

Rejecting free will has an additional downside. If there's no free will, you don't deserve praise for your accomplishments, you haven't earned or are entitled to anything. Dennett feels this—not only will the streets be overrun with rapists and murderers if we junk free will, but in addition, "no one would deserve to receive the prize they competed for in good faith and won." Oh, *that* worry, that your prizes will feel empty. In my experience, it's going to be plenty hard to convince people that a remorseless murderer doesn't deserve blame. But that's going to be dwarfed by the difficulty of convincing people that they themselves don't deserve to be praised if they've helped that old woman cross the street.* That problem with rejecting free will seems legit, if rarefied; we'll return to this.[6]

For me, the biggest problem with accepting that there's no free will takes the nefarious-neurosurgeon parable down a different path. The surgery is done, and the surgeon lies to the patient about no longer having free will. And rather than falling into mundane criminality, the patient falls into profound malaise, an enervation because of the pointlessness. In the short story "What's Expected of Us," Ted Chiang takes a cue from Libet, writing about a gizmo called the Predictor, with a button and a light. Whenever you press the button, the light goes on a second before. No matter what you do, no matter how much you try not to think about pressing the button, strategize to sneak up on it, the light comes on a second before you press the button. In the moment between the light coming on and your supposedly freely choosing to press the button, your future action is already a determined past. The result? People are hollowed out. "Some people, realizing that their choices don't matter, refuse to make any choices at all. Like a legion of Bartleby the Scriveners, they no longer engage in spontaneous action. Eventually, a third of those who play with a Predictor must be hospitalized because they won't feed themselves. The end state is akinetic mutism, a kind of waking coma."[7]

It's that yawning chasm where, amid "This happened because of what came before, which happened because of what came before that . . . ," there's no place for meaning or purpose. Which haunts philosophers, along with the rest of us. Ryan Lake of Clemson University writes that rejecting belief in free will would make sincere regret or apology impossible, robbing us of "an essential component of our relationships with others." Peter Tse writes, "I find [a leading incompatibilist's] denial of moral

*At the 2018 Harvard graduation, the poised, articulate student chosen to give a speech, Jin Park, showed that he understood turtles. Why was he there in that celebration of talents and accomplishments? Because, he explained, day after day, his undocumented immigrant father worked as a line cook in restaurants (that probably exploited the hell out of him, since he lacked papers), because his undocumented immigrant mother toiled endlessly, giving pedicures in beauty salons. "My talents are indistinguishable from their labors; they are one and the same."

responsibility a profoundly nihilistic view of human beings, their choices, and life in general." Philosopher Robert Bishop of Wheaton College, in dissecting Dennett's thinking, concludes that "he believes that the consoling perspective he offers is the only way for any of us to maintain a healthy, affirmative, outlook on life and remain meaningfully engaged in it." Life lived "as if," viewed through free will–colored glasses.[8]

This one looms over us. Evolution, chaos, emergence, have taken the most unexpected turns in us, producing biological machines that can know our machine-ness, and whose emotional responses to that knowledge feel real. *Are* real. Pain is painful. Happiness makes life frabjous. I try to ruthlessly hold myself to the implications of all this turtling, and sometimes I even succeed. But there is one tiny foothold of illogic that I can't overcome for even a millisecond, to my intellectual shame and personal gratitude. It is logically indefensible, ludicrous, meaningless to believe that something "good" can happen to a machine. Nonetheless, I am certain that it is good if people feel less pain and more happiness.

Despite these various downsides, I think that it is essential that we face our lack of free will. It may look now like we're heading for a major anticlimax for this book, one that is about as appealing as subsisting on locusts: "This is how the world works; suck it up." Sure, if you have a burn patient on the edge of death, probably hold off on telling them that their family didn't survive. But otherwise, it's usually good to go with the truth, especially about free will—faith can sustain, but nothing devastates as surely as the discovery that your deeply held faith has been misplaced all along. We claim we're rational beings, so go and prove it. Deal.

But "Toughen up, there's no free will" isn't remotely the point.

Maybe you're deflated by the realization that part of your success in life is due to the fact that your face has appealing features. Or that your praiseworthy self-discipline has much to do with how your cortex was constructed when you were a fetus. That someone loves you because of, say, how their oxytocin receptors work. That you and the other machines don't have meaning.

If these generate a malaise in you, it means one thing that trumps everything else—you are one of the lucky ones. You are privileged enough to have success in life that was not of your own doing, and to cloak yourself with myths of freely willed choices. Heck, it probably means that you've both found love *and* have clean running water. That your town wasn't once a prosperous place where people manufactured things but is now filled with shuttered factories and no jobs, that you didn't grow up in the sort of neighborhood where it was nearly impossible to "Just Say No" to drugs because there were so few healthy things to say yes to, that your mother wasn't working three jobs and barely making the rent when she was pregnant with you, that a pounding on the door isn't from ICE. That when you encounter a stranger, their insula and amygdala don't activate because you belong to an out-group. That when you are truly in need, you're not ignored.

If you are among these lucky very, *very* few, the ultimate implications of this book don't concern you.*

A LIBERATORY SCIENCE (WITHOUT TONGUE IN CHEEK)

A Case Study

In the process of working on this book, I spoke to a number of people who were involved in advocacy for people suffering from obesity. One told me about when she first learned about the hormone leptin.[†,9]

*And even phrased this way, this is a false dichotomy, making a distinction between the benighted few who can ignore all this and remain convinced that they deserve their superyacht and the unwashed majority who need to be convinced that it's not their fault that they don't own one. Every page really applies to all of us, because we are all destined at times to blame, be blamed, hate, be hated, feel entitled, and suffer the entitled.

†I'm making reference in the title to this section to the great intellectual hoax known as the Sokal affair. Physicist Alan Sokal of NYU and University College London got fed up with the

As background, leptin is the poster child of the "It's a biological disease, not a measure of your lack of self-discipline" insight, regulating fat storage throughout the body and, most significant, telling your hypothalamus when you've eaten enough. Abnormally low levels of leptin signaling* produce an abnormally low capacity for feeling satiated, resulting in severe obesity, beginning in childhood. This individual turned out to carry a leptin mutation; inspection of a family picture album suggested that it had been there for generations.

Mutation puts us in the world of medical exotica. Regular ol' unmutated leptin and its receptor genes come in various flavors, differing in the efficiency with which they function. Same for the literally hundreds of other genes implicated in regulating body mass index (BMI). Of course, environment also plays a major role. Just to home in on one of our familiar

---

intellectual emptiness, agitprop, and toeing of party line in a lot of postmodernist thinking. He thus wrote a paper that (a) agreed that physics and math are guilty of the sins of various antiprogressive -isms; (b) confessed that the supposed "truths" of science, as well as the supposed existence of "physical reality," are mere social constructs; (c) fawningly cited leading postmodernists; and (d) was packed with science gibberish. It was submitted to and duly published in 1996 by *Social Text*, a leading postmodernist cultural studies journal, as "Transgressing the Boundaries: Toward a Transformative Hermeneutics of Quantum Gravity." The hoax was then revealed. Massive brouhaha, conferences of postmodernists condemning his "bad faith," Jacques Derrida calling him "sad," and so on. I thought the paper was glorious, a hilarious parody of postmodernist cant (e.g., "the content of any science is profoundly constrained by the language within which its discourses are formulated; and mainstream Western physical science has, since Galileo, been formulated in the language of mathematics. But *whose* mathematics?"). With tongue firmly in cheek, Sokal proclaimed the paper's goal to be fostering a "liberatory science" that would be freed from the tyranny of "absolute truth" and "objective reality." Thus, in the present case, I'm noting "without tongue in cheek," because I'm going to argue that science dumping the concept of free will is truly liberating.

(The Sokal affair was seized upon by the likes of Rush Limbaugh as an exposé of the Left's intellectual fraudulence, with Sokal embraced as some sort of right-wing scourge. This infuriated me, as Sokal had walked the walk as a leftist—for example, in the 1980s, he left his cushy academic post to teach math in Nicaragua during the Sandinista revolution. Furthermore, anything the Right had to say about truth ended with "alternative facts" in Trump's first week in office. As an aside, in college, Sokal lived down the hall, two years ahead of me and thus too intimidating of a big kid to talk to; his brilliance, wonderful eccentricity, and willingness to call BS were already legendary.)

*Nitpicky aside: Why say "too little leptin signaling" rather than just "too little leptin"? *Signaling* is a broader term, reflecting that a problem can be at the level of the amount of a messenger (e.g., a hormone or neurotransmitter) or with the sensitivity of cells to the messenger (e.g., abnormal levels/function of receptors for the messenger). Sometimes the radio station is screwed up, sometimes it's the radio in your kitchen. (Do people still have radios?)

outposts, the womb, your lifelong propensity toward obesity is influenced by whether you were undernourished as a fetus, whether your pregnant mother smoked, drank, or took illicit drugs, even by the gut bacteria she transferred to your fetal gut.* Some of the precise genes that would have been epigenetically modified in your fetal pancreas and fat cells have even been identified. And as usual, different versions of genes interact differently with different environments. One gene variant increases obesity risk, but only when coupled with your mother having smoked during pregnancy. The impact of a variant of another gene is stronger in urban dwellers than in rural. Some variants increase the risk of obesity depending on your gender, race, or ethnicity, depending on whether you exercise (in other words, a genetics of why exercise melts away fat in some people but not others), depending on the specifics of your diet, whether you drink, and so on. On a larger scale, be of low socioeconomic status, or live in a place where you're surrounded by inequality (on the levels of countries, states, and cities), and the same diet is more likely to make you obese.[10]

Collectively, these genes and gene/environment interactions regulate every nook and cranny of biology, are relevant to everything from the avidity with which a newborn nurses to why two adults with the same elevated BMI have different risks of adult-onset diabetes.

Let's look again at that table from chapter 4:

"Biological stuff"	*Do you have grit?*
Having destructive sexual urges	*Do you resist acting upon them?*

*An extraordinary, famed example is the Dutch Hunger Winter, when the occupying Nazis cut off the food supply in the Netherlands in the winter of 1944–45, and twenty to forty thousand Dutch starved to death. If you were a fetus then, with you and your mother severely deprived of nutrients and calories, epigenetic changes produced a lifelong thrifty metabolism, a body voraciously adept at storing calories. Be one of those fetuses, and sixty years later, you had a dramatically increased risk of obesity, metabolic syndrome, diabetes, and, as we've seen, schizophrenia.

Being a natural marathoner	*Do you fight through the pain?*
Not being all that bright	*Do you triumph by studying extra hard?*
Having a proclivity toward alcoholism	*Do you order ginger ale instead?*
Having a beautiful face	*Do you resist concluding that you're entitled to people being nice to you because of it?*

Lots of the effects we're considering come from the left side of the table, the features of your biology that you were just handed by luck. Some of them concern how efficiently your gut absorbs nutrients versus flushing them down the toilet; how readily fat is stored or mobilized; whether you tend to accumulate fat in your butt or abdomen (the former's healthier); whether stress hormones strengthen that propensity. Great news: you can still be judgmental—life's caprices may bless some people and curse others as to their natural attributes, you say . . . but what really matters is your self-discipline when playing the hand you were dealt.

But some of these genetic effects are harder to categorize, as to which side of the table they should be placed. For example, genes code for types of taste receptors in your tongue. Hmm, is this merely a biological attribute such that even though food might taste better to you than to others, you are still expected to resist gluttony? Or is it possible for food to taste so good that it cannot be resisted?* Hormones like leptin that signal whether you feel full generate some similar difficulties in categorizing.

And then there are genetic effects related to obesity that are squarely on that right side of the chart, the world where we're judged for the backbone and character we bring to our natural attributes. The genetics of how

*And the world of processed foods involves scientists trying to achieve that state with whatever food their boss sells.

many dopamine neurons you formed, mediating anticipation and reward. The genetics of how much pictures of appealing food activate those neurons when you're dieting. How intensely stress produces cravings for high-carb/high-fat foods, how aversive hunger feels. And of course, how readily your frontal cortex regulates parts of the hypothalamus relevant to hunger, bringing in the ever-present issue of willpower. Once again, both sides of the chart are made of the same biology.

This scientific truth has had zero impact on the general public. Encouraging studies show that the average levels of implicit, unconscious bias against people as a function of their race, age, or sexual orientation have all decreased significantly over the last decade. But not implicit biases against obese individuals. They've gotten worse. Significantly, it's there among medical students, particularly among those who are thin, White, and male. Even your average obese individual shows implicit antiobesity biases, unconsciously associating obesity with laziness; this sort of self-loathing is rare among stigmatized groups. And this self-loathing has a price; for example, for people with the same diet and BMI, internalizing an antiobesity bias triples the chances of metabolic disease.* Throw in the explicit biases, and we have the world in which the obese are discriminated against when it comes to jobs, housing, health care (and one where stigma typically worsens obesity, rather than magically generating successful willpower).[11]

In other words, a realm in which people's lives are ruined, where they are blamed for biology over which they had no control. And what happened when the person I was speaking with fully grasped the implications of what a leptin mutation means? "It was the start of my no longer thinking of myself as a fat pig, of being my own worst bully."

*Really, for the same BMI? Of course. More self-loathing, more secretion of stress hormones resulting in more preferential storage of fat in the gut (among other downsides), more of an increase in metabolic and cardiovascular disease risk.

OVER AND OVER AND OVER

Everywhere you look, there's that pain and self-loathing, staining all of life, about traits that are manifestations of biology. "I find myself beating myself up at times, wondering why I can't get my shit together, wondering if these disorders say something about my character," writes Sam about his bipolar disorder. "Over the years, I started to assume I was just lazy. Instead of thinking there might be something wrong biologically, I assumed it was all my fault. And, every time I'd resolve to be better at being attentive in class, or neat or diligent about homework, I'd inevitably fail," wrote Arielle about her ADHD. "I called myself evil, cold, weird," said Marianne about her autism spectrum disorder.[*,12]

Again and again, the same voice, in domains where blame is as absurd as deciding that you were responsible for your height. Oh, but then there's blame even there: "My mom (5ft 6) and my dad (6ft 1) constantly yell at me for being short saying I'm not active enough and don't sleep enough," writes one unnamed person. And Manas, living in India at the intersection of issues of height and of the societal obsession with shades of brownness writes: "I grew taller than everyone at home because I had an active lifestyle. I might be tall but I am darker than the rest of the people at home. That goes to show that we win in some areas but lose in others," the deep misattributed pain made clear when *because* appears.[13]

Then there's the learning about someone's own different-ness. "[I was] so liberated knowing that there is a name for what I am experiencing," writes Kat about her bipolar disorder; Erin about her borderline personality disorder: "My struggles with mental ill health were validated." Sam, about his mood disorder: The discovery that "your first diet or binge didn't 'cause' your eating disorder. Your first cut didn't 'cause' your depression." Michelle writes about her ADHD, "Everything fell into place. I wasn't crap because I found [tax] returns painful, blurted out stuff and was messy.

*Then there's the ghastly quora.com/Is-it-my-fault-my-husband-hits-me.

I wasn't crap at all. I have a neuro difference." Marianne about her autism: "I wished only that I hadn't lost so much of my life hating myself."[14]

And all the while, chaoticism teaches us that "being normal" is an impossibility, that it ultimately just means that you have the same sorts of abnormalities that are accepted as out of our control that everyone else has. Hey, it's normal that you can't cause objects to levitate.

Then there's the liberation of understanding that what you mistook as the consequences of different choices could be nothing more than a butterfly flapping its wings. I once spent a day teaching some incarcerated men about the brain. Afterward, one guy asked me, "My brother and I grew up in the same house. He's the vice president of a bank; how'd I wind up like this?" We talked, figured out a likely explanation for his brother—by whatever hiccup of chance, his motor cortex and visual cortex gave him great hand-eye coordination, and he happened to be spotted playing pickup basketball by the right person . . . who got him a scholarship to the fancy prep school on the other side of the tracks that groomed him into the ruling class.

Then there's one of the deepest sources of pain. I once lectured at an elementary school about other primates. Afterward, a deeply homely child asked if baboons cared if you're not pretty. As *Wicked*'s ostracized green-skinned Elphaba sings about a boy who could make someone feel loved and desired, she concludes, "He could be that boy. But I'm not that girl." And every time someone less attractive is less likely to be hired, to get a raise, to be voted for, to be exonerated by a jury, an implicit belief is being expressed that lack of beauty on the outside and lack of beauty on the inside go hand in hand.

Naturally, sexuality comes into this too. In 1991, the superb neuroscientist Simon LeVay at the Salk Institute rocked the world with front-page news. LeVay, gay and still reeling from the death of the love of his life from AIDS, had discovered a part of the brain that differed structurally depending on whether you loved people of your gender or the opposite one. Sexual orientation as a biological trait—a release from the cesspool of a pastor whose church would picket funerals with signs saying GOD HATES

FAGS, from medieval conversion therapy. As Lady Gaga sings, "God makes no mistakes, I'm on the right track, baby, I was born this way." For the lucky, this was no news, something they'd known all along. For the less fortunate, there was release from the belief that they could have, should have, chosen to love differently than they did. Or the revelation could have been among those on the outside—parents writing to LeVay about being freed from the likes of "If only I hadn't encouraged him to pursue arts camp instead of basketball camp, he wouldn't have turned out gay."[15]

Blame shows up as well concerning fertility, where a woman's lack of reproductive potency can prompt a doctor to grossly exaggerate the effects of stress on fertility ("You're too uptight," "You're too type A"), where psychoanalytic toxins still fester ("The problem is your ambivalence about having a child"), where blame is heaped on lifestyle choices ("You wouldn't have had the abortion that left scar tissue in your uterus if you hadn't slept around and been careless"). Where, as studies show, infertility can be as psychiatrically debilitating as cancer.[16]

A particularly pernicious consequence of misplaced belief in captaining your own ship comes with the work of Duke University epidemiologist Sherman James. He described a personality style that he called "John Henryism," named for the American folk hero, the railroad construction worker who drove steel with unmatched strength; challenged by his boss to compete against a new machine doing the same, he vowed that no machine was going to keep him down, battled and defeated it . . . only to then drop dead from exhaustion. The John Henryism profile is one of someone who feels like they can take on any challenge if they apply themselves enough, endorsing statements on a questionnaire like "When things don't go the way I want them to, that just makes me work even harder" or "I've always felt that I could make of my life pretty much what I wanted to make of it." Well, what's wrong with that? It sounds like a good, healthy locus of control. Unless, like John Henry, you were an African American blue-collar worker or sharecropper, where this attributional style results in a greatly increased risk of cardiovascular disease. It's a pathogenic belief that with enough effort, you could overcome a racist system guaranteed to keep you down.* A fatal belief that you should be able to control the uncontrollable.[17]

There's our nation with its cult of meritocracy that judges your worth by your IQ and your number of degrees. A nation that spews bilge about equal economic potential while, as of 2021, the top 1 percent has 32 percent of the wealth, and the bottom half less than 3 percent, where you can find an advice column headlined "It's Not Your Fault if You Are Born Poor, but It's Your Fault if You Die Poor," which goes on to say that if that was your lamentable outcome, "I'll say you're a wasted sperm."[18]

Having a neuropsychiatric disorder, having been born into a poor family, having the wrong face or skin color, having the wrong ovaries, loving the wrong gender. Not being smart enough, beautiful enough, successful enough, extroverted enough, lovable enough. Hatred, loathing, disappointment, the have-nots persuaded to believe that they deserve to be where they are because of the blemish on their face or their brain. All wrapped in the lie of a just world.

In 1911, the poet Morris Rosenfeld wrote the song "Where I Rest," at a time when it was the immigrant Italians, Irish, Poles, and Jews who were exploited in the worst jobs, worked to death or burned to death in sweatshops.† It always brings me to tears, provides one metaphor for the lives of the unlucky:[19]

*James doesn't see the same among higher-SES African Americans or among Whites at all.

†The words for "Mayn Rue-Plats" were in Yiddish, at a time when it was the language of socialist firebrands on the Lower East Side of New York rather than ultraorthodox ayatollahs. Rosenfeld wrote it in response to the Triangle Shirtwaist Factory fire in March 1911, in which 146 sweatshop workers—almost all immigrants, almost all women, some as young as fourteen—died because the owners had locked an exit, believing that otherwise, workers would sneak out the back way with stolen clothes. A jury found the owners liable for wrongful death, forcing them to pay all of seventy-five dollars compensation to each family of the dead, while the owners themselves received more than sixty thousand dollars for the loss of their factory. Seventeen months later, one of them was found to have once again locked the exits in his new factory and was levied the minimum fine of twenty dollars. One hundred two years later, the Rana Plaza building in Dhaka, Bangladesh, collapsed, killing *1,134* sweatshop workers inside. Cracks had been discovered in the

Where I Rest

Look not for me in nature's greenery
You will not find me there, I fear.
Where lives are wasted by machinery
That is where I rest, my dear.

Look not for me where birds are singing
Enchanting songs find not my ear.
For in my slavery, chains a-ringing
Is the music I do hear.

Not where the streams of life are flowing
I draw not from these fountains clear.
But where we reap what greed is sowing
Hungry teeth and falling tears.

But if your heart does love me truly
Join it with mine and hold me near.
Then will this world of toil and cruelty
Die in birth of Eden here.*

It is the events of one second before to a million years before that determine whether your life and loves unfold next to bubbling streams or machines choking you with sooty smoke. Whether at graduation ceremonies you wear the cap and gown or bag the garbage. Whether the thing you are viewed as deserving is a long life of fulfillment or a long prison sentence.

There is no justifiable "deserve." The only possible moral conclusion is that you are no more entitled to have your needs and desires met than is any other human. That there is no human who is less worthy than you to

---

building the day before, resulting in its evacuation; the owners informed workers that anyone not back on the job the next day would be docked a month's pay.

*Translation by Daniel Kahn.

have their well-being considered.* You may think otherwise, because you can't conceive of the threads of causality beneath the surface that made you you, because you have the luxury of deciding that effort and self-discipline aren't made of biology, because you have surrounded yourself with people who think the same. But this is where the science has taken us.

And we need to accept the absurdity of hating any person for anything they've done; ultimately, that hatred is sadder than hating the sky for storming, hating the earth when it quakes, hating a virus because it's good at getting into lung cells. This is where the science has brought us as well.

Not everyone agrees; they suggest that the science that has filled these pages is about the statistical properties of populations, unable to predict enough about the individual. They suggest that we don't know enough yet. But we know that every step higher in an Adverse Childhood Experience score increases the odds of adult antisocial behavior by about 35 percent; given that, we already know enough. We know that your life expectancy will vary by thirty years depending on the country you're born in,† twenty years depending on the American family into which you happen to be born; we already know enough. And we already know enough, because we understand that the biology of frontocortical function explains why at life's junctures, some people consistently make the wrong decision. We already know enough to understand that the endless people whose lives are less fortunate than ours don't implicitly "deserve" to be invisible. Ninety-nine percent of the time I can't remotely achieve this mindset, but there is nothing to do but try, because it will be freeing.

Those in the future will marvel at what we didn't yet know. There will be scholars opining about why in the course of a few decades around the start of the third millennium, most Americans stopped opposing gay marriage. History majors will struggle on final exams to remember whether it

*I have been made aware that this bears some resemblance to the Buddhist concept of "unselfing." I have absolutely nothing useful to say about Buddhism beyond that.

†As of 2022, eighty-five years in Japan, fifty-five in the Central African Republic.

was the nineteenth, twentieth, or twenty-first century when people began to understand epigenetics. They will view us as being as ignorant as we view the goitered peasants who thought Satan caused seizures. That borders on the inevitable. But it need not be inevitable that they also view us as heartless.

Acknowledgments

I've been very lucky in my life, something which I certainly did not earn (see the previous ~400 pages for details). In my book-writing realm, that good fortune has included having wonderful, generous colleagues and friends who, along with my family, have provided feedback (sometimes in the form of conversations going back decades, and/or who have read sections of this book, amid any mistakes being my own). These include:

Peter Alces, William and Mary Law School

David Barash, University of Washington

Alessandro Bartolomucci, University of Minnesota

Robert Bishop, Wheaton College

Sean Carroll, Johns Hopkins University

Gregg Caruso, State University of New York

Jerry Coyne, University of Chicago

Paul Ehrlich, Stanford University

Hank Greely, Stanford University

Josh Greene, Harvard University

Daniel Greenwood, Hofstra University School of Law, cofounder of the "Third-Floor Holmes Hall Ethics of Free Will and Determinism" lecture series almost half a century ago

Sam Harris

Robin Hiesinger, Free University of Berlin

Jim Kahn, University of California, San Francisco

Neil Levy, Oxford University

Liqun Luo, Stanford University

Rickard Sjoberg, Umea University, Sweden

The late Bruce Waller, Youngstown State University.

Thanks as well to Bhupendra Madhiwalla, Tom Mendosa, Raul Rivers, and Harlen Tanenbaum.

I am now many books into having Katinka Matson as my book agent, and many years into having Steven Barclay as my speaking agent—deep thanks to you both for your friendship and for always having my back.

At Penguin Random House, I thank Hilary Roberts for her careful read and suggestions as copy editor. I warmly thank Mia Council for overseeing the process of getting this book to print, and for providing some truly insightful feedback. Most of all, I thank Scott Moyers, my editor for this and my previous book; your help has been such that at every juncture where my writing/thinking/self-confidence has stalled, my automatic first thought now is "What would Scott say?"

I closed my lab about a decade ago. Typically, lab scientists ending research at what was a relatively young age do so to become the dean of something or other, or an editor of a science journal. As such, my bidding farewell to pipetting in order to sit at home writing is unorthodox; I am grateful to Stanford University for the intellectual freedom that it gifts its faculty, and to the two chairs of my department during this period—Martha Cyert and Time Stearns—and the late, truly beloved Bob Simoni.

And heck, while we're at it, thanks to Tony Fauci for battling the Forces of Darkness. And good going Malala.

Thanks to our twelve-pound Havanese, Kupenda, and our eighty-five-pound Golden Retriever, Safi. The former has taught me how social status is more about social intelligence than muscle mass, passing his days terrorizing the helpless, hapless latter. And to the primate members of my family—Benjamin and Rachel, who bring me joy beyond measure, and to Lisa, my everything.

Appendix

NEUROSCIENCE 101

Consider two different scenarios.

First: Think back to when you hit puberty. You'd been primed by a parent or teacher about what to expect. You woke up with a funny feeling, found your jammies alarmingly soiled. You excitedly woke up your parents, who, along with my family, got tearful; they took embarrassing pictures, a sheep was slaughtered in your honor, you were carried through town in a sedan chair while neighbors chanted in an ancient language. This was a big deal.

But be honest—would your life be so different if those endocrine changes had instead occurred twenty-four hours later?

Second scenario: Emerging from a store, you are unexpectedly chased by a lion. As part of the stress response, your brain increases your heart rate and blood pressure, dilates blood vessels in your leg muscles, which are now frantically working, sharpens sensory processing to produce a tunnel vision of concentration.

And how would things have turned out if your brain took twenty-four hours to send those commands? Dead meat.

That's what makes the brain special. Hit puberty tomorrow instead of today? So what? Make some antibodies in an hour instead of now? Rarely fatal. Same for delaying depositing calcium in your bones. But much of what the nervous system is about is encapsulated in the frequent question in this book: What happened one second before? Incredible speed.

The nervous system is about contrasts, unambiguous extremes, having something or having nothing to say, maximizing signal-to-noise ratios. And this is demanding and expensive.*

ONE NEURON AT A TIME

The basic cell type of the nervous system, what we typically call a "brain cell," is the neuron. The hundred billion or so in our brains communicate with each other, forming complex circuits. In addition, there are "glia" cells, which do a lot of gofering—providing structural support and insulation for neurons, storing energy for them, helping to mop up neuronal damage.

Naturally, this neuron/glia comparison is all wrong. There are about ten glial cells for every neuron, coming in various subtypes. They greatly influence how neurons speak to each other, and also form glial networks that communicate completely differently from neurons. So glia are important. Nonetheless, to make this primer more manageable, I'm going to be very neuron-centric.

Part of what makes the nervous system so distinctive is how distinctive neurons are as cells. Cells are usually small, self-contained entities—consider red blood cells, which are round little discs.

Neurons, in contrast, are highly asymmetrical, elongated beasts, typically with processes sticking out all over the place. Consider this drawing of a single neuron seen under a microscope in the early twentieth century by one of the patriarchs of the field, Santiago Ramón y Cajal:

*Which, among other things, is why the nervous system is so vulnerable to injury. Someone has a cardiac arrest. Their heart stops for a few minutes before it is shocked into beating again, and during those few minutes, the entire body is deprived of blood, oxygen, and glucose. And at the end of those few minutes of "hypoxia-ischemia," every cell in the body is miserable and queasy. Yet it is preferentially brain cells (and a consistent subset of them) that are now destined to die over the next few days.

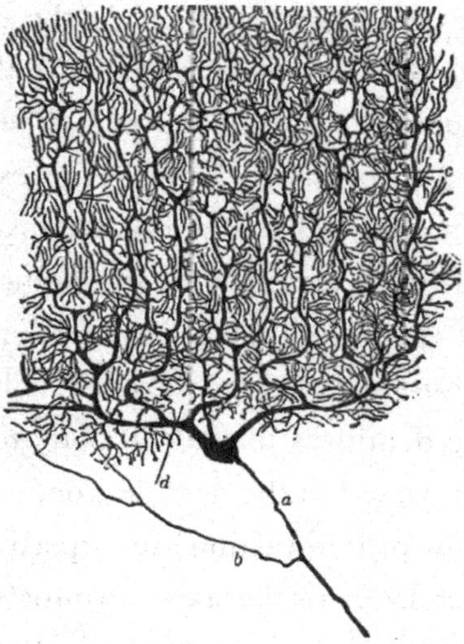

It's like the branches of a manic tree, explaining the jargon that this is a highly "arborized" neuron (a point explored at length in chapter 7, concerning how those arbors form in the first place).

Many neurons are also outlandishly large. A zillion red blood cells fit on the proverbial period at the end of this sentence. In contrast, there are single neurons in the spinal cord that send out projection cables that are many feet long. There are spinal cord neurons in blue whales that are half the length of a basketball court.

Now for the subparts of a neuron, the key to understanding its function.

What neurons do is talk to each other, cause each other to get excited. At one end of a neuron are its metaphorical ears, specialized processes that

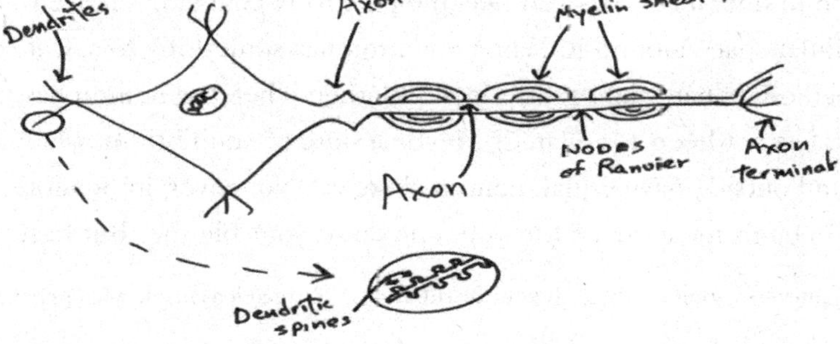

receive information from another neuron. At the other end are the processes that are the mouth, that communicate with the next neuron in line.

The ears, the inputs, are called dendrites. The output begins with a single long cable called an axon, which then ramifies into axonal endings—these axon terminals are the mouths (ignore the myelin sheath for the moment). The axon terminals connect to the spines on the branches of dendrites of the next neuron in line. Thus, a neuron's dendritic ears are informed that the neuron behind it is excited. The flow of information then sweeps from the dendrites to the cell body to the axon to the axon terminals, and is then passed to the next neuron.

Let's translate "flow of information" into quasi chemistry. What actually goes from the dendrites to the axon terminals? A wave of electrical excitation. Inside the neuron are various positively and negatively charged ions. Just outside the neuron's membrane are other positively and negatively charged ions. When a neuron has gotten an exciting signal from the previous neuron at a spine on a dendritic branch, channels in the membrane in that spine open, allowing various ions to flow in, others to flow out, and the net result is that the inside of the end of that dendrite becomes more positively charged. The charge spreads toward the axon terminal, where it is passed to the next neuron. That's it for the chemistry.

Two gigantically important details:

The Resting Potential

So when a neuron has gotten a hugely excitatory message from the previous neuron in line, its insides can become positively charged, relative to the extracellular space around it. When a neuron has something to say, it screams its head off. What might things look like then when the neuron has nothing to say, has not been stimulated? Maybe a state of equilibrium, where the inside and outside have equal, neutral charges.* No, never, impossible. That's good enough for some cell in your spleen or your big toe. But back to that

*For chemists: in other words, so that the distribution of charged ions inside and out balance each other.

critical issue, that neurons are all about contrasts. When a neuron has nothing to say, it isn't some passive state of things just trickling down to zero. Instead, it's an active process. An active, intentional, forceful, muscular, sweaty process. Instead of the "I have nothing to say" state being one of default charge neutrality, the inside of the neuron is *negatively* charged.

You couldn't ask for a more dramatic contrast: I have nothing to say = inside of the neuron is negatively charged. I have something to say = inside is positive. No neuron ever confuses the two. The internally negative state is called the resting potential. The excited state is called the action potential. And why is generating this dramatic resting potential such an active process? Because neurons have to work like crazy, using various pumps in their membranes, to push some positively charged ions out, to keep some negatively charged ones in, all in order to generate that negative internal resting state. Along comes an excitatory signal; channels open, and oceans of ions rush this way and that to generate the excitatory positive internal charge. And when that wave of excitation has passed, the channels close and the pumps have to get everything back to where they started, regenerating that negative resting potential. Remarkably, neurons spend nearly half their energy on the pumps that generate the resting potential. It doesn't come cheap to generate dramatic contrasts between having nothing to say and having some exciting news.

Now that we understand resting potentials and action potentials, on to the other gigantically important detail.

That's Not What Action Potentials Really Are

What I've just outlined is that a single dendritic spine receives an excitatory signal from the previous neuron (i.e., the previous neuron has had an action potential); this generates an action potential in that spine, which spreads the axonal branch that it is on, on toward the cell body, over it, and on to the axon and the axon terminals, and passes the signal to the next neuron in line. Not true.

Instead: The neuron is sitting there with nothing to say, which is to say that it's displaying a resting potential; all of its insides are negatively charged. Along comes an excitatory signal at one dendritic spine on one dendritic branch, emanating from the axon terminal of the previous neuron in line. As a result, channels open and ions flow in and out in that one spine. But not enough to make the entire insides of the neuron positively charged. Simply a little less negative just inside that spine. Just to attach some numbers here that don't matter in the slightest, things shift from the resting potential charge being around –70 millivolts (mV) to around –60 mV. Then the channels close. That little hiccup of becoming less negative* spreads farther to nearby spines on that branch of dendrite. The pumps have started working, pumping ions back to where they were in the first place. So at that dendritic spine, the charge went from –70 mV to –60 mV. But a little bit down the branch, things then go from –70 to –65 mV. Farther down, –70 to –69. In other words, that excitatory signal dissipates. You've taken a nice smooth, calm lake, in its resting state, and tossed a little pebble in. It causes a bit of a ripple right there, which spreads outward, getting smaller in its magnitude, until it dissipates not far from where the pebble hit. And miles away, at the lake's axonal end, that ripple of excitation has had no effect whatsoever.

In other words, if a single dendritic spine is excited, that's not enough to pass the excitation down to the axonal end and on to the next neuron. How does a message ever get passed on? Back to that wonderful drawing of a neuron by Cajal.

All those bifurcating dendritic branches are dotted with spines. And in order to get sufficient excitation to sweep from the dendritic end of the neuron to the axonal end, you have to have summation—the same spine must be stimulated repeatedly and rapidly and/or, more commonly, a bunch of the spines must be stimulated at once. You can't get a wave, rather than just a ripple, unless you've thrown in a lot of pebbles.

At the base of the axon, where it emerges from the cell body, is a

*Jargon: that little bit of "depolarization."

specialized part (called the axon hillock). If all those summated dendritic inputs produce enough of a ripple to move the resting potential around the hillock from –70 mV to about –40 mV, a threshold is passed. And once that happens, all hell breaks loose. A different class of channels opens in the membrane of the hillock, which allows a massive migration of ions, producing, finally, a positive charge (about +30 mV). In other words, an action potential. Which then opens up those same types of channels in the next smidgen of axonal membrane, regenerating the action potential there, and then the next, and the next, all the way down to the axon terminals.

From an informational standpoint, a neuron has two different types of signaling systems. From the dendritic spines to the axon hillock, it's an analog signal, with gradations of signals that dissipate over space and time. And from the axon hillock on to the axon terminals, it's a digital system with all-or-none signaling that regenerates down the length of the axon.

Let's throw in some imaginary numbers, in order to appreciate the significance of this. Let's suppose an average neuron has about one hundred dendritic spines and about one hundred axon terminals. What are the implications of this in the context of that analog/digital feature of neurons?

Sometimes nothing interesting. Consider neuron A, which, as just introduced, has one hundred axon terminals. Each one of those connects to one of the one hundred dendritic spines of the next neuron in line, neuron B. Neuron A has an action potential, which propagates down to all of its hundred axon terminals, which excites all one hundred dendritic spines in neuron B. The threshold at the axon hillock of neuron B requires fifty of the spines to get excited around the same time in order to generate an action potential; thus, with all one hundred of the spines firing, neuron B is guaranteed to get an action potential and is going to pass on neuron A's message.

Now instead, neuron A projects half of its axon terminals to neuron B and half to neuron C. It has an action potential; does that guarantee one in neurons B and C? Each of those neurons' axon hillocks has that threshold of needing a signal from fifty pebbles at once, in which case they have action potentials—neuron A has caused action potentials in two downstream neurons, has dramatically influenced the function of two neurons.

Now instead, neuron A evenly distributes its axon terminals among ten different target neurons, neurons B through K. Is its action potential going to produce action potentials in the target neurons? No way—continuing our example, the ten dendritic spines' worth of pebbles in each target neuron is way below the threshold of fifty pebbles.

So what will ever cause an action potential in, say, neuron K, which only has ten of its dendritic spines getting excitatory signals from neuron A? Well, what's going on with its other ninety dendritic spines? In this scenario, they're getting inputs from other neurons—nine of them, with ten inputs from each. In other words, any given neuron integrates the inputs from all the neurons projecting to it. And out of this comes a rule: *the more neurons that neuron A projects to, by definition, the more neurons it can influence; however, the more neurons it projects to, the smaller its average influence will be upon each of those target neurons.* There's a trade-off.

This doesn't matter in the spinal cord, where one neuron typically sends all its projections to the next one in line. But in the brain, one neuron will disperse its projections to scads of other ones and receive inputs from scads of other ones, with each neuron's axon hillock determining whether its threshold is reached and an action potential generated. The brain is wired in these networks of divergent and convergent signaling.

Now to put in a flabbergasting real number: your average neuron has about *ten thousand to fifty thousand* dendritic spines and about the same number of axon terminals. Factor in a hundred billion neurons, and you see why brains, rather than kidneys, write good poetry.

Just for completeness, here are a couple of final facts that should be ignored if this has already been more than you wanted. Neurons have some additional tricks, at the end of an action potential, to enhance the contrast between nothing-to-say and something-to-say even more, a means of ending the action potential really fast and dramatically—something called delayed rectification and another thing called the hyperpolarized refractory period. Another minor detail from that diagram above: a type of glial cell wraps around an axon, forming a layer of insulation called a myelin sheath; this "myelination" causes the action potential to shoot down the axon faster.

And one final detail of great future importance: the threshold of the axon hillock can change over time, thus changing the neuron's excitability. What things change thresholds? Hormones, nutritional state, experience, and other factors that fill this book.

We've now made it from one end of a neuron to the other. How exactly does a neuron with an action potential communicate its excitation to the next neuron in line?

TWO NEURONS AT A TIME: SYNAPTIC COMMUNICATION

Suppose an action potential triggered in neuron A has swept down to all those tens of thousands of axon terminals. How is this excitation passed on to the next neuron(s)?

The Defeat of the Synctitium-ites

If you were your average nineteenth-century neuroscientist, the answer was easy. Their explanation would be that a fetal brain is made up of huge numbers of separate neurons that slowly grow their dendritic and axonal processes. And eventually, the axon terminals of one neuron reach and touch the dendritic spines of the next neuron(s), and they merge, forming a continuous membrane between the two cells. From all those separate fetal neurons, the mature brain forms this continuous, vastly complex net of one single superneuron, called a "synctitium." Thus excitation readily flows from one neuron to the next because they aren't really separate neurons.

Late in the nineteenth century, an alternative view emerged, namely that each neuron remained an independent unit, and that the axon terminals of one neuron didn't actually touch the dendritic spines of the next. Instead, there's a tiny gap between the two. This notion was called the neuron doctrine.

The adherents to the synctitium school were arrogant as hell and even knew how to spell "synctitium," so they weren't shy in saying that they thought that the neuron doctrine was asinine. Show me the gaps between axon terminals and dendritic spines, they demanded of these heretics, and tell me how excitation jumps from one neuron to the next.

And then in 1873, it all got solved by the Italian neuroscientist Camillo Golgi, who invented a technique for staining brain tissue in a novel fashion. And the aforementioned Cajal used this "Golgi stain" to stain all the processes, all the branches and branchlets and twigs of the dendrites and axon terminals of single neurons. Crucially, the stain didn't spread from one neuron to the next. There wasn't a continuous merged net of a single superneuron. Individual neurons are discrete entities. The neuron doctrine–ers vanquished the synctitium-ites.*

Hooray, case closed; there are indeed micro-microscopic gaps between axon terminals and dendritic spines; these gaps are called synapses (which weren't directly visualized until the invention of electron microscopy in the 1950s, putting the last nail in the synctitial coffin). But there's still that problem of how excitation propagates from one neuron to the next, leaping across the synapse.

The answer, whose pursuit dominated neuroscience in the middle half of the twentieth century, is that the electrical excitation doesn't leap across the synapse. Instead, it gets translated into a different type of signal.

---

*Ironic footnote: Cajal was the chief exponent of the neuron doctrine. And the leading voice in favor of synctitiums? Golgi; the technique he invented showed that he was wrong. He apparently moped the entire way to Stockholm to receive his Nobel Prize in 1906—shared with Cajal. The two loathed each other, didn't even speak. In his Nobel address, Cajal managed to muster the good manners to praise Golgi. Golgi, in his, attacked Cajal and the neuron doctrine; dickhead.

Neurotransmitters

Sitting inside each axon terminal, tethered to the membrane, are little balloons, called vesicles, filled with many copies of a chemical messenger. Along comes the action potential that initiated at the very start of the axon, at that neuron's axon hillock. It sweeps over the terminal and triggers the release of those chemical messengers into the synapse. Which they float across, reaching the dendritic spine on the other side, where they excite the neuron. These chemical messengers are called neurotransmitters.

How do neurotransmitters, released from the "presynaptic" side of the synapse, cause excitation in the "postsynaptic" dendritic spine? Sitting on the membrane of the spine are receptors for the neurotransmitter. Time to introduce one of the great clichés of biology. The neurotransmitter molecule has a distinctive shape (with each copy of the molecule having the same). The receptor has a binding pocket of a distinctive shape that is perfectly complementary to the shape of the neurotransmitter. And thus the neurotransmitter—cliché time—fits into the receptor like a key into a lock. No other molecule fits snugly into that receptor; the neurotransmitter molecule won't fit snugly into any other type of receptor. Neurotransmitter binds to receptor, which triggers those channels to open, and the currents of ionic excitation begin in the dendritic spine.

This describes "transsynaptic" communication with neurotransmitters. Except for one detail: What happens to the neurotransmitter molecules after they bind to the receptors? They don't bind forever—remember that action potentials occur on the order of a few thousandths of a second. Instead, they float off the receptors, at which point the neurotransmitters have to be cleaned up. This occurs in one of two ways. First, for the ecologically minded synapse, there are "reuptake pumps" in the membrane of the axon terminal. They take up the neurotransmitters and recycle them, putting them back into those secretory vesicles to be used again.*

*More with the keys in locks—the reuptake pumps have a shape that is complementary to the shape of the neurotransmitter, so that the latter is the only thing taken back up into the axon terminal.

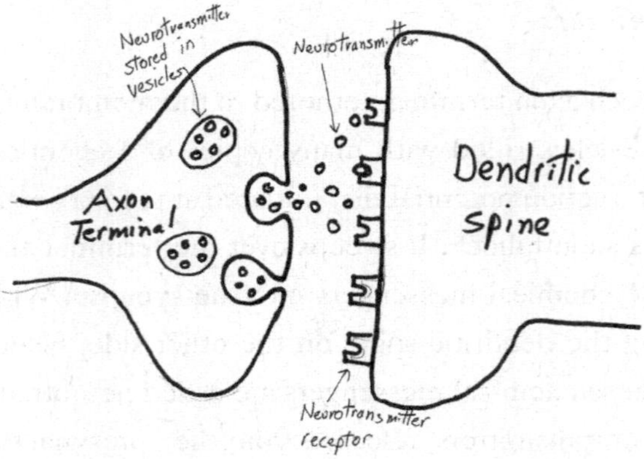

The second option is for the neurotransmitter to be degraded in the synapse by an enzyme, with the breakdown products flushed out to sea (i.e., the extracellular environment, and from there on to the cerebrospinal fluid, the bloodstream, and eventually the bladder).

These housekeeping steps are hugely important. Suppose you want to increase the amount of neurotransmitter signaling across a synapse. Let's translate that into the excitation terms of the previous section—you want to increase excitability across the synapse, such that an action potential in the presynaptic neuron has more of an oomph in the postsynaptic neuron, which is to say it has an increased likelihood of causing an action potential in that second neuron. You could increase the amount of neurotransmitter released—the presynaptic neuron yells louder. Or you could increase the amount of receptor on the dendritic spine—the postsynaptic neuron is listening more acutely.

But as another possibility, you could decrease the activity of the reuptake pump. As a result, less of the neurotransmitter is removed from the synapse. Thus, it sticks around longer and binds to the receptors repeatedly, amplifying the signal. Or as the conceptual equivalent, you could decrease the activity of the degradative enzyme; less neurotransmitter is broken down, and more sticks around longer in the synapse, having an enhanced effect. As we'll see, some of the most interesting findings that help explain individual differences in the behaviors that concern us in this book relate to the amounts of neurotransmitter made and released, and the amounts and functioning of the receptors, reuptake pumps, and degradative enzymes.

Types of Neurotransmitters

What is this mythic neurotransmitter molecule, released by action potentials from the axon terminals of all of the hundred billion neurons? Here's where things get complicated, because there is more than one type of neurotransmitter.

Why more than one? The same thing happens in every synapse, which is that the neurotransmitter binds to its key-in-a-lock receptor and triggers the opening of various channels that allow the ions to flow and makes the inside of the spine a bit less negatively charged.

One reason is that different neurotransmitters depolarize to different extents—in other words, some have more excitatory effects than others—and for different durations. This allows for a lot more complexity in information being passed from one neuron to the next.

And now to double the size of our palette, there are some neurotransmitters that don't depolarize, don't increase the likelihood of the next neuron in line having an action potential. They do the opposite—they "hyperpolarize" the spine, opening different types of channels that make the resting potential even more negative (e.g., shifting from –70 mV to –80 mV). In other words, there are such things as *inhibitory* neurotransmitters. You can see how that has just made things more complicated—a neuron with its ten thousand to fifty thousand dendritic spines is getting excitatory inputs of differing magnitudes from various neurons, getting inhibitory ones from other neurons, integrating all of this at the axon hillock.

Thus, there are lots of different classes of neurotransmitters, each binding to a unique receptor site that is complementary to its shape. Are there a bunch of different types of neurotransmitters in each axon terminal, so that an action potential triggers the release of a whole orchestration of signaling? Here is where we invoke Dale's principle, named for Henry Dale, one of the grand pooh-bahs of the field, who in the 1930s proposed a rule whose veracity forms the very core of each neuroscientist's sense of well-being: an action potential releases the same type of neurotransmitter from all of the axon terminals of a neuron. As such, there will be a distinctive neurochemical profile to a particular neuron: Oh, that neuron is a neurotransmitter A–type neuron. And what that also means is that the neurons that it talks to have neurotransmitter A receptors on their dendritic spines.*

There are dozens of neurotransmitters that have been identified. Some of the most renowned: serotonin, norepinephrine, dopamine, acetylcholine, glutamate (the most excitatory neurotransmitter in the brain), and GABA (the most inhibitory). It's at this point that medical students are tortured with all the multisyllabic details of how each neurotransmitter is synthesized—its precursor, the intermediate forms the precursor is converted to until finally arriving at the real thing, the painfully long names of the various enzymes that catalyze the syntheses. Amid that, there are some pretty simple rules built around three points:

a. You do not ever want to find yourself running for your life from a lion and, oopsies, the neurons that tell your muscles to run fast go off-line because they've run out of neurotransmitter. Commensurate with that, neurotransmitters are made from precursors that are plentiful; often, they are simple dietary constituents. Serotonin and dopamine, for example, are made from the dietary amino acids tryptophan and tyrosine, respectively. Acetylcholine is made from dietary choline and lecithin.†

*What that also implies is that if a neuron is receiving axonal projections to five thousand of its spines from a neurotransmitter A–releasing neuron and five thousand from a neurotransmitter B–releasing one, it expresses different receptors on those two populations of its spines.

†Whoa, does that mean that you can regulate the amounts of neurotransmitters with your diet? People got very excited about this possibility in my student days. For the most part, though, this has been a bust—for example, if you were so deprived of proteins that contain tyrosine that you can't make enough dopamine, you'd already be dead for lots of reasons.

b. A neuron can potentially have dozens of action potentials a second. Each involves restocking of the vesicles with more neurotransmitter, releasing it, mopping up afterward. Given that, you do not want your neurotransmitters to be huge, complex, ornate molecules, each of which requires generations of stonemasons to construct. Instead, they are all made in a small number of steps from their precursors. They're cheap and easy to make. For example, it only takes two simple synthetic steps to turn tyrosine into dopamine.

c. Finally, to complete this pattern of neurotransmitter synthesis as cheap and easy, generate multiple neurotransmitters from the same precursor. In neurons that use dopamine as the neurotransmitter, for example, there are two enzymes that do those two construction steps. Meanwhile, in norepinephrine-releasing neurons, there's an additional enzyme that converts the dopamine to norepinephrine.

Cheap, cheap, cheap. Which makes sense. Nothing becomes obsolete faster than a neurotransmitter after it has done its postsynaptic thing. Yesterday's newspaper is useful today only for house-training puppies. A final point that will be of huge relevance to come: just as the threshold of the axon hillock can change over time in response to experience, nearly every facet of the nuts and bolts of neurotransmitter-ology can be changed by experience as well.

Neuropharmacology

As these neurotransmitterology insights emerged, this allowed scientists to begin to understand how various "neuroactive" and "psychoactive" drugs and medicines work.

Broadly, such drugs fall into two categories: those that increase signaling across a particular type of synapse and those that decrease it. We already saw some of the strategies for increasing signaling: (a) administer a drug that stimulates more synthesis of the neurotransmitter (for example, by administering the precursor or using a drug that increases the activity of the enzymes that synthesize the neurotransmitter; as an example, Parkinson's disease involves a loss of dopamine in one brain region, and a bulwark of treatment is to boost dopamine levels by administering the drug L-DOPA, which is the immediate precursor of dopamine); (b) administer a synthetic version of the neurotransmitter or a drug that is structurally close enough to the real thing to fool the receptors (psilocybin, for example, is structurally similar to serotonin and activates a subtype of its receptors); (c) stimulate the postsynaptic neuron to make more receptors (fine in theory, not easily done); (d) inhibit degradative enzymes so that more of the neurotransmitter sticks around in the synapse; (e) inhibit the reuptake of the neurotransmitter, prolonging its effects in the synapse (the modern antidepressant of choice, Prozac, does exactly that in serotonin synapses and thus is often referred to as an "SSRI," a selective serotonin reuptake inhibitor).*

Meanwhile, a pharmacopeia of drugs is available to decrease signaling across synapses, and you can see what their underlying mechanisms are going to include—blocking the synthesis of a neurotransmitter, blocking its release, blocking its access to its receptor, and so on. Fun example: Acetylcholine stimulates your diaphragm to contract. Curare, the poison used in darts by indigenous people in the Amazon, blocks acetylcholine receptors. You stop breathing.

---

*So, if SSRIs boost serotonin signaling and lessen symptoms of depression, the cause of depression must be too little serotonin. Well, maybe not. (A) A paucity of serotonin may be the cause of only some subtypes of depression—SSRIs most certainly don't help everyone and to varying extents; (B) for other subtypes, serotonin shortage may be one of the contributing causes, or even completely irrelevant; (C) just because more serotonin signaling equals less depression, that doesn't necessarily mean that the initial problem was too little serotonin—after all, just because duct tape can cure a leaking pipe doesn't mean that the leak was initially caused by a shortage of duct tape; (D) despite the "selective" part of the SSRI acronym, the drugs are actually not perfectly selective and effect other neurotransmitters as well, meaning that these others may be relevant rather than serotonin; (E) despite what SSRIs do to serotonin signaling, it is possible that the problem is too *much* serotonin—this can arise through a scenario that is so multilayered that it leaves my students gasping for air; (F) even more stuff. As such, a controversy is now raging as to whether the "serotonin hypothesis" (i.e., that depression is caused by too little serotonin) has been oversold. Which seems likely.

MORE THAN TWO NEURONS AT A TIME

We have now triumphantly reached the point of thinking about three neurons at a time. And within not too many pages, we will have gone wild and considered even more than three. The purpose of this section is to see how circuits of neurons work, the intermediate step before examining what entire regions of the brain have to do with our behaviors. As such, the examples here were chosen merely to give a flavor of how things work at this level. Having some understanding of the building blocks of circuits like these is enormously important for chapter 12's focus on how circuits in the brain can change in response to experience.

Neuromodulation

Consider the following diagram below:

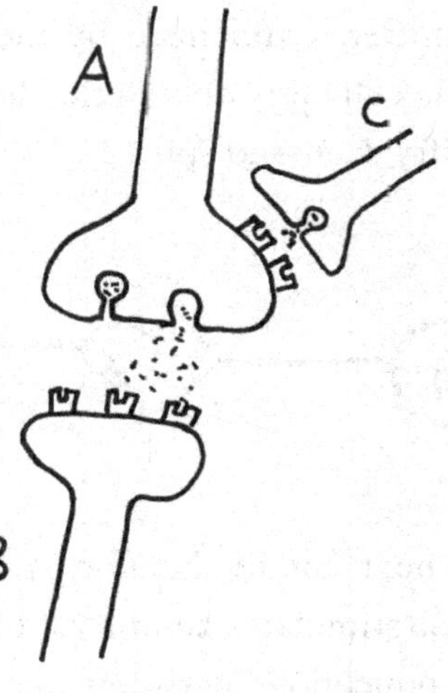

The axon terminal of neuron A forms a synapse with the dendritic spine of postsynaptic neuron B and releases an excitatory neurotransmitter. The usual. Meanwhile, neuron C sends an axon terminal projection on to neuron A. But not to a normal place, a dendritic spine. Instead, its axon terminal synapses onto the axon terminal of neuron A.

What's up with this? Neuron C releases the inhibitory neurotransmitter GABA, which floats across that "axo-axonic" synapse and binds to receptors on that side of neuron A's axon terminal. And its inhibitory effect (i.e., making that –70 mV resting potential even more negative) snuffs out any action potential hurtling down that branch of the axon, keeps it from getting to the very end and releasing neurotransmitter; thus, rather than directly influencing neuron B, neuron C is altering the ability of neuron A to influence B. In the jargon of the field, neuron C is playing a "neuromodulatory" role in this circuit.

Sharpening a Signal over Time and Space

Now for a new type of circuitry. To accommodate this, I'm using a simpler way of representing neurons. As diagrammed, neuron A sends all of its ten thousand to fifty thousand axonal projections to neuron B and releases an excitatory neurotransmitter, symbolized by the plus sign. The circle in neuron B represents the cell body plus all the dendritic branches that contain ten thousand to fifty thousand spines:

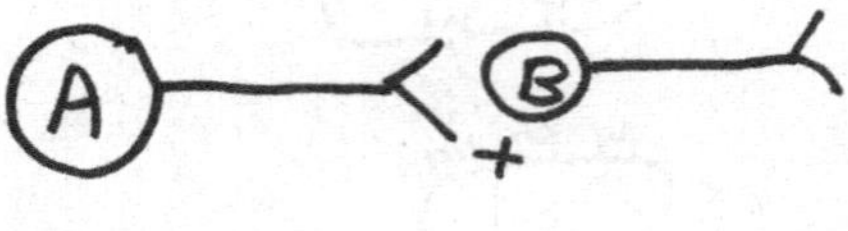

Now consider the next circuit. Neuron A stimulates neuron B, the usual. In addition, it also stimulates neuron C. This is routine, with neuron A splitting its axonal projections between the two target cells, exciting both. And what does neuron C do? It sends an inhibitory projection back on to neuron A, forming a negative feedback loop. Back to the brain loving contrasts, energetically screaming its head off when it has something to say, and energetically being silent otherwise. This is a more macro level of the same. Neuron A fires off a series of action potentials. What better way to energetically communicate when it's all over than for it to become majorly silent, thanks to the inhibitory feedback loop? It's a means of sharpening a signal over time.* And note that neuron A can "determine" how powerful that negative feedback signal will be by how many of the ten thousand axon terminals it shunts toward neuron C instead of B.

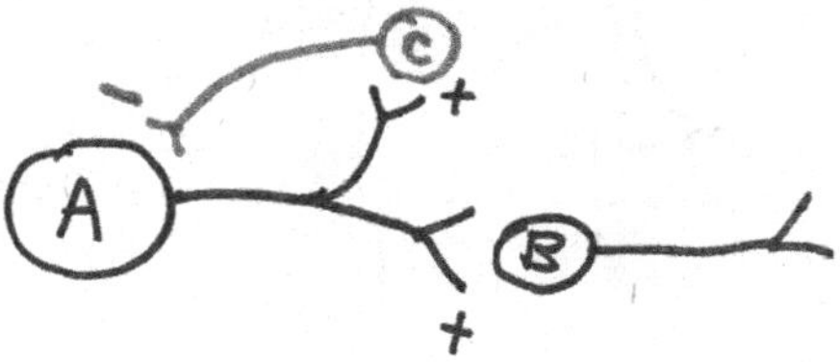

Such "temporal sharpening" of a signal can be accomplished in another way:

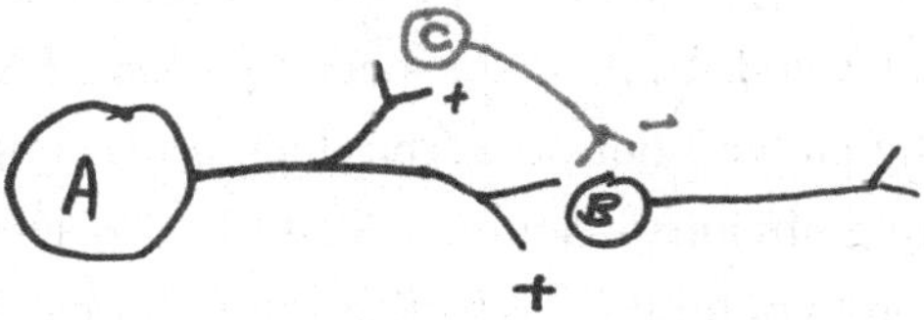

Neuron A stimulates B and C. Neuron C sends an inhibitory signal on to neuron B that will arrive after B starts getting stimulated (since the A/C/B loop is two synaptic steps, versus one for A/B). Result? Sharpening a signal with "feed-forward inhibition."

Now for the other type of sharpening of a signal, of increasing the

*And this makes sense only after introducing an additional fact. Thanks to random, probabilistic hiccups with the ion channels now and then neurons will occasionally have a random, spontaneous action potential from out of nowhere (which is looked at in depth in chapter 10 when considering what quantum indeterminacy has to do with brain function [*psst*—not much]). So neuron A intentionally fires off ten action potentials, followed soon after by two random ones. That might make it hard to tell if neuron A meant to yell ten, eleven, or twelve times. By calibrating the circuit so that the inhibitory feedback signal shows up right after the tenth action potential, the two random ones afterward are prevented, and it is easier to tell what neuron A meant. The signal has been sharpened by damping the noise.

signal-to-noise ratio. Consider this six-neuron circuit, where neuron A stimulates B, C stimulates D, and E stimulates F:

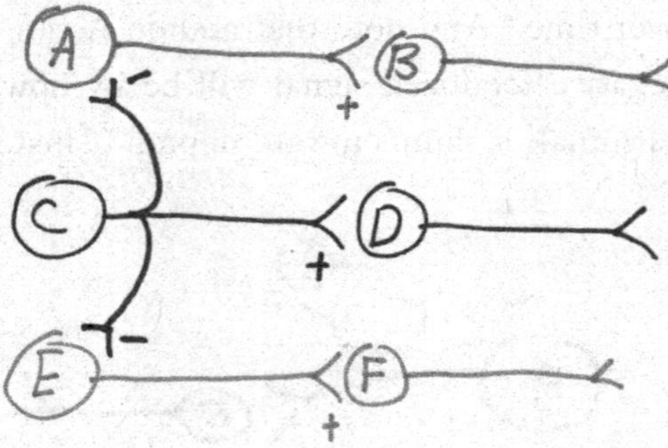

Neuron C sends an excitatory projection on to neuron D. But in addition, neuron C's axon sends collateral inhibitory projections on to neurons A and E.* Thus, if neuron C is stimulated, it both stimulates neuron D *and* silences neurons A and E. With such "lateral inhibition," C screams its head off while A and E become especially silent. It's a means of sharpening a spatial signal (and note that the diagram is simplified, in that I've omitted something obvious—neurons A and E also send inhibitory collateral projections on to neuron C, as well as the neurons on the other sides of them in this imaginary two-dimensional network).

Lateral inhibition like this is ubiquitous in sensory systems. Shine a tiny dot of light onto an eye. Wait, was that photoreceptor neuron A, C, or E that just got stimulated? Thanks to lateral inhibition, it is clearer that it was C. Ditto in tactile systems, allowing you to tell that it was this smidgen of skin that was just touched, not a little this way or that. Or in the ears, telling you that the tone you are hearing is an A, not an A-sharp or A-flat.†

*Thanks to the wisdom of Dale, we know that the same neurotransmitter(s) is coming out of every axon terminal of neuron C. In other words, the same neurotransmitter can be excitatory at some synapses and inhibitory at others. This is determined by what type of ion channel the receptor is coupled to in the dendritic spine.

†Similar circuitry is also seen in the olfactory system, which has always puzzled me. What's just lateral to the smell of an orange? The smell of a tangerine?

Thus, what we've seen is another example of contrast enhancement in the nervous system. What is the significance of the fact that the silent state of a neuron is negatively charged, rather than a neutral zero millivolts? A way of sharpening a signal within a neuron. Feedback, feed-forward, and lateral inhibition with these sorts of collateral projections? A way of sharpening a signal over space and time within a circuit.

Two Different Types of Pain

This next circuit encompasses some of the elements just introduced and explains why there are, broadly, two different types of pain. I love this circuit because it is just so elegant:

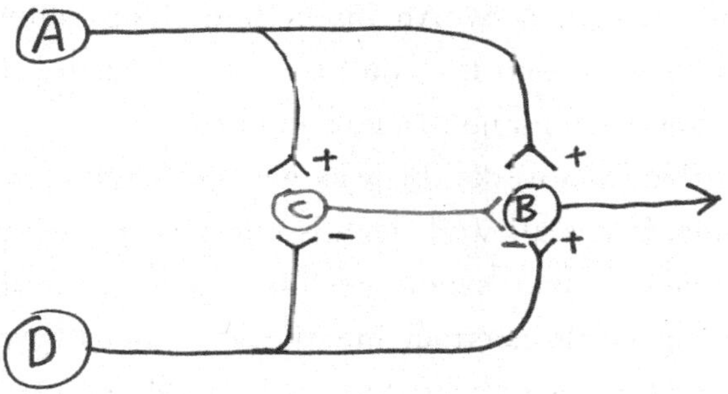

Neuron A's dendrites sit just below the surface of the skin, and the neuron has an action potential in response to a painful stimulus. Neuron A then stimulates neuron B, which projects up the spinal cord, letting you know that something painful just happened. But neuron A also stimulates neuron C, which inhibits B. This is one of our feed-forward inhibitory circuits. Result? Neuron B fires briefly and then is silenced, and you perceive this as a sharp pain—you've been poked with a needle.

Meanwhile, there's neuron D, whose dendrites are in the same general area of the skin and respond to a different type of painful stimulus. As before, neuron D excites neuron B, message is sent up to the brain. But it also sends projections to neuron C, where it *inhibits* it. Result? When neuron D is activated by a pain signal, it inhibits the ability of neuron C to inhibit neuron B. And you perceive it as a throbbing, continuous pain, like a burn or abrasion. Importantly, this is reinforced further by the fact that action potentials travel down the axon of neuron D much slower than in neuron A (having to do with that myelin that I mentioned earlier—details aren't important). So the pain in neuron A's world is not only transient but immediate. Pain in the neuron D branch is not only long-lasting but has a slower onset.

The two types of fibers can interact, and we often intentionally force them to. Suppose that you have some sort of continuous, throbbing pain—say, an insect bite. How can you stop the throbbing? Briefly stimulate the fast fiber. This adds to the pain for an instant, but by stimulating neuron C, you shut the system down for a while. And that is precisely what we often do in circumstances. An insect bite throbs unbearably, and we scratch hard right around it to dull the pain. And the slow-chronic-pain pathway is shut down for up to a few minutes.

The fact that pain works this way has important clinical implications. For one thing, it has allowed scientists to design treatments for people with severe chronic pain syndromes (for example, certain types of back injury). Implant a little electrode into the fast-pain pathway and attach it to a stimulator on the person's hip; too much throbbing pain, buzz the stimulator, and after a brief, sharp pain, the chronic throbbing is turned off for a while; works wonders in many cases.

Thus we have a circuit that encompasses a temporal sharpening mechanism, introduces the double negative of inhibiting inhibitors, and is just all-around cool. And one of the biggest reasons why I love it is that it was first proposed in 1965 by these great neurobiologists Ronald Melzack and Patrick Wall. It was merely proposed as a theoretical model ("No one has ever seen this sort of wiring, but we propose that it's got to look something like this, given how pain works"). And subsequent work showed that's exactly how this part of the nervous system is wired.

Circuitry built on these sorts of elements is extremely important as well in chapter 12 in explaining how we generalize, form categories—

where you look at a picture and say, "I can't tell you who the artist is, but this is by *one of those* Impressionist painters," or where you're thinking of "one of those" presidents between Lincoln and Teddy Roosevelt, or "one of those" dogs that herd sheep.

ONE MORE ROUND OF SCALING UP

A neuron, two neurons, a neuronal circuit. We're ready now, as a last step, to scale up to the level of thousands, hundreds of thousands, of neurons at once. Look up an image of a liver sliced through in cross section and viewed through a microscope. It's just a homogeneous field of cells, an undifferentiated carpet; if you've seen one part, you've seen it all. Boring.

In contrast, the brain is anything but that, showing a huge amount of internal organization.

In other words, the cell bodies of neurons that have related functions are clumped together in particular regions of the brain, and the axons that they send to other parts of the brain are organized into these projection cables. What all this means, crucially, is that *different parts of the brain do different things.* All the regions of the brain have names (usually multisyllabic and derived from Greek or Latin), as do the subregions and the sub-subregions. Moreover, each talks to a consistent collection of other regions (i.e., sends axons to them) and is talked to by a consistent collection (i.e., receives axonal projections from them). Which part of the brain is talking to which other part tells you a lot about function. For example, neurons that receive information that your body temperature has risen send projections to neurons that regulate sweating and activate them at such times. And just to show how complicated this all gets, if you're around someone who is, well, sufficiently hot that your body *feels* warmer, those *same* neurons will activate projections that they have to neurons that cause your gonads to get all giggly and tongue-tied.

You can go crazy studying all the details of connections between different brain regions, as I've seen tragically in the case of many a neuroanatomist who relishes all these details. For our purposes, there are some key points:

—Each particular region contains millions of neurons. Some familiar names on this level of analysis: hypothalamus, cerebellum, cortex, hippocampus.

—Some regions have very distinct and compact subregions, and each is referred to as a "nucleus." (This is confusing, as the part of every cell that contains the DNA is also called the nucleus. What can you do?) Some probably totally unfamiliar names, just as examples: the basal nucleus of Meynert, the supraoptic nucleus of the hypothalamus, the charmingly named inferior olive nucleus.

—As described, the cell bodies of the neurons with related functions are clumped together in their particular region or nucleus and send their axonal projections off in the same direction, merging together into a cable (aka a "fiber tract").

—Back to that myelin wrapping around axons that helps action potentials propagate faster. Myelin tends to be white, sufficiently so that the fiber tract cables in the brain look white. Thus, they're generically referred to as "white matter." The clusters where the (unmyelinated) neuronal cell bodies are clumped together are "gray matter."

Enough with the primer. Back to the book.

Notes

1. TURTLES ALL THE WAY DOWN

1. For a review of experimental philosophy, see: J. Knobe et al., "Experimental Philosophy," *Annual Review of Psychology* 63 (2012): 81. Also see: David Bourget and David Chalmers, eds., "The 2020 PhilPapers Survey," 2020, survey2020.philpeople.org/survey/results/all.

   Free will belief in children across cultures: Gopnik and Kushnir's work: T. Kushnir et al., "Developing Intuitions about Free Will between Ages Four and Six," *Cognition* 138 (2015): 79; N. Chernyak, C. Kang, and T. Kushnir, "The Cultural Roots of Free Will Beliefs: How Singaporean and U.S. Children Judge and Explain Possibilities for Action in Interpersonal Contexts," *Developmental Psychology* 55 (2019): 866; N. Chernyak et al., "A Comparison of American and Nepalese Children's Concepts of Freedom of Choice and Social Constrain," *Cognitive Science* 37 (2013): 1343; A. Wente et al., "How Universal Are Free Will Beliefs? Cultural Differences in Chinese and U.S. 4- and 6-Year-Olds," *Child Development* 87 (2016): 666.

   Belief in free will is widespread cross-culturally, but not universal: D. Wisniewski, R. Deutschland, and J.-D. Haynes, "Free Will Beliefs Are Better Predicted by Dualism Than Determinism Beliefs across Different Cultures," *PLoS One* 14 (2019): e0221617; R. Berniunasa et al., "The Weirdness of Belief in Free Will," *Consciousness and Cognition* 87 (2021): 103054; H. Sarkissian et al., "Is Belief in Free Will a Cultural Universal?," *Mind and Language* 25 (2021): 346.

   Driving study: E. Awad et al., "Drivers Are Blamed More Than Their Automated Cars When Both Make Mistakes," *Nature Human Behaviour*, 4 (2020): 134.

2. L. Egan, P. Bloom, and L. Santos, "Choice-Induced Preferences in the Absence of Choice: Evidence from a Blind Two Choice Paradigm with Young Children and Capuchin Monkeys," *Journal of Experimental and Social Psychology* 46 (2010): 204.

3. Footnote (p. 6): For overviews of their ideas, see: G. Strawson, "The Impossibility of Moral Responsibility," *Philosophical Studies* 75 (1994): 5; D. Pereboom, *Living without Free Will* (Cambridge University Press, 2001); G. Caruso, *Rejecting Retributivism: Free Will, Punishment, and Criminal Justice* (Cambridge University Press, 2021); N. Levy, *Hard Luck: How Luck Undermines Free Will and Moral Responsibility* (Oxford University Press, 2011); and S. Harris, *Free Will* (Simon & Schuster, 2012).

   For a somewhat different take, but in a similar spirit, see B. Waller, *Against Moral Responsibility* (MIT Press, 2011).

Similar broad rejection of free will comes in the writings of scientists such as evolutionary biologist Jerry Coyne of the University of Chicago, psychologist/neuroscientists Jonathan Cohen of Princeton University, Josh Greene of Harvard University, and Paul Glimcher of NYU, and molecular biology god the late Francis Crick.

A small number of legal scholars, such as Pete Alces of William & Mary Law School, break stride with the basic assumptions of their field in also rejecting the existence of free will.

4. M. Vargas, "Reconsidering Scientific Threats to Free Will," in *Moral Psychology*, vol. 4, *Free Will and Moral Responsibility*, ed. W. Sinnott-Armstrong (MIT Press, 2014).
5. R. Baumeister, "Constructing a Scientific Theory of Free Will," in *Moral Psychology*, vol. 4, *Free Will and Moral Responsibility*, ed. W. Sinnott-Armstrong (MIT Press, 2014).
6. A. Mele, "Free Will and Substance Dualism: The Real Scientific Threat to Free Will?," in *Moral Psychology*, vol. 4, *Free Will and Moral Responsibility*, ed. W. Sinnott-Armstrong.
7. R. Nisbett and T. Wilson, "Telling More Than We Can Know: Verbal Reports on Mental Processes," *Psychological Review* 84 (1977): 231.

2. THE FINAL THREE MINUTES OF A MOVIE

1. Footnote: J. McHugh and P. Mackowiak, "Death in the White House: President William Henry Harrison's Atypical Pneumonia," *Clinical Infectious Diseases* 59 (2014): 990. Harrison's doctor treated him with an array of medications, which probably hastened his death. There was opium, which, as is known to opium addicts, causes major constipation, allowing the typhoid bacteria to linger longer, dividing. He was also given carbonated alkali, which probably impaired the ability of stomach acids to kill the bacteria. And just for good measure and for no clear reason, he was also given considerable amounts of mercury, which is neurotoxic. McHugh and Mackowiak convincingly suggest that enteric disease from contaminated water made James Polk seriously ill while president and killed Zachary Taylor while in office.
2. Libet published his initial data in B. Libet et al., "Time of Conscious Intention to Act in Relation to Onset of Cerebral Activity (Readiness-Potential): The Unconscious Initiation of a Freely Voluntary Act," *Brain: A Journal of Neurology* 106 (1983): 623; "Infamous": E. Nahmias, "Intuitions about Free Will, Determinism, and Bypassing," in *The Oxford Handbook of Free Will*, 2nd ed., ed. R. Kane (Oxford University Press, 2011).
3. P. Sanford et al., "Libet's Intention Reports Are Invalid: A Replication of Dominik et al. (2017)," *Consciousness and Cognition* 77 (2020): 102836. This paper was in response to an earlier one: T. Dominik et al., "Libet's Experiment: Questioning the Validity of Measuring the Urge to Move," *Consciousness and Cognition* 49 (2017): 255. Media accounts of the Libet experiment: E. Racine et al., "Media Portrayal of a Landmark Neuroscience Experiment on Free Will," *Science Engineering Ethics* 23 (2007): 989.
4. P. Haggard, "Decision Time for Free Will," *Neuron* 69 (2011): 404; P. Haggard and M. Eimer, "On the Relation between Brain Potentials and the Awareness of Voluntary Movements," *Experimental Brain Research* 126 (1999): 128.
5. J.-D. Haynes, "The Neural Code for Intentions in the Human Brain," in *Bioprediction, Biomarkers, and Bad Behavior*, ed. I. Singh and W. Sinnott-Armstrong (Oxford University Press, 2013); S. Bode and J. Haynes, "Decoding Sequential Stages of Task Preparation in the Human Brain," *Neuroimage* 45 (2009): 606; S. Bode et al., "Tracking the Unconscious Generation of Free Decisions Using Ultra-high Field fMRI," *PLoS One* 6, no. 6 (2011): e21612; C. Soon et al., "Unconscious Determinants of Free Decisions in the Human Brain," *Nature Neuroscience* 11 (2008): 543. The SMA as a gateway (footnote): R. Sjöberg, "Free Will and Neurosurgical Resections of the Supplementary Motor Area: A Critical Review," *Acta Neurochirgica* 163 (2021): 1229.
6. I. Fried, R. Mukamel, and G. Kreiman, "Internally Generated Preactivation of Single Neurons in Human Medial Frontal Cortex Predicts Volition," *Neuron* 69 (2011): 548; I. Fried,

"Neurons as Will and Representation," *Nature Reviews Neuroscience* 23 (2022): 104; H. Gelbard-Sagiv et al., "Internally Generated Reactivation of Single Neurons in Human Hippocampus during Free Recall," *Science* 322 (2008): 96.

7. Bell ringing delayed: W. Banks and E. Isham, "We Infer Rather Than Perceive the Moment We Decided to Act," *Psychological Science* 20 (2009): 17. Effect of happiness on readiness potential: D. Rigoni, J. Demanet, and G. Sartori, "Happiness in Action: The Impact of Positive Affect on the Time of the Conscious Intention to Act," *Frontiers in Psychology* 6 (2015): 1307. Also see H. Lau et al., "Attention to Intention," *Science* 303 (2004): 1208.
8. M. Desmurget et al., "Movement Intention after Parietal Cortex Stimulation in Humans," *Science* 324 (2009): 811.
9. Anarchic hand syndrome: C. Marchetti and S. Della Sala, "Disentangling the Alien and Anarchic Hand," *Cognitive Neuropsychiatry* 3 (1998): 191; S. Della Sala, C. Marchetti, and H. Spinnler, "Right-Sided Anarchic (Alien) Hand: A Longitudinal Study," *Neuropsychologia* 29 (1991): 1113.
10. Transcranial magnetic stimulation: J. Brasil-Neto et al., "Focal Transcranial Magnetic Stimulation and Response Bias in a Forced-Choice Task," *Journal of Neurology, Neurosurgery and Psychiatry* 55 (1992): 964. Magicians: A. Pailhes and G. Kuhn, "Mind Control Tricks: Magicians' Forcing and Free Will," *Trends in Cognitive Sciences* 25 (2021): 338; H. Kelley, "Magic Tricks: The Management of Causal Attributions," in *Perspectives on Attribution Research and Theory: The Bielefeld Symposium*, ed. D. Gorlitz (Ballinger, 1980).

    Footnote: D. Knoch et al., "Diminishing Reciprocal Fairness by Disrupting the Right Prefrontal Cortex," *Science* 314 (2006): 829.
11. D. Wegner, *The Illusion of Conscious Will* (MIT Press, 2002).
12. Footnote (p. 26): P. Tse, "Two Types of Libertarian Free Will Are Realized in the Human Brain," in *Neuroexistentialism*, ed. G. Caruso (Oxford University Press, 2017).
13. Libet's overview: B. Libet, "Unconscious Cerebral Initiative and the Role of Conscious Will in Voluntary Action," *Behavioral and Brain Sciences* 8 (1985): 529. Criticisms of the Libet study: R. Doty, "The Time Course of Conscious Processing: Vetoes by the Uninformed?," *Behavioral and Brain Sciences* 8 (1985): 541; C. Wood, "Pardon, Your Dualism Is Showing," *Behavioral and Brain Sciences* 8 (1985): 557; G. Wasserman, "Neural/Mental Chronometry and Chronotheology," *Behavioral and Brain Sciences* 8 (1985): 556.
14. M. Vargas, "Reconsidering Scientific Threats to Free Will," in *Moral Psychology*, vol. 4, *Free Will and Moral Responsibility*, ed. W. Sinnott-Armstrong (MIT Press, 2014).
15. Both viewpoints in K. Smith, "Taking Aim at Free Will," *Nature* 477 (2011): 23.
16. Driving simulation: O. Perez et al., "Preconscious Prediction of a Driver's Decision Using Intracranial Recordings," *Journal of Cognitive Neuroscience* 27 (2015): 1492. Bungee jumping: Nann et al., "To Jump or Not to Jump—the Bereitschaftspotential Required to Jump into 192-Meter Abyss," *Science Reports* 9 (2019): 2243.
17. U. Maoz et al., "Neural Precursors of Decisions That Matter—an ERP Study of Deliberate and Arbitrary Choice," *eLife* 8 (2019): e39787. For the quote, see Daniel Dennett, "Is Free Will an Illusion? What Can Cognitive Science Tell Us?," Santa Fe Institute, May 14, 2014, YouTube video, 1:21:19, youtube.com/watch?v=wGPIzSe5cAU&t=3890s, around 41:00.
18. This and related studies are discussed in Haynes, "Neural Code for Intentions."
19. O. Bai et al., "Prediction of Human Voluntary Movement Before It Occurs," *Clinical Neurophysiology* 122 (2011): 364.
20. Nearly forty years after Libet: A. Schurger et al., "What Is the Readiness Potential?," *Trends in Cognitive Science* 25 (2010): 558. Urge versus decision: S. Pockett and S. Purdy, "Are Voluntary Movements Initiated Preconsciously? The Relationships between Readiness Potentials, Urges and Decisions," in *Conscious Will and Responsibility: A Tribute to Benjamin Libet*, ed. W. Sinnott-Armstrong and L. Nadel (Oxford University Press, 2020). The Gazzaniga quote comes from M. Gazzaniga, "On Determinism and Human Responsibility," in *Neuroexistentialism*, ed. G. Caruso (Oxford University Press, 2017).

21. The Mele quote is from A. Mele, *Free: Why Science Hasn't Disproved Free Will* (Oxford University Press, 2014), 32. Roskies is quoted in K. Smith, "Taking Aim at Free Will," *Nature* 477 (2011): 2, on page 24.
22. New insights about comas (from the footnote): A. Owen et al., "Detecting Awareness in the Vegetative State," *Science* 313 (2006): 1402; M. Monti et al., "Willful Modulation of Brain Activity in Disorders of Consciousness," *New England Journal of Medicine* 362 (2010): 579.
23. M. Shadlen and A. Roskies, "The Neurobiology of Decision-Making and Responsibility: Reconciling Mechanism and Mindedness," *Frontiers in Neuroscience* 6 (2012), doi.org/10.3389/fnins.2012.00056.
24. A. Schlegel et al., "Hypnotizing Libet: Readiness Potentials with Non-conscious Volition," *Consciousness and Cognition* 33 (2015): 196.
25. Caruso explores this idea in a number of publications, most recently in his excellent G. Caruso, *Rejecting Retributivism: Free Will, Punishment, and Criminal Justice* (Cambridge University Press, 2021). To me, at least, issues of whether preconsciousness and consciousness can exist simultaneously take us into the philosophical underbrush. For true aficionados, this brings up the frothy but influential ideas of philosopher Jaegwon Kim of Brown University. If I understand it: (a) assume that conscious mental states, while the emergent product of underlying physical properties (i.e., thingies like molecules and neurons), are different from them; (b) something like a behavior cannot be caused by both a mental state and its underlying physical bases (which came to be called Kim's "causal exclusion principle"); (c) physical events (like pushing a button or moving your tongue and larynx to tell your generals to start a war) are caused by prior physical events. So mental states don't cause behaviors. I guess this is kind of interesting. Well, maybe not, because in my view, mental states and their underlying physical/neurobiological bases can't be separated—they're just two different conceptual entry points to considering the same processes. More of this in later chapters. Some of his papers: J. Kim, "Concepts of Supervenience," *Philosophy and Phenomenological Research* 45 (1984): 153; J. Kim, "Making Sense of Emergence," *Philosophical Studies* 95 (1995): 3.
26. E. Nahmias, "Intuitions about Free Will, Determinism, and Bypassing," in *The Oxford Handbook of Free Will*, 2nd ed., ed. R. Kane (New York: Oxford University Press, 2011).
27. Do it or not study: E. Filevich, S. Kuhn, and P. Haggard, "There Is No Free Won't: Antecedent Brain Activity Predicts Decisions to Inhibit," *PLoS One* 8, no. 2 (2013): e53053. Brain-computer interface study: M. Schultze-Kraft et al., "The Point of No Return in Vetoing Self-Initiated Movements," *Proceedings of the National Academy of Sciences of the United States of America* 113 (2016): 1080.
28. Footnote: Libet's first report of his findings: Libet et al., "Time of Conscious Intention to Act." His 1985 discussion of it is found in Libet, "Unconscious Cerebral Initiative."
29. Gambling study: D. Campbell-Meiklejohn et al., "Knowing When to Stop: The Brain Mechanisms of Chasing Losses," *Biological Psychiatry* 63 (2008): 293. Alcohol on board: Y. Liu et al., "'Free Won't' after a Beer or Two: Chronic and Acute Effects of Alcohol on Neural and Behavioral Indices of Intentional Inhibition," *BMC Psychology* 8 (2020): 2. Kids versus adults: M. Schel, K. Ridderinkhof, and E. Crone, "Choosing Not to Act: Neural Bases of the Development of Intentional Inhibition," *Developmental Cognitive Neuroscience* 10 (2014): 93.
30. "Freedom arises from": B. Brembs, "Towards a Scientific Concept of Free Will as a Biological Trait: Spontaneous Actions and Decision-Making in Invertebrates," *Proceedings of the Royal Society B: Biological Sciences* 278 (2011): 930; the paper approaches the topic from the very unorthodox (and interesting) angle of examining decision-making in insects. Mele quote: Mele, *Free*, 32.
31. N. Levy, *Hard Luck: How Luck Undermines Free Will and Moral Responsibility* (Oxford University Press, 2011).
32. Footnote: H. Frankfurt, "Alternate Possibilities and Moral Responsibility," *Journal of Philosophy* 66 (1969): 829.

33. H. Frankfurt, "Three Concepts of Free Action," *Aristotelian Society Proceedings, Supplementary Volumes* 49 (1975): 113, quote on p. 122; M. Shadlen and A. Roskies, "The Neurobiology of Decision-making and Responsibility: Reconciling Mechanism and Mindedness," *Frontiers in Neuroscience* 23 April (2012): 1, quote is on p. 10.

    Footnote: Sjöberg, "Free Will and Neurosurgical Resections."
34. D. Dennett, *Freedom Evolves* (Penguin, 2004); the quote comes from p. 276. Dennett also expresses these ideas in a wide variety of his other books, e.g., D. Dennett, *Elbow Room: The Varieties of Free Will Worth Wanting* (MIT Press, 1984); D. Dennett, *Freedom Evolves* (Viking, 2003); his lectures, e.g., Dennett, "Is Free Will an Illusion?"; and his debates, e.g., D. Dennett and G. Caruso, *Just Deserts: Debating Free Will* (Polity, 2021);
35. N. Levy, "Luck and History-Sensitive Compatibilism," *Philosophical Quarterly* 59 (2009): 237, the quote comes from p. 244; D. Dennett, "Review of 'Against Moral Responsibility,'" in *Naturalism*, https://www.naturalism.org/resources/book-reviews/dennett-review-of-against-moral-responsibility.

    As a theme in this chapter, people have been arguing about Libetian issues for forty years, and the references cited barely scratched the surface of really interesting takes on these issues. Others include: G. Gomes, "The Timing of Conscious Experience: A Critical Review and Reinterpretation of Libet's Research," *Consciousness and Cognition* 7 (1998): 559; A. Batthyany, "Mental Causation and Free Will after Libet and Soon: Reclaiming Conscious Agency," in *Irreducibly Conscious: Selected Papers on Consciousness*, ed. A. Batthyany and A. Elitzur (Universitäts-Verlag Winter, 2009); A. Lavazza, "Free Will and Neuroscience: From Explaining Freedom Away to New Ways of Operationalizing and Measuring It," *Frontiers in Human Neuroscience* 10 (2016): 262; C. Frith, S. Blakemore, and D. Wolpert, "Abnormalities in the Awareness and Control of Action," *Philosophical Transactions of the Royal Society B: Biological Sciences* 355 (2000): 1404; A. Guggisberg and A. Mottaz, "Timing and Awareness of Movement Decisions: Does Consciousness Really Come Too Late?," *Frontiers of Human Neuroscience* 7 (2013), doi.org/10.3389/fnhum.2013.00385; T. Bayne, "Neural Decoding and Human Freedom," in *Moral Psychology* vol. 4, *Free Will and Moral Responsibility*, ed. W. Sinnott-Armstrong (MIT Press, 2014).

3. WHERE DOES INTENT COME FROM?

1. Implicit bias and shootings: J. Correll et al., "Across the Thin Blue Line: Police Officers and Racial Bias in the Decision to Shoot," *Journal of Personality and Social Psychology* 92 (2007): 1006; J. Correll et al., "The Police Officer's Dilemma: Using Ethnicity to Disambiguate Potentially Threatening Individuals," *Journal of Personality and Social Psychology* 83 (2002): 1314. For an excellent overview of the entire field, see J. Eberhardt, *Biased: Uncovering the Hidden Prejudice That Shapes What We See, Think, and Do* (Viking, 2019).
2. Implicit effects of disgust: D. Pizarro, Y. Inbar, and C. Helion, "On Disgust and Moral Judgment," *Emotion Review* 3 (2011): 267; T. Adams, P. Stewart, and J. Blanchard, "Disgust and the Politics of Sex: Exposure to a Disgusting Odorant Increases Politically Conservative Views on Sex and Decreases Support for Gay Marriage," *PLoS One* 9 (2014): e95572; Y. Inbar, D. Pizarro, and P. Bloom, "Disgusting Smells Cause Decreased Liking of Gay Men," *Emotion* 12 (2012): 23; J. Terrizzi, N. Shook, and W. Ventis, "Disgust: A Predictor of Social Conservatism and Prejudicial Attitudes Toward Homosexuals," *Personality and Individual Differences* 49 (2010): 587.
3. More disgust: S. Tsao and D. McKay, "Behavioral Avoidance Tests and Disgust in Contamination Fears: Distinctions from Trait Anxiety," *Behavioral Research Therapeutics* 42 (2004): 207; B. Olatunji, B. Puncochar, and R. Cox, "Effects of Experienced Disgust on Morally-Relevant Judgments," *PLoS One* 11 (2016): e0160357.
4. And more disgust: H. Chapman and A. Anderson, "Things Rank and Gross in Nature: A Review and Synthesis of Moral Disgust," *Psychological Bulletin* 139 (2013): 300; P. Rozin et al.,

"The CAD Triad Hypothesis: A Mapping between Three Moral Emotions (Contempt, Anger, Disgust) and Three Moral Codes (Community, Autonomy, Divinity)," *Journal of Personality and Social Psychology* 76 (1999): 574. The insula, when activated by aversive emotional states, talking to the amygdala: D. Gehrlach et al., "Aversive State Processing in the Posterior Insular Cortex," *Nature Neuroscience* 22 (2019): 1424.

5. Implicit effects of sweet tastes: M. Schaefer et al., "Sweet Taste Experience Improves Prosocial Intentions and Attractiveness Ratings," *Psychological Research* 85 (2021): 1724; B. Meier et al., "Sweet Taste Preferences and Experiences Predict Prosocial Inferences, Personalities, and Behaviors," *Psychological Sciences* 102 (2012): 163.
6. Confusing beauty and moral goodness: Q. Cheng et al., "Neural Correlates of Moral Goodness and Moral Beauty Judgments," *Brain Research* 1726 (2020): 146534; T. Tsukiura and R. Cabeza, "Shared Brain Activity for Aesthetic and Moral Judgments: Implications for the Beauty-Is-Good Stereotype," *Social Cognitive and Affective Neuroscience* 6 (2011): 138; X. Cui et al., "Different Influences of Facial Attractiveness on Judgments of Moral Beauty and Moral Goodness," *Science Reports* 9 (2019): 12152; T. Wang et al., "Is Moral Beauty Different from Facial Beauty? Evidence from an fMRI Study," *Social Cognitive and Affective Neuroscience* 10 (2015): 814; Q. Luo et al., "The Neural Correlates of Integrated Aesthetics between Moral and Facial Beauty," *Science Reports* 9 (2019): 1980; C. Ferrari et al., "The Dorsomedial Prefrontal Cortex Mediates the Interaction between Moral and Aesthetic Valuation: A TMS Study on the Beauty-Is-Good Stereotype," *Social Cognitive and Affective Neuroscience* 12 (2017): 707.

   Then there's an irresistible study showing that botanists choose to spend their careers studying prettier flowers (blue ones, taller ones): M. Adamo et al., "Plant Scientists' Research Attention Is Skewed towards Colourful, Conspicuous and Broadly Distributed Flowers," *Nature Plants* 7 (2021): 574. To the best of my knowledge, I did not choose to devote thirty-three summers to studying wild baboons because I thought they were as pretty as a picture.
7. The initial study that introduced the term "Macbeth effect": C. Zhong and K. Lijenquist, "Washing Away Your Sins: Threatened Morality and Physical Cleansing," *Science* 313 (2006): 1454.

   Additional behavior studies of the Macbeth effect: S. W. Lee and N. Schwarz, "Dirty Hands and Dirty Mouths: Embodiment of the Moral-Purity Metaphor Is Specific to the Motor Modality Involved in Moral Transgression," *Psychological Sciences* 21 (2010): 1423; E. Kalanthroff, C. Aslan, and R. Dar, "Washing Away Your Sins Will Set Your Mind Free: Physical Cleansing Modulates the Effect of Threatened Morality on Executive Control," *Cognition and Emotion* 31 (2017): 185; S. Schnall, J. Benton, and S. Harvey, "With a Clean Conscience: Cleanliness Reduces the Severity of Moral Judgments," *Psychological Sciences* 19 (2008): 1219; K. Kaspar, V. Krapp, and P. Konig, "Hand Washing Induces a Clean Slate Effect in Moral Judgments: A Pupillometry and Eye-Tracking Study," *Scientific Reports* 5 (2015): 10471.

   Brain imaging studies of the Macbeth effect: C. Denke et al., "Lying and the Subsequent Desire for Toothpaste: Activity in the Somatosensory Cortex Predicts Embodiment of the Moral-Purity Metaphor," *Cerebral Cortex* 26 (2016): 477; M. Schaefer et al., "Dirty Deeds and Dirty Bodies: Embodiment of the Macbeth Effect Is Mapped Topographically onto the Somatosensory Cortex," *Scientific Reports* 6 (2015): 18051.

   For a study suggesting that this linkage may not be universal: E. Gámez, J. M. Díaz, and H. Marrero, "The Uncertain Universality of the Macbeth Effect with a Spanish Sample," *Spanish Journal of Psychology* 14 (2011): 156.

   Finally, one study showing that among university students, social science majors are more vulnerable to the Macbeth effect than are engineering students: M. Schaefer, "Morality and Soap in Engineers and Social Scientists: The Macbeth Effect Interacts with Professions," *Psychological Research* 83 (2019): 1304.
8. Ginger and moral disgust: J. Tracy, C. Steckler, and G. Heltzel, "The Physiological Basis of Psychological Disgust and Moral Judgments," *Journal of Personality and Social Psychology:*

Attitudes and Social Cognition 116 (2019): 15. An interesting paper showing that disgust influences moral judgments less concerning distant events, and that this is probably mediated by a psychological framing where it is someone else, rather than you, who has to directly interact with the disgusting stimulus: M. van Dijke et al., "So Gross and Yet So Far Away: Psychological Distance Moderates the Effect of Disgust on Moral Judgment," *Social Psychological and Personality Science* 9 (2018): 689.

9. The original study of judges: S. Danziger, J. Levav, and L. Avnaim-Pesso, "Extraneous Factors in Judicial Decisions," *Proceedings of the National Academy of Science of the United States of America* 108 (2011): 6889. This study was challenged by some other researchers, suggesting that the finding is an artifact of poor study design; in my opinion, the original authors effectively rebutted these charges. See notes 28 and 29 in chapter 4 for details about this.

   More on the topic: L. Aaroe and M. Petersen, "Hunger Games: Fluctuations in Blood Glucose Levels Influence Support for Social Welfare," *Psychological Sciences* 24 (2013): 2550.

   A connection between hunger for food and for money: B. Briers et al., "Hungry for Money: The Desire for Caloric Resources Increases the Desire for Financial Resources and Vice Versa," *Psychological Sciences* 17 (2006): 939.

   Some circumstance where the connection is demonstrable only in some domains: J. Hausser et al., "Acute Hunger Does Not Always Undermine Prosociality," *Nature Communications* 10 (2019): 4733; S. Fraser and D. Nettle, "Hunger Affects Social Decisions in a Multi-round Public Goods Game but Not a Single-Shot Ultimatum Game," *Adaptive Human Behavior* 6 (2020): 334; I. Harel and T. Kogut, "Visceral Needs and Donation Decisions: Do People Identify with Suffering or with Relief?," *Journal of Experimental and Social Psychology* 56 (2015): 24.

   As is so often the case, the suggestion that this phenomenon is influenced by culture: E. Rantapuska et al., "Does Short-Term Hunger Increase Trust and Trustworthiness in a High Trust Society?," *Frontiers of Psychology* 8 (2017): 1944.

10. For more details about this general topic, see chapter 3 in R. Sapolsky, *Behave: The Biology of Humans at Our Best and Worst* (Penguin Press, 2017).

11. The classic study demonstrating that testosterone does not generate aggression de novo but, instead, amplifies preexisting social learning about aggression: A. Dixson and J. Herbert, "Testosterone, Aggressive Behavior and Dominance Rank in Captive Adult Male Talapoin Monkeys (*Miopithecus talapoin*)," *Physiology and Behavior* 18 (1977): 539.

   How some of the behavioral effects of testosterone arise from their effects in the brain: K. Kendrick and R. Drewett, "Testosterone Reduces Refractory Period of Stria Terminalis Neurons in the Rat Brain," *Science* 204 (1979): 877; K. Kendrick, "Inputs to Testosterone-Sensitive Stria Terminalis Neurones in the Rat Brain and the Effects of Castration," *Journal of Physiology* 323 (1982): 437; K. Kendrick, "The Effect of Castration on Stria Terminalis Neurone Absolute Refractory Periods Using Different Antidromic Stimulation Loci," *Brain Research* 248 (1982): 174; K. Kendrick, "Electrophysiological Effects of Testosterone on the Medial Preoptic-Anterior Hypothalamus of the Rat," *Journal of Endocrinology* 96 (1983): 35; E. Hermans, N. Ramsey, and J. van Honk, "Exogenous Testosterone Enhances Responsiveness to Social Threat in the Neural Circuitry of Social Aggression in Humans," *Biological Psychiatry* 63 (2008): 263.

   In 1990, the ethologist John Wingfield of the University of California at Davis, along with colleagues, published an immensely influential paper about the nature of testosterone's effects on aggression. Their "Challenge Hypothesis" posits that not only does testosterone not cause aggression, it doesn't uniformly just amplify preexisting social tendencies toward aggression either. Instead, at times when an organism is challenged for social status, testosterone amplifies whatever behaviors are needed to maintain status. Well, that doesn't seem like much of an elaboration—if you're a male baboon whose rank is being challenged, aggression is what you need to maintain status. But when it comes to humans, there are greater subtleties, because status can be maintained in different ways. For example, in an economic game where status is accrued through generous economic offers, testosterone increases such generosity. See: J. Wingfield et al., "The 'Challenge Hypothesis': Theoretical Implications for Patterns of Testosterone Secretion, Mating Systems, and Breeding Strategies," *American Naturalist* 136 (1990): 829. The hypothesis helps explain a wide range of testosterone-dependent behaviors: J. Wingfield "The Challenge Hypothesis: Where It Began and Relevance to Humans," *Hormones and Behavior* 92 (2017): 9. Also see: J. Archer, "Testosterone and Human Aggression: An Evaluation of the Challenge Hypothesis," *Neuroscience and Biobehavioral Reviews* 30 (2006): 319.

12. Papers concerning the behavioral and neurobiological underpinnings of testosterone make people reactive to perceived threat: E. Hermans, N. Ramsey, and J. van Honk, "Exogenous Testosterone Enhances Responsiveness to Social Threat in the Neural Circuitry of Social Aggression in Humans," *Biological Psychiatry* 63 (2008): 263; J. van Honk et al., "A Single Administration of Testosterone Induces Cardiac Accelerative Responses to Angry Faces in Healthy Young Women," *Behavioral Neuroscience* 115 (2001): 238; N. Wright et al., "Testosterone Disrupts Human Collaboration by Increasing Egocentric Choices," *Proceedings of the Royal Society B: Biological Sciences* 279 (2012): 2275; P. Mehta and J. Beer, "Neural Mechanisms of the Testosterone-Aggression Relation: The Role of Orbitofrontal Cortex," *Journal of Cognitive Neuroscience* 22 (2010): 2357; G. van Wingen et al., "Testosterone Reduces Amygdala-Orbitofrontal Cortex Coupling," *Psychoneuroendocrinology* 35 (2010): 105; P. Bos et al., "The Neural Mechanisms by Which Testosterone Acts on Interpersonal Trust," *Neuroimage* 2 (2012): 730.

13. Some studies exploring sources of individual differences in the workings of the testicular system: C. Laube, R. Lorenz, and L. van den Bos, "Pubertal Testosterone Correlates with Adolescent Impatience and Dorsal Striatal Activity," *Development and Cognitive Neuroscience* 42 (2020): 100749; B. Mohr et al., "Normal, Bound and Nonbound Testosterone Levels in Normally Ageing Men: Results from the Massachusetts Male Ageing Study," *Clinical Endocrinology* 62 (2005): 64; W. Bremner, M. Vitiello, and P. Prinz, "Loss of Circadian Rhythmicity in Blood Testosterone Levels with Aging in Normal Men," *Journal of Clinical Endocrinology and Metabolism* 56 (1983): 1278; S. Beyenburg et al., "Androgen Receptor mRNA Expression in the Human Hippocampus," *Neuroscience Letters* 294 (2000): 25.

14. Some good general reviews, see: R. Feldman, "Oxytocin and Social Affiliation in Humans," *Hormones and Behavior* 61 (2012): 380; Z. Donaldson and L. Young, "Oxytocin, Vasopressin, and the Neurogenetics of Sociality," *Science* 322 (2008): 900; P. S. Churchland and P. Winkielman, "Modulating Social Behavior with Oxytocin: How Does It Work? What Does It Mean?," *Hormones and Behavior* 61 (2012): 392.

    Papers related to differences in the oxytocin system comparing monogamous and polygamous rodents: L. Young et al., "Increased Affiliative Response to Vasopressin in Mice Expressing the V_{1a} Receptor from a Monogamous Vole," *Nature* 400 (1999): 766; M. Lim et al., "Enhanced Partner Preference in a Promiscuous Species by Manipulating the Expression of a Single Gene," *Nature* 429 (2004): 754.

    Papers related to differences in the oxytocin system comparing monogamous and polygamous nonhuman primates: A. Smith et al., "Manipulation of the Oxytocin System Alters Social Behavior and Attraction in Pair-Bonding Primates, *Callithrix penicillata*," *Hormones and Behavior* 57 (2010): 255; M. Jarcho et al., "Intranasal VP Affects Pair Bonding and Peripheral Gene Expression in Male *Callicebus cupreus*," *Genes, Brain and Behavior* 10 (2011): 375; C. Snowdon et al., "Variation in Oxytocin Is Related to Variation in Affiliative Behavior in Monogamous, Pairbonded Tamarins," *Hormones and Behavior* 58 (2010): 614.

    The neurobiology underlying these oxytocin effects: The hypothalamic pathway that differs by sex: N. Scott et al., "A Sexually Dimorphic Hypothalamic Circuit Controls Maternal Care and Oxytocin Secretion," *Nature* 525 (2016): 519. For an example of oxytocin working in the insular cortex to modify social interactions, see M. Carter-Rogers et al., "Insular Cortex Mediates Approach and Avoidance Response to Social Affective Stimuli," *Nature*

Neuroscience 21 (2018): 404. Likewise for oxytocin working in the amygdala: Y. Liu et al., "Oxytocin Modulates Social Value Representations in the Amygdala," *Nature Neuroscience* 22 (2019): 633; J. Wahis et al., "Astrocytes Mediate the Effect of Oxytocin in the Central Amygdala on Neuronal Activity and Affective States in Rodents," *Nature Neuroscience* 24 (2021): 529.

Oxytocin and parenting, including paternal behavior: O. Bosch and I. Neumann, "Both Oxytocin and Vasopressin Are Mediators of Maternal Care and Aggression in Rodents: From Central Release to Sites of Action," *Hormones and Behavior* 61 (2012): 293; Y. Kozorovitskiy et al., "Fatherhood Affects Dendritic Spines and Vasopressin V1a Receptors in the Primate Prefrontal Cortex," *Nature Neuroscience* 9 (2006): 1094; Z. Wang, C. Ferris, and G. De Vries "Role of Septal Vasopressin Innervation in Paternal Behavior in Prairie Voles," *Proceedings of the National Academy of Sciences of the United States of America* 91 (1994): 400.

Genetic and epigenetic differences mediating individual differences in oxytocin sensitivity: Marsh et al., "The Influence of Oxytocin Administration on Responses to Infants and Potential Moderation by OXTR Genotype," *Psychopharmacology* (Berlin) 224 (2012): 469; M. J. Bakermans-Kranenburg and M. H. van Ijzendoorn, "Oxytocin Receptor (OXTR) and Serotonin Transporter (5-HTT) Genes Associated with Observed Parenting," *Social Cognitive and Affective Neuroscience* 3 (2008): 128; E. Hammock and L. Young, "Microsatellite Instability Generates Diversity in Brain and Sociobehavioral Traits," *Science* 308 (2005): 1630.

Annals of totally irresistible findings: M. Nagasawa et al., "Oxytocin-Gaze Positive Loop and the Coevolution of Human-Dog Bonds," *Science* 348 (2015): 333. When a dog and its human gaze into each other's eyes, both secrete oxytocin; administer oxytocin to one of the two and they gaze longer—eliciting more oxytocin secretion in the other. In other words, a hormonal system central to parental behavior and pair-bonding that is at least a hundred million years old has, in the last thirty thousand years, been co-opted for human/wolf interactions.

15. Oxytocin effects on fear and anxiety: M. Yoshida et al., "Evidence That Oxytocin Exerts Anxiolytic Effects via Oxytocin Receptor Expressed in Serotonergic Neurons in Mice," *Journal of Neuroscience* 29 (2009): 2259. Oxytocin working in the amygdala: D. Viviani et al., "Oxytocin Selectively Gates Fear Responses through Distinct Outputs from the Central Nucleus," *Science* 333 (2011): 104; H. Knobloch et al., "Evoked Axonal Oxytocin Release in the Central Amygdala Attenuates Fear Response," *Neuron* 73 (2012): 553; "Oxytocin Attenuates Amygdala Responses to Emotional Faces Regardless of Valence," *Biological Psychiatry* 62 (2007): 1187; P. Kirsch et al., "Oxytocin Modulates Neural Circuitry for Social Cognition and Fear in Humans," *Journal of Neuroscience* 25 (2005): 11489; I. Labuschagne et al., "Oxytocin Attenuates Amygdala Reactivity to Fear in Generalized Social Anxiety Disorder," *Neuropsychopharmacology* 35 (2010): 2403.

Oxytocin blunting the stress-response: M. Heinrichs et al., "Social Support and Oxytocin Interact to Suppress Cortisol and Subjective Responses to Psychosocial Stress," *Biological Psychiatry* 54 (2003): 1389.

Oxytocin effects on empathy, trust, and cooperation: S. Rodrigues et al., "Oxytocin Receptor Genetic Variation Relates to Empathy and Stress Reactivity in Humans," *Proceedings of the National Academy of Sciences of the United States of America* 106 (2009): 21437; M. Kosfeld et al., "Oxytocin Increases Trust in Humans," *Nature* 435 (2005): 673; A. Damasio, "Brain Trust," *Nature* 435 (2005): 571; S. Israel et al., "The Oxytocin Receptor (OXTR) Contributes to Prosocial Fund Allocations in the Dictator Game and the Social Value Orientations Task," *Public Library of Science One* 4 (2009): e5535; P. Zak, R. Kurzban, and W. Matzner, "Oxytocin Is Associated with Human Trustworthiness," *Hormones and Behavior* 48 (2005): 522; T. Baumgartner et al., "Oxytocin Shapes the Neural Circuitry of Trust and Trust Adaptation in Humans," *Neuron* 58 (2008): 639; J. Filling et al., "Effects of Intranasal Oxytocin and Vasopressin on Cooperative Behavior and Associated Brain Activity in Men,"

Psychoneuroendocrinology 37 (2012): 447; A. Theodoridou et al., "Oxytocin and Social Perception: Oxytocin Increases Perceived Facial Trustworthiness and Attractiveness," *Hormones and Behavior* 56 (2009): 128. A failure of replication: C. Apicella et al., "No Association between Oxytocin Receptor (OXTR) Gene Polymorphisms and Experimentally Elicited Social Preferences," *Public Library of Science One* 5 (2010): e11153.

Oxytocin effects on aggression: M. Dhakar et al., "Heightened Aggressive Behavior in Mice with Lifelong versus Postweaning Knockout of the Oxytocin Receptor," *Hormones and Behavior* 62 (2012): 86; J. Winslow et al., "Infant Vocalization, Adult Aggression, and Fear Behavior of an Oxytocin Null Mutant Mouse," *Hormones and Behavior* 37 (2005): 145.

16. C. De Dreu, "Oxytocin Modulates Cooperation within and Competition between Groups: An Integrative Review and Research Agenda," *Hormones and Behavior* 61 (2012): 419; C. De Dreu et al., "The Neuropeptide Oxytocin Regulates Parochial Altruism in Intergroup Conflict among Humans," *Science* 328 (2011): 1408; C. De Dreu et al., "Oxytocin Promotes Human Ethnocentrism," *Proceedings of the National Academy of Sciences of the United States of America* 108 (2011): 1262.
17. K. Parker et al., "Preliminary Evidence That Plasma Oxytocin Levels Are Elevated in Major Depression," *Psychiatry Research* 178 (2010): 359; S. Freeman et al., "Effect of Age and Autism Spectrum Disorder on Oxytocin Receptor Density in the Human Basal Forebrain and Midbrain," *Translational Psychiatry* 8 (2018): 257.
18. R. Sapolsky, "Stress and the Brain: Individual Variability and the Inverted-U," *Nature Neuroscience* 25 (2015): 1344.
19. Effects of stress and stress hormones on the amygdala: J. Rosenkranz, E. Venheim, and M. Padival, "Chronic Stress Causes Amygdala Hyperexcitability in Rodents," *Biological Psychiatry* 67 (2010): 1128; S. Duvarci and D. Pare, "Glucocorticoids Enhance the Excitability of Principal Basolateral Amygdala Neurons," *Journal of Neuroscience* 27 (2007) 4482; A. Kavushansky and G. Richter-Levin, "Effects of Stress and Corticosterone on Activity and Plasticity in the Amygdala," *Journal of Neuroscience Research* 84 (2006): 1580; P. Rodríguez Manzanares et al., "Previous Stress Facilitates Fear Memory, Attenuates GABAergic Inhibition, and Increases Synaptic Plasticity in the Rat Basolateral Amygdala," *Journal of Neuroscience* 25 (2005): 8725.

Effects of stress and stress hormones on interactions between the amygdala and hippocampus: A. Kavushansky et al., "Activity and Plasticity in the CA1, the Dentate Gyrus, and the Amygdala Following Controllable Versus Uncontrollable Water Stress," *Hippocampus* 16 (2006): 35; H. Lakshminarasimhan and S. Chattarji, "Stress Leads to Contrasting Effects on the Levels of Brain Derived Neurotrophic Factor in the Hippocampus and Amygdala," *Public Library of Science One* 7 (2012): e30481; S. Ghosh, T. Laxmi, and S. Chattarji, "Functional Connectivity from the Amygdala to the Hippocampus Grows Stronger after Stress," *Journal of Neuroscience* 33 (2013): 7234.

20. Behavioral effects of stress and stress hormones: S. Preston et al., "Effects of Anticipatory Stress on Decision-Making in a Gambling Task," *Behavioral Neuroscience* 121 (2007): 257; P. Putman et al., "Exogenous Cortisol Acutely Influences Motivated Decision Making in Healthy Young Men," *Psychopharmacology* 208 (2010): 257; P. Putman, E. Hermans, and J. van Honk, "Cortisol Administration Acutely Reduces Threat-Selective Spatial Attention in Healthy Young Men," *Physiology and Behavior* 99 (2010): 294; K. Starcke et al., "Anticipatory Stress Influences Decision Making under Explicit Risk Conditions," *Behavioral Neuroscience* 122 (2008): 1352.

Sex differences and stress/stress hormone effects: R. van den Bos, M. Harteveld, and H. Stoop, "Stress and Decision-Making in Humans: Performance Is Related to Cortisol Reactivity, Albeit Differently in Men and Women," *Psychoneuroendocrinology* 34 (2009): 1449; N. Lighthall, M. Mather, and M. Gorlick, "Acute Stress Increases Sex Differences in Risk Seeking in the Balloon Analogue Risk Task," *Public Library of Science One* 4 (2009): e6002; N. Lighthall et al., "Gender Differences in Reward-Related Decision Processing under Stress," *Social Cognitive and Affective Neuroscience* 7 (2012): 476.

Stress and stress hormone effects on aggression: D. Hayden-Hixson and C. Ferris, "Steroid-Specific Regulation of Agonistic Responding in the Anterior Hypothalamus of Male Hamsters," *Physiology and Behavior* 50 (1991): 793; A. Poole and P. Brain, "Effects of Adrenalectomy and Treatments with ACTH and Glucocorticoids on Isolation-Induced Aggressive Behavior in Male Albino Mice," *Progress in Brain Research* 41 (1974): 465; E. Mikics, B. Barsy, and J. Haller, "The Effect of Glucocorticoids on Aggressiveness in Established Colonies of Rats," *Psychoneuroendocrinology* 32 (2007): 160; R. Böhnke et al., "Exogenous Cortisol Enhances Aggressive Behavior in Females, but Not in Males," *Psychoneuroendocrinology* 35 (2010): 1034; K. Bertsch et al., "Exogenous Cortisol Facilitates Responses to Social Threat under High Provocation," *Hormones and Behavior* 59 (2011): 428.

Stress and stress hormone effects on moral decision-making: K. Starcke, C. Polzer, and O. Wolf, "Does Everyday Stress Alter Moral Decision-Making?," *Psychoneuroendocrinology* 36 (2011): 210; F. Youssef, K. Dookeeram, and V. Basdeo, "Stress Alters Personal Moral Decision Making," *Psychoneuroendocrinology* 37 (2012): 491.

21. For more details about this general topic, see chapter 4 in Sapolsky, *Behave*.
22. Footnote: For a great history of the (re)discovery of adult neurogenesis, see M. Specter, "How the Songs of Canaries Upset a Fundamental Principle of Science," *New Yorker*, July 23, 2001.

The behavioral consequences of adult neurogenesis: G. Kempermann, "What Is Adult Hippocampal Neurogenesis Good For?," *Frontiers of Neuroscience* 16 (2022), doi.org/10.3389/fnins.2022.852680; Y. Li, Y. Luo, and Z. Chen, "Hypothalamic Modulation of Adult Hippocampal Neurogenesis in Mice Confers Activity-Dependent Regulation of Memory and Anxiety-Like Behavior," *Nature Neuroscience* 25 (2022): 630; D. Seib et al., "Hippocampal Neurogenesis Promotes Preference for Future Rewards," *Molecular Psychiatry* 26 (2021): 6317; C. Anacker et al., "Hippocampal Neurogenesis Confers Stress Resilience by Inhibiting the Ventral Dentate Gyrus," *Nature* 559 (2018): 98.

Amid all this fascination, experience is also changing the consequential birth of those less flashy glial cells in the adult brain: A. Delgado et al., "Release of Stem Cells from Quiescence Reveals Gliogenic Domains in the Adult Mouse Brain," *Science* 372 (2021): 1205.

The debate as to how much adult neurogenesis actually occurs in humans: S. Sorrells et al., "Human Hippocampal Neurogenesis Drops Sharply in Children to Undetectable Levels in Adults," *Nature* 555 (2018): 377. For a retort: M. Baldrini et al., "Human Hippocampal Neurogenesis Persists throughout Aging," *Cell Stem Cell* 22 (2018): 589. For an opinion piece with a similar viewpoint: G. Kempermann, F. Gage, and L. Aigner, "Human Neurogenesis: Evidence and Remaining Questions," *Cell Stem Cell* 23 (2018): 25. Then a vote for the revolutionaries: S. Ranade, "Single-Nucleus Sequencing Finds No Adult Hippocampal Neurogenesis in Humans," *Nature Neuroscience* 25 (2022): 2.

23. R. Hamilton et al., "Alexia for Braille Following Filateral Occipital Stroke in an Early Blind Woman," *Neuroreport* 11 (2000): 237; E. Striem-Amit et al., "Reading with Sounds: Sensory Substitution Selectively Activates the Visual Word Form Area in the Blind," *Neuron* 76 (2012): 640; A. Pascual-Leone, "Reorganization of Cortical Motor Outputs in the Acquisition of New Motor Skills," in *Recent Advances in Clinical Neurophysiology*, ed. J. Kinura and H. Shibasaki (Elsevier Science, 1996), pp. 304–8.
24. S. Rodrigues, J. LeDoux, and R. Sapolsky, "The Influence of Stress Hormones on Fear Circuitry," *Annual Review of Neuroscience* 32 (2009): 289.
25. For a general review, see: B. Leuner and E. Gould, "Structural Plasticity and Hippocampal Function," *Annual Review of Psychology* 61 (2010): 111.

Effects of stress on hippocampal structure: A. Magarinos and B. McEwen, "Stress-Induced Atrophy of Apical Dendrites of Hippocampal CA3c Neurons: Involvement of Glucocorticoid Secretion and Excitatory Amino Acid Receptors," *Neuroscience* 69 (1995): 89; A. Magarinos et al., "Chronic Psychosocial Stress Causes Apical Dendritic Atrophy of Hippocampal CA3 Pyramidal Neurons in Subordinate Tree Shrews," *Journal of Neuroscience* 16

(1996): 3534; B. Eadie, V. Redila, and B. Christie, "Voluntary Exercise Alters the Cytoarchitecture of the Adult Dentate Gyrus by Increasing Cellular Proliferation, Dendritic Complexity, and Spine Density," *Journal of Comparative Neurology* 486 (2005): 39; A. Vyas et al., "Chronic Stress Induces Contrasting Patterns of Dendritic Remodeling in Hippocampal and Amygdaloid Neurons," *Journal of Neuroscience* 22 (2002): 6810.

Neuroplasticity related to depression: P. Videbach and B. Revnkilde, "Hippocampal Volume and Depression: A Meta-analysis of MRI Studies," *American Journal of Psychiatry* 161 (2004): 1957; L. Gerritsen et al., "Childhood Maltreatment Modifies the Relationship of Depression with Hippocampal Volume," *Psychological Medicine* 45 (2015): 3517.

Effects of exercise and stimulation on neuroplasticity: J. Firth et al., "Effect of Aerobic Exercise on Hippocampal Volume in Humans: A Systematic Review and Meta-analysis," *Neuroimage* 166 (2018): 230; G. Clemenson, W. Deng, and F. Gage, "Environmental Enrichment and Neurogenesis: From Mice to Humans," *Current Opinion in Behavioral Sciences* 4 (2015): 56.

Estrogen and neuroplasticity: B. McEwen, "Estrogen Actions throughout the Brain," *Recent Progress in Hormone Research* 57 (2002): 357; N. Lisofsky et al., "Hippocampal Volume and Functional Connectivity Changes during the Female Menstrual Cycle," *Neuroimage* 118 (2015): 154; K. Albert et al., "Estrogen Enhances Hippocampal Gray-Matter Volume in Young and Older Postmenopausal Women: A Prospective Dose-Response Study," *Neurobiology of Aging* 56 (2017): 1.

26. N. Brebe et al., "Pair-Bonding, Fatherhood, and the Role of Testosterone: A Meta-analytic Review," *Neuroscience & Biobehavioral Reviews* 98 (2019): 221; Y. Ulrich-Lai et al., "Chronic Stress Induces Adrenal Hyperplasia and Hypertrophy in a Subregion-Specific Manner," *American Journal of Physiology: Endocrinology and Metabolism* 291 (2006): E965.
27. J. Foster, "Modulating Brain Function with Microbiota," *Science* 376 (2022): 936; J. Cryan and S. Mazmanian, "Microbiota-Brain Axis: Context and Causality," *Science* 376 (2022): 938. Also: C. Chu et al., "The Microbiota Regulate Neuronal Function and Fear Extinction Learning," *Nature* 574 (2019): 543. A great example of events over the course of weeks to months changing behavior without conscious awareness can be found in S. Mousa, "Building Social Cohesion between Christians and Muslims through Soccer in Post-ISIS Iraq," *Science* 369 (2020): 866. Soccer teams in a league were experimentally composed of either solely Christian players or a mixture of the two religions (without players being aware of this intentional design as part of a study). Spending a season playing with Muslim teammates made Christian players far more affiliative with their Muslim teammates on the field—without changing overtly stated attitudes about Muslims.
28. For more details about this general topic, see chapter 5 in Sapolsky, *Behave*.
29. A. Caballero, R. Granbeerg, and K. Tseng, "Mechanisms Contributing to Prefrontal Cortex Maturation during Adolescence," *Neuroscience & Biobehavioral Reviews* 70 (2016): 4; K. Delevich et al., "Coming of Age in the Frontal Cortex: The Role of Puberty in Cortical Maturation," *Seminars in Cell & Developmental Biology* 118 (2021): 64. Chronically disrupting sleep in adolescent mice changes the workings of the dopamine reward system in adulthood, and not in a good direction; in other words, our mothers were right when urging us to resist the teenage pull toward crazy sleeping hours: W. Bian et al., "Adolescent Sleep Shapes Social Novelty Preference in Mice," *Nature Neuroscience* 25 (2022): 912.
30. E. Sowell et al., "Mapping Continued Brain Growth and Gray Matter Density Reduction in Dorsal Frontal Cortex: Inverse Relationships during Postadolescent Brain Maturation," *Journal of Neuroscience* 21 (2021): 8819; J. Giedd, "The Teen Brain: Insights from Neuroimaging," *Journal of Adolescent Health* 42 (2008): 335.
31. Footnote: C. González-Acosta et al., "von Economo Neurons in the Human Medial Frontopolar Cortex," *Frontiers in Neuroanatomy* 12 (2018), doi.org/10.3389/fnana.2018.00064; R. Hodge, J. Miller, and E. Lein, "Transcriptomic Evidence That von Economo Neurons Are

Regionally Specialized Extratelencephalic-Projecting Excitatory Neurons," *Nature Communications* 11 (2020): 1172.

32. For more details about this general topic, as well as specifics about the evolution of delayed frontal cortical maturation, see chapter 6 in Sapolsky, *Behave*.
33. For a good introduction to Kohlberg's truly monumental work, see D. Garz, *Lawrence Kohlberg: An Introduction* (Barbra Budrich, 2009).
34. D. Baumrind, "Child Care Practices Anteceding Three Patterns of Preschool Behavior," *Genetic Psychology Monographs* 75 (1967): 43; E. Maccoby and J. Martin, "Socialization in the Context of the Family: Parent-Child Interaction," in *Handbook of Child Psychology*, ed. P. Mussen (Wiley, 1983).
35. J. R. Harris, *The Nurture Assumption: Why Children Turn Out the Way They Do* (Free Press, 1998).
36. W. Wei, J. Lu, and L. Wang, "Regional Ambient Temperature Is Associated with Human Personality," *Nature Human Behaviour* 1 (2017): 890; R. McCrae et al. "Climatic Warmth and National Wealth: Some Culture-Level Determinants of National Character Stereotypes," *European Journal of Personality* 21 (2007): 953; G. Hofsteded and R. McCrae, "Personality and Culture Revisited: Linking Traits and Dimensions of Culture," *Cross-Cultural Research* 38 (2004): 52.
37. I. Weaver et al., "Epigenetic Programming by Maternal Behavior," *Nature Neuroscience* 7 (2004): 847. For an example of early-life stress causing epigenetic changes in the function of the adult brain all the way down to gene regulation in individual neurons, see H. Kronman et al., "Long-Term Behavioral and Cell-Type-Specific Molecular Effects of Early Life Stress Are Mediated by H3K79me2 Dynamics in Medium Spiny Neurons," *Nature Neuroscience* 24 (2021): 667. One would think that the adverse effects of, say, low socioeconomic status in childhood would occur as a result of brain development being delayed. Instead, the problem is that the early-life stress *accelerates* maturation of the brain, meaning that the window for brain construction being sculpted by experience closes earlier: U. Tooley, D. Bassett, and P. Mackay, "Environmental Influences on the Pace of Brain Development," *Nature Reviews Neuroscience* 22 (2021): 372.
38. D. Francis et al., "Nongenomic Transmission Across Generations of Maternal Behavior and Stress Responses in the Rat," *Science* 286 (1999): 1155; N. Provencal et al., "The Signature of Maternal Rearing in the Methylome in Rhesus Macaque Prefrontal Cortex and T Cells," *Journal of Neuroscience* 32 (2012): 15626. Among wild baboons, having a low dominance rank shortens the life expectancy not only of a female but of the next generation as well: M. Zipple et al., "Intergenerational Effects of Early Adversity on Survival in Wild Baboons," *eLife* 8 (2019): e47433.
39. The concept of Adverse Childhood Experiences was pioneered by Vincent Felitti of Kaiser Permanente San Diego/UCSD and Robert Anda of the CDC. See, for example: V. Felitti et al., "Relationship of Childhood Abuse and Household Dysfunction to Many of the Leading Causes of Death in Adults: The Adverse Childhood Experiences (ACE) Study," *American Journal of Preventative Medicine* 14 (1998): 245. Their original focus was on the relationship between ACE score and adult health. For example, see V. Felitti, "The Relation between Adverse Childhood Experiences and Adult Health: Turning Gold into Lead," *Permanente Journal* 6 (2002): 44. Their findings were replicated widely and expanded upon. See, for example: K. Hughes et al., "The Effect of Multiple Adverse Childhood Experiences on Health: A Systematic Review and Meta-analysis," *Lancet Public Health* 2 (2017): e356; K. Petruccelli, J. Davis, and T. Berman, "Adverse Childhood Experiences and Associated Health Outcomes: A Systematic Review and Meta-analysis," *Child Abuse & Neglect* 97 (2019): 104127. Extensive research then began to focus on the relationship between ACE score and adult violence and antisocial behavior. See these publications (from which the 35 percent increase estimate was generated): T. Moffitt et al., "A Gradient of Childhood Self-Control

Predicts Health, Wealth, and Public Safety," *Proceedings of the National Academy of Sciences of the United States of America* 108 (2011): 2693; J. Reavis et al., "Adverse Childhood Experiences and Adult Criminality: How Long Must We Live Before We Possess Our Own Lives?," *Permanente Journal* 17 (2013): 44; J. Craig et al., "A Little Early Risk Goes a Long Bad Way: Adverse Childhood Experiences and Life-Course Offending in the Cambridge Study," *Journal of Criminal Justice* 53 (2017): 34; J. Stinson et al., "Adverse Childhood Experiences and the Onset of Aggression and Criminality in a Forensic Inpatient Sample," *International Journal of Forensic Mental Health* 20 (2021): 374; L. Dutin et al., "Criminal History and Adverse Childhood Experiences in Relation to Recidivism and Social Functioning in Multi-problem Young Adults," *Criminal Justice and Behavior* 48, no. 5 (2021): 637; B. Fox et al., "Trauma Changes Everything: Examining the Relationship between Adverse Childhood Experiences and Serious, Violent and Chronic Juvenile Offenders," *Child Abuse & Neglect* 46 (2015): 163; M. Baglivio et al., "The Relationship between Adverse Childhood Experiences (ACE) and Juvenile Offending Trajectories in a Juvenile Offender Sample," *Journal of Criminal Justice* 43 (2015): 229. For good reviews, see: M. Baglivio, "On Cumulative Childhood Traumatic Exposure and Violence/Aggression: The Implications of Adverse Childhood Experiences (ACE)," in *Cambridge Handbook of Violent Behavior and Aggression*, 2nd ed., ed. A. Vazsonyi, D. Flannery, and M. DeLisi (Cambridge University Press, 2018), p. 467; G. Graf et al., "Adverse Childhood Experiences and Justice System Contact: A Systematic Review," *Pediatrics* 147 (2021): e2020021030.

40. The "relative age effect" is considered at length in both M. Gladwell, *Outliers: The Story of Success* (Little Brown, 2008), and S. Levitt and S. Dubner, *Superfreakonomics: Global Cooling, Patriotic Prostitutes, and Why Suicide Bombers Should Buy Life Insurance* (William Morrow, 2009). For more explorations of the phenomenon, see: E. Dhuey and S. Lipscomb, "What Makes a Leader? Relative Age and High School Leadership," *Economic Educational Review* 27 (2008): 173; D. Lawlor et al., "Season of Birth and Childhood Intelligence: Findings from the Aberdeen Children of the 1950s Cohort Study," *British Journal of Educational Psychology* 76 (2006): 481; A. Thompson, R. Barnsley, and J. Battle, "The Relative Age Effect and the Development of Self-Esteem," *Educational Research* 46 (2004): 313.
41. For more detail about this general topic, see chapter 7 in Sapolsky, *Behave*.
42. T. Roseboom et al., "Hungry in the Womb: What Are the Consequences? Lessons from the Dutch Famine," *Maturitas* 70 (2011): 141; B. Horsthemke, "A Critical View on Transgenerational Epigenetic Inheritance in Humans," *Nature Communications* 9 (2018): 2973; B. Van den Bergh et al., "Prenatal Developmental Origins of Behavior and Mental Health: The Influence of Maternal Stress in Pregnancy," *Neuroscience and Biobehavioral Reviews* 117 (2020): 26; F. Gomes, X. Zhu, and A. Grace, "Stress during Critical Periods of Development and Risk for Schizophrenia," *Schizophrenia Research* 213 (2019): 107; A. Brown and E. Susser, "Prenatal Nutritional Deficiency and Risk of Adult Schizophrenia," *Schizophrenia Bulletin* 3 (2008): 1054; D. St. Clair et al., "Rates of Adult Schizophrenia Following Prenatal Exposure to the Chinese Famine of 1959–1961," *Journal of the American Medical Association* 294 (2005): 557. This entire topic has been subsumed under the concept of "origins of adult disease," pioneered by David Barker at the University of Southampton in the UK. See, for example: D. Barker et al., "Fetal Origins of Adult Disease: Strength of Effects and Biological Basis," *International Journal of Epidemiology* 31 (2002): 1235. For a skeptical read of this entire literature, with the conclusion that the magnitude of the effects are generally overblown, see S. Richardson, *The Maternal Imprint: The Contested Science of Maternal-Fetal Effects* (University of Chicago Press, 2021).
43. For more details about this general topic, see chapter 7 in Sapolsky, *Behave*.
44. J. Bacque-Cazenave et al., "Serotonin in Animal Cognition and Behavior," *Journal of Molecular Science* 21 (2020): 1649; E. Coccaro et al., "Serotonin and Impulsive Aggression," *CNS Spectrum* 20 (2015): 295; J. Siegel and M. Crockett, "How Serotonin Shapes Moral Judgment

and Behavior," *Annals of the New York Academy of Sciences* 1299 (2013): 42; J. Palacios, "Serotonin Receptors in Brain Revisited," *Brain Research* 1645 (2016): 46.

45. J. Liu et al., "Tyrosine Hydroxylase Gene Polymorphisms Contribute to Opioid Dependence and Addiction by Affecting Promoter Region Function," *Neuromolecular Medicine* 22 (2020): 391.
46. M. Bakermans-Kranenburg and M. van Ijzendoorn, "Differential Susceptibility to Rearing Environment Depending on Dopamine-Related Genes: New Evidence and a Meta-analysis," *Development and Psychopathology* 23 (2011): 39; M. Sweitzer et al., "Polymorphic Variation in the Dopamine D4 Receptor Predicts Delay Discounting as a Function of Childhood Socioeconomic Status: Evidence for Differential Susceptibility," *Social Cognitive and Affective Neuroscience* 8 (2013): 499; N. Perroud et al., "COMT but Not Serotonin-Related Genes Modulates the Influence of Childhood Abuse on Anger Traits," *Genes Brain and Behavior* 9 (2010): 193; S. Lee et al., "Association of Maternal Dopamine Transporter Genotype with Negative Parenting: Evidence for Gene x Environment Interaction with Child Disruptive Behavior," *Molecular Psychiatry* 15 (2010): 548. For a nice example of some of these same gene/upbringing patterns in other primates, see M. Champoux et al., "Serotonin Transporter Gene Polymorphism, Differential Early Rearing, and Behavior in Rhesus Monkey Neonates," *Molecular Psychiatry* 7 (2002): 1058. It is worth noting that there have been controversies over the years regarding some of these gene/upbringing interactions in humans, with one side arguing that they are not reliable and consistently replicated, with others arguing that these relationships are robust when only considering the studies that were actually done well. For example, see: M. Wankerl et al., "Current Developments and Controversies: Does the Serotonin Transporter Gene-Linked Polymorphic Region (5-HTTLPR) Modulate the Association Between Stress and Depression?," *Current Opinion in Psychiatry* 23 (2010): 582.
47. E. Lein et al., "Genome-wide Atlas of Gene Expression in the Adult Mouse Brain," *Nature* 445 (2007): 168; Y. Jin et al., "Architecture of Polymorphisms in the Human Genome Reveals Functionally Important and Positively Selected Variants in Immune Response and Drug Transporter Genes," *Human Genomics* 12 (2018): 43.
48. For more detail about this general topic, see chapter 8 in Sapolsky, *Behave.*
49. Cross-cultural differences: H. Markus and S. Kitayama, "Culture and Self: Implications for Cognition, Emotion, and Motivation," *Psychological Review* 98 (1991): 224; A. Cuddy et al., "Stereotype Content Model across Cultures: Towards Universal Similarities and Some Differences," *British Journal of Social Psychology* 48 (2009): 1; R. Nisbett, *The Geography of Thought: How Asians and Westerners Think Differently . . . and Why* (Free Press, 2003).

    Neural bases of some of these differences: S. Kitayama and A. Uskul, "Culture, Mind, and the Brain: Current Evidence and Future Directions," *Annual Review of Psychology* 62 (2011): 419; B. Park et al., "Neural Evidence for Cultural Differences in the Valuation of Positive Facial Expressions," *Social Cognitive and Affective Neuroscience* 11 (2015): 243; B. Cheon et al., "Cultural Influences on Neural Basis of Intergroup Empathy," *Neuroimage* 57 (2011): 642.

    Cross-cultural differences in shame versus guilt: H. Katchadourian, *Guilt: The Bite of Conscience* (Stanford General Books, 2011); J. Jacquet, *Is Shame Necessary? New Uses for an Old Tool* (Pantheon, 2015).
50. T. Hedden et al., "Cultural Influences on Neural Substrates of Attentional Control," *Psychological Science* 19 (2008): 12; S. Han and G. Northoff, "Culture-Sensitive Neural Substrates of Human Cognition: A Transcultural Neuroimaging Approach," *Nature Reviews Neuroscience* 9 (20008): 646; T. Masuda and R. E. Nisbett, "Attending Holistically vs. Analytically: Comparing the Context Sensitivity of Japanese and Americans," *Journal of Personality and Social Psychology* 81 (2001): 922; J. Chiao, "Cultural Neuroscience: A Once and Future Discipline," *Progress in Brain Research* 178 (2009): 287.
51. K. Zhang and H. Changsha, *World Heritage in China* (Press of South China University of Technology, 2006).

52. T. Talhelm et al., "Large-Scale Psychological Differences within China Explained by Rice versus Wheat Agriculture," *Science* 344 (2014): 603; T. Talhelm, X. Zhang, and S. Oishi, "Moving Chairs in Starbucks: Observational Studies Find Rice-Wheat Cultural Differences in Daily Life in China," *Science Advances* 4 (2018), DOI:10.1126/sciadv.aap8469.
53. Footnote: The genetics of cross-cultural differences: H. Harpending and G. Cochran, "In Our Genes," *Proceedings of the National Academy of Sciences of the United States of America* 99 (2002): 10.

    Specific papers in the area: Y. Ding et al., "Evidence of Positive Selection Acting at the Human Dopamine Receptor D4 Gene Locus," *Proceedings of the National Academy of Sciences of the United States of America* 99 (2002): 309; F. Chang et al., "The World-wide Distribution of Allele Frequencies at the Human Dopamine D4 Receptor Locus," *Human Genetics* 98 (1996): 891; K. Kidd et al., "An Historical Perspective on 'The World-wide Distribution of Allele Frequencies at the Human Dopamine D4 Receptor Locus,'" *Human Genetics* 133 (2014): 431; C. Chen et al., "Population Migration and the Variation of Dopamine D4 Receptor (DRD4) Allele Frequencies around the Globe," *Evolution and Human Behavior* 20 (1999): 309.

    For a nontechnical introduction to this topic, see R. Sapolsky, "Are the Desert People Winning?," *Discover*, August 2005, 38.
54. Footnote: M. Fleisher, *Kuria Cattle Raiders: Violence and Vigilantism on the Tanzania/Kenya Frontier* (University of Michigan Press, 2000); M. Fleisher, "'War Is Good for Thieving!': The Symbiosis of Crime and Warfare among the Kuria of Tanzania," *Africa* 72 (200): 1. In these tensions, I root for my Maasai, of course; Maasai/Kuria tensions have been going on for a long, long time but thanks to the arbitrariness of what some European colonials did in the last century, when the two groups fight, it counts as an international conflict; R. McMahon, *Homicide in Pre-famine and Famine Ireland* (Liverpool University Press, 2013); R. Nisbett and D. Cohen, *Culture of Honor: The Psychology of Violence in the South* (Westview Press, 1996); B. Wyatt-Brown, *Southern Honor: Ethics and Behavior in the Old South* (Oxford University Press, 1982). Theory about the origins of the southern culture of honor among pastoralists in the British Isles: D. Fischer, *Albion's Seed* (Oxford University Press, 1989).
55. Footnote: E. Van de Vliert, "The Global Ecology of Differentiation between Us and Them," *Nature Human Behaviour* 4 (2020): 270.

    Second footnote: F. Lederbogen et al., "City Living and Urban Upbringing Affect Neural Social Stress Processing in Humans," *Nature* 474 (2011): 498; D. Kennedy and R. Adolphs, "Stress and the City," *Nature* 474 (2011): 452; A. Abbott, "City Living Marks the Brain," *Nature* 474 (2011): 429; M. Gelfand et al., "Differences between Tight and Loose Cultures: A 33-Nation Study," *Science* 332 (2011): 1100.
56. Footnote: K. Hill and R. Boyd, "Behavioral Convergence in Humans and Animals," *Science* 371 (2021): 235; T. Barsbai, D. Lukas, and A. Pondorfer, "Local Convergence of Behavior across Species," *Science* 371 (2021): 292. For more details about this general topic, see chapter 9 in Sapolsky, *Behave.*
57. For an overview of this topic, see chapter 10 in Sapolsky, *Behave.*
58. P. Alces, *Trialectic: The Confluence of Law, Neuroscience, and Morality* (University of Chicago Press, 2023). P. Tse, "Two Types of Libertarian Free Will Are Realized in the Human Brain," in *Neuroexistentialism: Meaning, Morals, and Purpose in the Age of Neuroscience*, ed. G. Caruso and O. Flanagan (Oxford University Press, 2018).
59. N. Levy, *Hard Luck: How Luck Undermines Free Will and Moral Responsibility* (Oxford University Press, 2015), quote from p. 87.

    Footnote (p. 83): The poignance of this reality is wonderfully summarized by a quote from Charles Johnson's short story "China," *The Penguin Book of the American Short Story*, ed. J. Freeman (Penguin Press, 2021), p. 92: "'I can only be what I've been?' This he asked softly, but his voice trembled." I thank Mia Council for pointing this out.

4. WILLING WILLPOWER: THE MYTH OF GRIT

1. N. Levy, "Luck and History-Sensitive Compatibilism," *Philosophical Quarterly* 59 (2009): 237, quote from p. 242.
2. G. Caruso and D. Dennett, "Just Deserts," *Aeon*, https://aeon.co/essays/on-free-will-daniel-dennett-and-gregg-caruso-go-head-to-head.
3. R. Kane, "Free Will, Mechanism and Determinism," in *Moral Psychology*, vol. 4, *Free Will and Moral Responsibility*, ed. W. Sinnott-Armstrong (MIT Press, 2014), the quote comes from p. 130; M. Shadlen and A. Roskies, "The Neurobiology of Decision-Making and Responsibility: Reconciling Mechanism and Mindedness," *Frontiers of Neuroscience* 6 (2012), doi.org/10.3389/fnins.2012.00056.
4. S. Spence, *The Actor's Brain: Exploring the Cognitive Neuroscience of Free Will* (Oxford University Press, 2009).
5. P. Tse, "Two Types of Libertarian Free Will Are Realized in the Human Brain," in *Neuroexistentialism: Meaning, Morals and Purpose in the Age of Neuroscience*, ed. G. Caruso and O. Flanagan (Oxford University Press, 2013).
6. A. Roskies, "Can Neuroscience Resolve Issues about Free Will?," *Moral Psychology*, vol. 4, *Free Will and Moral Responsibility*, ed. W. Sinnott-Armstrong (MIT Press, 2014), the quote comes from p. 116; M. Gazzaniga, "Mental Life and Responsibility in Real Time with a Determined Brain," in *Moral Psychology*, vol. 4: *Free Will and Moral Responsibility*, ed. W. Sinnott-Armstrong (MIT Press, 2014), 59.
7. Families losing fortunes: C. Hill, "Here's Why 90% of Rich People Squander Their Fortunes," *MarketWatch*, April 23, 2017, marketwatch.com/story/heres-why-90-of-rich-people-squander-their-fortunes-2017-04-23.

   Footnote: J. White and G. Batty, "Intelligence across Childhood in Relation to Illegal Drug Use in Adulthood: 1970 British Cohort Study," *Journal of Epidemiology and Community Health* 66 (2012): 767.
8. J. Cantor, "Do Pedophiles Deserve Sympathy?" CNN, June 21, 2012.
9. Footnote: Z. Goldberger, "Music of the Left Hemisphere: Exploring the Neurobiology of Absolute Pitch," *Yale Journal of Biology and Medicine* 74 (2001): 323.
10. K. Semendeferi et al., "Humans and Great Apes Share a Large Frontal Cortex," *Nature Neuroscience* 5 (2002): 272; P. Schoenemann, "Evolution of the Size and Functional Areas of the Human Brain," *Annual Review of Anthropology* 35 (2006): 379. In addition, depending on the way it is measured, the human PFC is proportionately greater in size and/or more densely and complexly wired than in any other primate: J. Rilling and T. Insel, "The Primate Neocortex in Comparative Perspective Using MRI," *Journal of Human Evolution* 37 (1999): 191; R. Barton and C. Venditti, "Human Frontal Lobes Are Not Relatively Large," *Proceedings of the National Academy of Sciences of the United States of America* 110 (2013): 9001. Embedded in all these findings is the challenge of figuring out what precisely is the equivalent of the human frontal cortex in, say, a lab rat; see M. Carlen, "What Constitutes the Prefrontal Cortex?," *Science* 358 (2017): 478.
11. E. Miller and J. Cohen, "An Integrative Theory of Prefrontal Cortex Function," *Annual Review of Neuroscience* 24 (2001): 167; L. Gao et al., "Single-Neuron Projectome of Mouse Prefrontal Cortex," *Nature Neuroscience* 25 (2022): 515; V. Mante et al., "Context-Dependent Computation by Recurrent Dynamics in Prefrontal Cortex," *Nature* 503 (2013): 78. Some more examples of frontal cortical involvement in task switching: S. Bunge, "How We Use Rules to Select Actions: A Review of Evidence from Cognitive Neuroscience," *Cognitive, Affective & Behavioral Neuroscience* 4 (2004): 564; E. Crone et al., "Evidence for Separable Neural Processes Underlying Flexible Rule Use," *Cerebral Cortex* 16 (2005): 475.
12. R. Dunbar, "The Social Brain Meets Neuroimaging," *Trends in Cognitive Sciences* 16 (2011): 101; P. Lewis et al., "Ventromedial Prefrontal Volume Predicts Understanding of Others and Social Network Size," *Neuroimage* 57 (2011): 1624; K. Bickart et al., "Intrinsic Amygdala–

Cortical Functional Connectivity Predicts Social Network Size in Humans," *Journal of Neuroscience* 32 (2012): 14729; R. Kanai et al., "Online Social Network Size Is Reflected in Human Brain Structure," *Proceedings of the Royal Society B: Biological Sciences* 279 (2012): 1327; J. Sallet et al., "Social Network Size Affects Neural Circuits in Macaques," *Science* 334 (2011): 697.

13. J. Kubota, M. Banaji, and E. Phelps, "The Neuroscience of Race," *Nature Neuroscience* 15 (2012): 940.

    Footnote: Reviewed in J. Eberhardt, *Biased: Uncovering the Hidden Prejudice That Shapes What We See, Think, and Do* (Viking, 2019).

14. N. Eisenberger, M. Lieberman, and K. Williams, "Does Rejection Hurt? An FMRI Study of Social Exclusion," *Science* 302 (2003): 290; N. Eisenberger, "The Pain of Social Disconnection: Examining the Shared Neural Underpinnings of Physical and Social Pain," *Nature Reviews Neuroscience* 3 (2012): 421; C. Masten, N. Eisenberger, and L. Borofsky, "Neural Correlates of Social Exclusion during Adolescence: Understanding the Distress of Peer Rejection," *Social Cognitive and Affective Neuroscience* 4 (2009): 143. For an interesting study of the gene regulation in the prefrontal cortex mediating resilience during stress, see: Z. Lorsch et al., "Stress Resilience Is Promoted by a Zfp189-Driven Transcriptional Network in Prefrontal Cortex," *Nature Neuroscience* 22 (2019): 1413.

15. Neurobiology of fear: C. Herry et al., "Switching On and Off Fear by Distinct Neuronal Circuits," *Nature* 454 (2008): 600; S. Maren and G. Quirk, "Neuronal Signaling of Fear Memory," *Nature Reviews Neuroscience* 5 (2004): 844; S. Rodrigues, R. Sapolsky, and J. LeDoux, "The Influence of Stress Hormones on Fear Circuitry," *Annual Review of Neuroscience* 32 (2009): 289; O. Klavir et al., "Manipulating Fear Associations via Optogenetic Modulation of Amygdala Inputs to Prefrontal Cortex," *Nature Neuroscience* 20 (2017): 836; S. Ciocchi et al., "Encoding of Conditioned Fear in Central Amygdala Inhibitory Circuits," *Nature* 468 (2010): 277; W. Haubensak et al., "Genetic Dissection of an Amygdala Microcircuit That Gates Conditioned Fear," *Nature* 468 (2010): 270.

    Neurobiology of extinguishing fear: M. Milad and G. Quirk, "Neurons in Medial Prefrontal Cortex Signal Memory for Fear Extinction," *Nature* 420 (2002): 70; E. Phelps et al., "Extinction Learning in Humans: Role of the Amygdala and vmPFC," *Neuron* 43 (2004): 897.

    Neurobiology of re-expressing conditioned fear: R. Marek et al., "Hippocampus-Driven Feed-Forward Inhibition of the Prefrontal Cortex Mediates Relapse of Extinguished Fear," *Nature Neuroscience* 21 (2018): 384.

16. J. Greene and J. Paxton, "Patterns of Neural Activity Associated with Honest and Dishonest Moral Decisions," *Proceedings of the National Academy of Sciences of the United States of America* 106 (2009): 12506. Also see his superb book: J. Greene, *Moral Tribes: Emotion, Reason, and the Gap between Us and Them* (Penguin Press, 2013).

17. H. Terra et al., "Prefrontal Cortical Projection Neurons Targeting Dorsomedial Striatum Control Behavioral Inhibition," *Current Biology* 30 (2020): 4188; S. de Kloet et al., "Bidirectional Regulation of Cognitive Control by Distinct Prefrontal Cortical Output Neurons to Thalamus and Striatum," *Nature Communications* 12 (2021): 1994.

18. Frontal disinhibition: R. Bonelli and J. Cummings, "Frontal-Subcortical Circuitry and Behavior," *Dialogues in Clinical Neuroscience* 9 (2007); E. Huey, "A Critical Review of Behavioral and Emotional Disinhibition," *Journal of Nervous and Mental Disease* 208 (2020): 344 (I'm proud to say that the author, a professor at Columbia University Medical School, was once a stellar member of my lab).

    Frontal damage and criminality: B. Miller and J. Llibre Guerra, "Frontotemporal Dementia," *Handbook of Clinical Neurology* 165 (2019): 33; M. Brower and B. Price, "Neuropsychiatry of Frontal Lobe Dysfunction in Violent and Criminal Behaviour: A Critical Review," *Neurology, Neurosurgery and Psychiatry* 71 (2001): 720; E. Shiroma, P. Ferguson, and E. Pickelsimer, "Prevalence of Traumatic Brain Injury in an Offender Population: A Meta-analysis," *Journal of Corrective Health Care* 16 (2010): 147.

Footnote: J. Allman et al., "The von Economo Neurons in the Frontoinsular and Anterior Cingulate Cortex," *Annals of the New York Academy of Sciences* 1225 (2011): 59; C. Butti et al., "von Economo Neurons: Clinical and Evolutionary Perspectives," *Cortex* 49 (2013): 312; H. Evrard et al., "von Economo Neurons in the Anterior Insula of the Macaque Monkey," *Neuron* 74 (2012): 482. For an appropriately skeptical critique of the linkage of empathy, mirror neurons and von Economo neurons see: G. Hickok, *The Myth of Mirror Neurons: The Real Neuroscience of Communication and Cognition* (Norton, 2014).

19. Y. Wang et al., "Neural Circuitry Underlying REM Sleep: A Review of the Literature and Current Concepts," *Progress in Neurobiology* 204 (2021): 102106; J. Greene et al., "An fMRI Investigation of Emotional Engagement in Moral Judgment," *Science* 293 (2001): 2105; J. Greene et al., "The Neural Bases of Cognitive Conflict and Control in Moral Judgment," *Neuron* 44 (2004): 389.
20. A. Barbey, M. Koenigs, and J. Grafman, "Dorsolateral Prefrontal Contributions to Human Intelligence," *Neuropsychologia* 51 (2013): 1361. For an overview of both the dlPFC and vmPFC, see Greene, *Moral Tribes.*
21. D. Knock et al., "Diminishing Reciprocal Fairness by Disrupting the Right Prefrontal Cortex," *Science* 314 (2006): 829; A. Bechara, "The Role of Emotion in Decision-Making: Evidence from Neurological Patients with Orbitofrontal Damage," *Brain and Cognition* 55 (2004): 30; A. Damasio, *The Feeling of What Happens: Body and Emotion in the Making of Consciousness* (Harcourt, 1999). These issues are also explored in L. Koban, P. Gianaros, and T. Wager, "The Self in Context: Brain Systems Linking Mental and Physical Health," *Nature Reviews Neuroscience* 22 (2021): 309.

    Footnote: E. Mas-Herrero, A. Dagher, and R. Zatorre, "Modulating Musical Reward Sensitivity Up and Down with Transcranial Magnetic Stimulation," *Nature Human Behaviour* 2 (2018); 27. See also J. Grahn, "Tuning the Brain to Musical Delight," *Nature Human Behaviour* 2 (2018): 17.
22. M. Koenigs et al., "Damage to the Prefrontal Cortex Increases Utilitarian Moral Judgments," *Nature* 446 (2007): 865; B. Thomas, K. Croft, and D. Tranel, "Harming Kin to Save Strangers: Further Evidence for Abnormally Utilitarian Moral Judgments after Ventromedial Prefrontal Damage," *Journal of Cognitive Neuroscience* 23 (2011): 2186; L. Young et al., "Damage to Ventromedial Prefrontal Cortex Impairs Judgment of Harmful Intent," *Neuron* 25 (2010): 845.
23. J. Saver and A. Damasio, "Preserved Access and Processing of Social Knowledge in a Patient with Acquired Sociopathy Due to Ventromedial Frontal Damage," *Neuropsychologia* 29 (1991): 1241; M. Donoso, A. Collins, and E. Koechlin, "Foundations of Human Reasoning in the Prefrontal Cortex," *Science* 344 (2014): 1481; T. Hare, "Exploiting and Exploring the Options," *Science* 344 (2014): 1446; T. Baumgartner et al., "Dorsolateral and Ventromedial Prefrontal Cortex Orchestrate Normative Choice," *Nature Neuroscience* 14 (2011): 1468; A. Bechara, "The Role of Emotion in Decision-Making: Evidence from Neurological Patients with Orbitofrontal Damage," *Brain and Cognition* 55 (2004): 30. Consequences of damage to the vmPFC: G. Moretto, M. Sellitto, and G. Pellegrino, "Investment and Repayment in a Trust Game after Centromedial Prefrontal Damage," *Frontiers of Human Neuroscience* 7 (2013): 593.
24. The PFC keeping track of long-lasting categorization rules: S. Reinert et al., "Mouse Prefrontal Cortex Represents Learned Rules for Categorization," *Nature* 593 (2021): 411. The PFC having to work hard continuously to keep track of an ongoing rule change can extend for weeks in rats (which is a long time for them): M. Chen et al., "Persistent Transcriptional Programmes Are Associated with Remote Memory," *Nature* 587 (2020): 437.

    "Cognitive load" has become highly controversial. The concepts of cognitive reserve and ego depletion were pioneered by social psychologist Roy Baumeister and colleagues: R. Baumeister and L. Newman, "Self-Regulation of Cognitive Inference and Decision Processes," *Personality and Social Psychology Bulletin* 20 (1994): 3; R. Baumeister, M. Muraven, and D.

Tice, "Ego Depletion: A Resource Model of Volition, Self-Regulation, and Controlled Processing," *Social Cognition* 18 (2000): 130; R. Baumeister et al., "Ego Depletion: Is the Active Self a Limited Resource?," *Journal of Personality and Social Psychology* 74 (1988): 1252. A number of studies, however, began to report problems with replicating the effect (e.g., L. Koppel et al., "No Effect of Ego Depletion on Risk Taking," *Science Reports* 9 [2019]: 9724). Amid that, others reported replications; see, for example: M. Hagger et al., "A Multilab Preregistered Replication of the Ego-Depletion Effect," *Perspectives on Psychological Science* 11 (2016): 546. Discussion of some of the possible sources of the confusion can be found in M. Friese et al., "Is Ego Depletion Real? An Analysis of Arguments," *Personality and Social Psychology Review* 23 (2019): 107. Baumeister and colleagues responded to the reported failures of replication with R. Baumeister and K. Vohs, "Misguided Effort with Elusive Implications," *Perspectives on Psychological Science* 11 (2016): 574. Meta-analyses of these studies have become so numerous—and produced conflicting conclusions as to whether the effect is for real—that there are now even meta-analyses of the meta-analyses: S. Harrison et al., "Exploring Strategies to Operationalize Cognitive Reserve: A Systematic Review of Reviews," *Journal of Clinical and Experimental Neuropsychology* 37 (2015): 253. I'm in no position to assess the debates surrounding the social psychology aspects of these studies, let alone those regarding data analysis; I'm on slightly more solid ground assessing the biological elements of these studies. As such, my relative outsider's read is that the effects are often for real but typically of considerably smaller magnitude than the early research suggested. This would certainly not be the first time this sort of revisionism has been necessary in science.

25. W. Hofmann, W. Rauch, and B. Gawronski, "And Deplete Us Not into Temptation: Automatic Attitudes, Dietary Restraint, and Self-Regulatory Resources as Determinants of Eating Behavior," *Journal of Experimental Social Psychology* 43 (2007): 497.
26. H. Kato, A. Jena, and Y. Tsugawa, "Patient Mortality after Surgery on the Surgeon's Birthday: Observational Study," *British Medical Journal* 371 (2020): m4381.
27. M. Kouchaki and I. Smith, "The Morning Morality Effect: The Influence of Time of Day on Unethical Behavior," *Psychological Sciences* 25 (2014): 95; F. Gino et al., "Unable to Resist Temptation: How Self-Control Depletion Promotes Unethical Behavior," *Organizational Behavior and Human Decision Processes* 115 (2011): 191–92; N. Mead et al., "Too Tired to Tell the Truth: Self-Control Resource Depletion and Dishonesty," *Journal of Experimental Social Psychology* 45 (2009): 594.

    These issues playing out in medical settings: T. Johnson et al., "The Impact of Cognitive Stressors in the Emergency Department on Physician Implicit Racial Bias," *Academy of Emergency Medicine* 23 (2016): 29; P. Trinh, D. Hoover, and F. Sonnenberg, "Time-of-Day Changes in Physician Clinical Decision Making: A Retrospective Study," *PLoS One* 16 (2021): e0257500; H. Nephrash and M. Barnett, "Association of Primary Care Clinic Appointment Time with Opioid Prescribing," *JAMA Open Network* 2 (2019): e1910373.
28. S. Danziger, J. Levav, and L. Avnaim-Pesso, "Extraneous Factors in Judicial Decisions," *Proceedings of the National Academy of Sciences of the United States of America* 108 (2011): 6889.

    Footnote: The hungry judge effect: K. Weinshall-Margel and J. Shapard, "Overlooked Factors in the Analysis of Parole Decisions," *Proceedings of the National Academy of Sciences of the United States of America* 108 (2011): E833. Also: A. Glöckner, "The Irrational Hungry Judge Effect Revisited: Simulations Reveal That the Magnitude of the Effect Is Overestimated," *Judgment and Decision Making* 11 (2016): 601. Additional studies: D. Hangartner, D. Kopp, and M. Siegenthaler, "Monitoring Hiring Discrimination through Online Recruitment Platforms," *Nature* 589 (2021): 572. See also P. Hunter, "Your Decisions Are What You Eat: Metabolic State Can Have a Serious Impact on Risk-Taking and Decision-Making in Humans and Animals," *EMBO Reports* 14 (2013): 505.

    Meanwhile, subsequent research has suggested a very different version of judicial decisions being influenced by implicit factors—on the average, judges give lighter sentences if it's the defendant's birthday that day. For example, in New Orleans courtrooms, there's about a 15 percent decrease in sentence length; tellingly, the effect is about twice as great if the judge and defendant are of the same race. Day before or after your birthday? No dice, has no effect. And even more telling but not surprising, no judge mentioned abstractions like birthdays in their judicial opinions. The paper's title aptly summarized how this is a case of conflicting values—"criminals should be punished" versus "we should be nice to people on their birthdays." D. Chen and P. Arnaud, "Clash of Norms: Judicial Leniency on Defendant Birthdays" *SSRN* (2020), ssrn.com/abstract=3203624 or http://dx.doi.org/10.2139/ssrn.3203624.

29. D. Kahneman, *Thinking, Fast and Slow* (Farrar, Straus and Giroux, 2013). Also, for insights into Kahneman's reasoning: H. Nohlen, F. van Harreveld, and W. Cunningham, "Social Evaluations under Conflict: Negative Judgments of Conflicting Information Are Easier Than Positive Judgments," *Social Cognitive and Affective Neuroscience* 14 (2019): 709.
30. Footnote: T. Baer and S. Schnall, "Quantifying the Cost of Decision Fatigue: Suboptimal Risk Decisions in Finance," *Royal Society Open Science* 5 (2021): 201059.
31. I. Beaulieu-Boire and A. Lang, "Behavioral Effects of Levodopa," *Movement Disorders* 30 (2015): 90.
32. L. R. Mujica-Parodi et al., "Chemosensory Cues to Conspecific Emotional Stress Activate Amygdala in Humans," *Public Library of Science One* 4, no. 7 (2009): e6415. Jaywalking: B. Pawlowski, R. Atwal, and R. Dunbar, "Sex Differences in Everyday Risk-Taking Behavior in Humans," *Evolutionary Psychology* 6 (2008): 29.

    Footnote: L. Chang et al., "The Face That Launched a Thousand Ships: The Mating-Warring Association in Men," *Personality and Social Psychology Bulletin* 37 (2011): 976; S. Ainsworth and J. Maner, "Sex Begets Violence: Mating Motives, Social Dominance, and Physical Aggression in Men," *Journal of Personality and Social Psychology* 103 (2012): 819; W. Iredale, M. van Vugt, and R. Dunbar, "Showing Off in Humans: Male Generosity as a Mating Signal," *Evolutionary Psychology* 6 (2008): 386; M. Van Vugt and W. Iredale, "Men Behaving Nicely: Public Goods as Peacock Tails," *British Journal of Psychology* 104 (2013): 3. Oh, those skateboarders: R. Ronay and W. von Hippel, "The Presence of an Attractive Woman Elevates Testosterone and Physical Risk Taking in Young Men," *Social Psychological and Personality Science* 1 (2010): 1.
33. J. Ferguson et al., "Oxytocin in the Medial Amygdala Is Essential for Social Recognition in the Mouse," *Journal of Neuroscience* 21 (2001): 8278; R. Griksiene and O. Ruksenas, "Effects of Hormonal Contraceptives on Mental Rotation and Verbal Fluency," *Psychoneuroendocrinology* 36 (2011): 1239–1248; R. Norbury et al., "Estrogen Therapy and Brain Muscarinic Receptor Density in Healthy Females: A SPET Study," *Hormones and Behavior* 5 (2007): 249.
34. Stress effects on the efficacy of frontal function: S. Qin et al., "Acute Psychological Stress Reduces Working Memory–Related Activity in the Dorsolateral Prefrontal Cortex," *Biological Psychiatry* 66 (2009): 25; L. Schwabe et al., "Simultaneous Glucocorticoid and Noradrenergic Activity Disrupts the Neural Basis of Goal-Directed Action in the Human Brain," *Journal of Neuroscience* 32 (2012): 10146; A. Arnsten, M. Wang, and C. Paspalas, "Neuromodulation of Thought: Flexibilities and Vulnerabilities in Prefrontal Cortical Network Synapses," *Neuron* 76 (2012): 223; A. Arnsten, "Stress Weakens Prefrontal Networks: Molecular Insults to Higher Cognition," *Nature Neuroscience* 18 (2015): 1376; E. Woo et al., "Chronic Stress Weakens Connectivity in the Prefrontal Cortex: Architectural and Molecular Changes," *Chronic Stress* 5 (2021), doi:24705470211029254.
35. Testosterone effects on the frontal cortex: P. Mehta and J. Beer, "Neural Mechanisms of the Testosterone-Aggression Relation: The Role of Orbitofrontal Cortex," *Journal of Cognitive Neuroscience* 22 (2010): 2357; E. Hermans et al., "Exogenous Testosterone Enhances Responsiveness to Social Threat in the Neural Circuitry of Social Aggression in Humans," *Biological Psychiatry* 63 (2008): 263; G. van Wingen et al., "Testosterone Reduces Amygdala-Orbitofrontal Cortex Coupling," *Psychoneuroendocrinology* 35 (2010): 105; I. Volman et al., "Endogenous Testosterone Modulates Prefrontal-Amygdala Connectivity during Social Emotional Behavior," *Cerebral Cortex* 21 (2011): 2282; P. Bos et al., "The Neural Mechanisms by Which

Testosterone Acts on Interpersonal Trust," *Neuroimage* 61 (2012): 730; P. Bos et al., "Testosterone Reduces Functional Connectivity during the 'Reading the Mind in the Eyes' Test," *Psychoneuroendocrinology* 68 (2016): 194; R. Handa, G. Hejnaa, and G. Murphy, "Androgen Inhibits Neurotransmitter Turnover in the Medial Prefrontal Cortex of the Rat Following Exposure to a Novel Environment," *Brain Research* 751 (1997): 131; T. Hajszan et al., "Effects of Androgens and Estradiol on Spine Synapse Formation in the Prefrontal Cortex of Normal and Testicular Feminization Mutant Male Rats," *Endocrinology* 148 (2007): 1963.

Oxytocin effects on frontal cortex: N. Ebner et al., "Oxytocin's Effect on Resting-State Functional Connectivity Varies by Age and Sex," *Psychoneuroendocrinology* 69 (2016): 50; S. Dodhia et al., "Modulation of Resting-State Amygdala-Frontal Functional Connectivity by Oxytocin in Generalized Social Anxiety Disorder," *Neuropsychopharmacology* 39 (2014): 2061.

Estrogen effects on the frontal cortex: R. Hill et al., "Estrogen Deficiency Results in Apoptosis in the Frontal Cortex of Adult Female Aromatase Knockout Mice," *Molecular and Cellular Neuroscience* 41 (2009): 1; R. Brinton et al., "Equilin, a Principal Component of the Estrogen Replacement Therapy Premarin, Increases the Growth of Cortical Neurons via an NMDA Receptor–Dependent Mechanism," *Experimental Neurology* 147 (1997): 211.

36. Effects of a variety of adverse experiences on the frontal cortex. Depression: E. Belleau, M. Treadway, and D. Pizzagalli, "The Impact of Stress and Major Depressive Disorder on Hippocampal and Medial Prefrontal Cortex Morphology," *Biological Psychiatry* 85 (2019): 443; F. Calabrese et al., "Neuronal Plasticity: A Link between Stress and Mood Disorders," *Psychoneuroendocrinology* 34, supp. 1 (2009): S208; S. Chiba et al., "Chronic Restraint Stress Causes Anxiety- and Depression-Like Behaviors, Downregulates Glucocorticoid Receptor Expression, and Attenuates Glutamate Release Induced by Brain-Derived Neurotrophic Factor in the Prefrontal Cortex," *Progress in Neuro-psychopharmacology and Biological Psychiatry* 39 (2012): 112; J. Radley et al., "Chronic Stress-Induced Alterations of Dendritic Spine Subtypes Predict Functional Decrements in an Hypothalamo-Pituitary-Adrenal-Inhibitory Prefrontal Circuit," *Journal of Neuroscience* 33 (2013): 14379.

Anxiety and PTSD: L. Mah, C. Szabuniewicz, and A. Fletcco, "Can Anxiety Damage the Brain?," *Current Opinions in Psychiatry* 29 (2016): 56; K. Moench and C. Wellman, "Stress-Induced Alterations in Prefrontal Dendritic Spines: Implications for Post-traumatic Stress Disorder," *Neuroscience Letters* 5 (2015): 601.

Social instability: M. Breach, K. Moench, and C. Wellman, "Social Instability in Adolescence Differentially Alters Dendritic Morphology in the Medial Prefrontal Cortex and Its Response to Stress in Adult Male and Female Rats," *Developmental Neurobiology* 79 (2019): 839.

37. Effects of alcohol and weed on the frontal cortex: C. Shields and C. Gremel, "Review of Orbitofrontal Cortex in Alcohol Dependence: A Disrupted Cognitive Map?," *Alcohol: Clinical and Experimental Research* 44 (2020): 1952; D. Eldreth, J. Matochik, and L. Cadet, "Abnormal Brain Activity in Prefrontal Brain Regions in Abstinent Marijuana Users," *Neuroimage* 23 (2004): 914; J. Quickfall and D. Crockford, "Brain Neuroimaging in Cannabis Use: A Review," *Journal of Neuropsychiatry and Clinical Neuroscience* 18 (2006): 318; V. Lorenzetti et al., "Does Regular Cannabis Use Affect Neuroanatomy? An Updated Systematic Review and Meta-analysis of Structural Neuroimaging Studies," *European Archives of Psychiatry and Clinical Neuroscience* 269 (2019): 59. Studies like these are nice validation of my decision as a fifteen-year-old to never drink or do drugs (and to stick with that resolution).

Exercise and the frontal cortex: D. Moore et al., "Interrelationships between Exercise, Functional Connectivity, and Cognition among Healthy Adults: A Systematic Review," *Psychophysiology* (2022): e14014; J. Graban, N. Hlavacova, and D. Jezova, "Increased Gene Expression of Selected Vesicular and Glial Glutamate Transporters in the Frontal Cortex in Rats Exposed to Voluntary Wheel Running," *Journal of Physiology and Pharmacology* 68

(2017): 709; M. Ceftis et al., "The Effect of Exercise on Memory and BDNF Signaling Is Dependent on Intensity," *Brain Structure and Function* 224 (2019): 1975.

Eating disorders and frontal cortex: B. Donnelly et al., "Neuroimaging in Bulimia Nervosa and Binge Eating Disorder: A Systematic Review," *Journal of Eating Disorders* 6 (2018): 3; V. Alfano et al., "Multimodal Neuroimaging in Anorexia Nervosa," *Journal of Neuroscience Research* 98 (2020): 2178.

And for a really interesting study, see: F. Lederbogen et al., "City Living and Urban Upbringing Affect Neural Social Stress Processing in Humans," *Nature* 474 (2011): 498.

38. E. Durand et al., "History of Traumatic Brain Injury in Prison Populations: A Systematic Review," *Annals of Physical Rehabilitation Medicine* 60 (2017): 95; E. Shiroma, P. Ferguson, and E. Pickelsimer, "Prevalence of Traumatic Brain Injury in an Offender Population: A Meta-analysis," *Journal of Corrective Health Care* 16 (2010): 147; M. Linden, M. Lohan, and J. Bates-Gaston, "Traumatic Brain Injury and Co-occurring Problems in Prison Populations: A Systematic Review," *Brain Injury* 30 (2016): 839; E. De Geus et al., "Acquired Brain Injury and Interventions in the Offender Population: A Systematic Review," *Frontiers of Psychiatry* 12 (2021): 658328.

Footnote: J. Pemment, "Psychopathy versus Sociopathy: Why the Distinction Has Become Crucial," *Aggression and Violent Behavior* 18 (2013): 458.

39. E. Pascoe and L. Smart Richman, "Perceived Discrimination and Health: A Meta-analytic Review," *Psychological Bulletin* 135 (2009): 531; U. Clark, E. Miller, and R. R. Hegde, "Experiences of Discrimination Are Associated with Greater Resting Amygdala Activity and Functional Connectivity," *Biological Psychiatry and Cognitive Neuroscience Neuroimaging* 3 (2018): 367; C. Masten, E. Telzer, and N. Eisenberger, "An FMRI Investigation of Attributing Negative Social Treatment to Racial Discrimination," *Journal of Cognitive Neuroscience* 23 (2011): 1042; N. Fani et al., "Association of Racial Discrimination with Neural Response to Threat in Black Women in the US Exposed to Trauma," *JAMA Psychiatry* 78 (2021): 1005.

40. Adolescent adversity: K. Yamamuro et al., "A Prefrontal-Paraventricular Thalamus Circuit Requires Juvenile Social Experience to Regulate Adult Sociability in Mice," *Nature Neuroscience* 23 (2020): 10; C. Drzewiecki et al., "Adolescent Stress during, but Not after, Pubertal Onset Impairs Indices of Prepulse Inhibition in Adult Rats," *Developmental Psychobiology* 63 (2021): 837; M. Breach, K. Moench, and C. Wellman, "Social Instability in Adolescence Differentially Alters Dendritic Morphology in the Medial Prefrontal Cortex and Its Response to Stress in Adult Male and Female Rats," *Developmental Neurobiology* 79 (2019): 839; M. Leussis et al., "The Enduring Effects of an Adolescent Social Stressor on Synaptic Density, Part II: Poststress Reversal of Synaptic Loss in the Cortex by Adinazolam and MK-801," *Synapse* 62 (2008): 185; K. Zimmermann, R. Richardson, and K. Baker, "Maturational Changes in Prefrontal and Amygdala Circuits in Adolescence: Implications for Understanding Fear Inhibition during a Vulnerable Period of Development," *Brain Science* 9 (2019): 65; L. Wise et al., "Long-Term Effects of Adolescent Exposure to Bisphenol A on Neuron and Glia Number in the Rat Prefrontal Cortex: Differences between the Sexes and Cell Type," *Neurotoxicology* 53 (2016): 186.

41. T. Koseki et al., "Exposure to Enriched Environments during Adolescence Prevents Abnormal Behaviours Associated with Histone Deacetylation in Phencyclidine-Treated Mice," *International Journal of Psychoneuropharmacology* 15 (2012): 1489; F. Sadegzadeh et al., "Effects of Exposure to Enriched Environment during Adolescence on Passive Avoidance Memory, Nociception, and Prefrontal BDNF Level in Adult Male and Female Rats," *Neuroscience Letters* 732 (2020): 135133; J. McCreary, Z. Erikson, and Y. Hao, "Environmental Intervention as a Therapy for Adverse Programming by Ancestral Stress," *Science Reports* 6 (2016): 37814.

42. Effects of childhood stress and trauma on the frontal cortex: C. Weems et al., "Post-traumatic Stress and Age Variation in Amygdala Volumes among Youth Exposed to Trauma," *Social Cognitive and Affective Neuroscience* 10 (2015): 1661; A. Garrett et al., "Longitudinal Changes

in Brain Function Associated with Symptom Improvement in Youth with PTSD," *Journal of Psychiatric Research* 114 (2019): 161; V. Carrion et al., "Reduced Hippocampal Activity in Youth with Posttraumatic Stress Symptoms: An fMRI Study," *Journal of Pediatric Psychology* 35 (2010): 559; V. Carrion et al., "Converging Evidence for Abnormalities of the Prefrontal Cortex and Evaluation of Midsagittal Structures in Pediatric Posttraumatic Stress Disorder: An MRI Study," *Psychiatry Research: Neuroimaging* 172 (2009): 226; K. Richert et al., "Regional Differences of the Prefrontal Cortex in Pediatric PTSD: An MRI Study," *Depression and Anxiety* 23 (2006): 17; A. Tomoda et al., "Reduced Prefrontal Cortical Gray Matter Volume in Young Adults Exposed to Harsh Corporal Punishment," *Neuroimage* 47 (2009): T66; A. Chocyk et al., "Impact of Early-Life Stress on the Medial Prefrontal Cortex Functions—a Search for the Pathomechanisms of Anxiety and Mood Disorders," *Pharmacology Reports* 65 (2013): 1462; A. Chocyk et al., "Early-Life Stress Affects the Structural and Functional Plasticity of the Medial Prefrontal Cortex in Adolescent Rats," *European Journal of Neuroscience* 38 (2013): 2089; A. Chocyk et al., "Early Life Stress Affects the Structural and Functional Plasticity in the Medial Prefrontal Cortex in Adolescent Rats," *European Journal of Neuroscience* 38 (2013): 2089 (note—this was the film in which the young Tom Hanks made his debut as the dorsolateral prefrontal cortex); M. Lopez et al., "The Social Ecology of Childhood and Early Life Adversity," *Pediatric Research* 89 (2021): 353; V. Carrion and S. Wong, "Can Traumatic Stress Alter the Brain? Understanding the Implications of Early Trauma on Brain Development and Learning," *Journal of Adolescent Health* 51 (2013): S23.

Effects of the neighborhood in which a child is developing: X. Zhang et al., "Childhood Urbanicity Interacts with Polygenic Risk for Depression to Affect Stress-Related Medial Prefrontal Function," *Translation Psychiatry* 11 (2021): 522; B. Ramphal et al., "Associations between Amygdala-Prefrontal Functional Connectivity and Age Depend on Neighborhood Socioeconomic Status," *Cerebral Cortex Communications* 1 (2020): tgaa033.

Mothering effects on frontocortical maturation: D. Liu et al., "Maternal Care, Hippocampal Glucocorticoid Receptors, and Hypothalamic-Pituitary-Adrenal Responses to Stress," *Science* 277 (1997); S. Uchida et al., "Maternal and Genetic Factors in Stress-Resilient and -Vulnerable Rats: A Cross-Fostering Study," *Brain Research* 1316 (2010): 43.

Amid this large, grim literature, there is an issue of whether this is a realm of pathology or adaptation. Major early-life adversity produces a brain that, in adulthood, is hyperreactive to threat and stress, has trouble turning off vigilance, is poor at long-term planning and gratification postponement, and so on. Does this constitute a case of a pathologically dysfunctional brain in adulthood? Or is it precisely the sort of brain you want (if this is what your childhood was like, better have this sort of brain in preparation for more of the same in adulthood)? This issue is considered in M. Teicher, J. Samson, and K. Ohashi, "The Effects of Childhood Maltreatment on Brain Structure, Function and Connectivity," *Nature Reviews Neuroscience* 17 (2016): 652.

43. D. Kirsch et al,, "Childhood Maltreatment, Prefrontal-Paralimbic Gray Matter Volume, and Substance Use in Young Adults and Interactions with Risk for Bipolar Disorder," *Science Reports* 11 (2021): 123; M. Monninger et al., "The Long-Term Impact of Early Life Stress on Orbitofrontal Cortical Thickness," *Cerebral Cortex* 30 (2020): 1307; A. Van Harmelen et al., "Hypoactive Medial Prefrontal Cortex Functioning in Adults Reporting Childhood Emotional Maltreatment," *Scan* 9 (2014): 2026; A. Van Harmelen et al., "Childhood Emotional Maltreatment Severity Is Associated with Dorsal Medial Prefrontal Cortex Responsivity to Social Exclusion in Young Adults," *PLoS One* 9 (2014): E85107; M. Underwood, M. Bakalian, and V. Johnson, "Less NMDA Receptor Binding in Dorsolateral Prefrontal Cortex and Anterior Cingulate Cortex Associated with Reported Early-Life Adversity but Not Suicide," *International Journal of Neuropsychopharmacology* 23 (2020): 311; R. Salokangas et al., "Effect of Childhood Physical Abuse on Social Anxiety Is Mediated via Reduced Frontal Lobe and Amygdala-Hippocampus Complex Volume in Adult Clinical High-Risk Subjects," *Schizophrenia Research* 22 (2021): 101; M. Kim et al., "A Link between Childhood Adversity and Trait

Anger Reflects Relative Activity of the Amygdala and Dorsolateral Prefrontal Cortex," *Biological Psychiatry Cognitive Neuroscience and Neuroimaging* 3 (2018): 644; T. Kraynak et al., "Retrospectively Reported Childhood Physical Abuse, Systemic Inflammation, and Resting Corticolimbic Connectivity in Midlife Adults," *Brain, Behavior and Immunity* 82 (2019): 203.

44. C. Hendrix, D. Dilks, and B. McKenna, "Maternal Childhood Adversity Associates with Frontoamygdala Connectivity in Neonates," *Biological Psychiatry, Cognitive Neuroscience and Neuroimaging* 6 (2021): 470.

45. M. Monninger, E. Kraaijenvanger, and T. Pollok, "The Long-Term Impact of Early Life Stress on Orbitofrontal Cortical Thickness," *Cerebral Cortex* 30 (2020): 1307; N. Bush et al., "Kindergarten Stressors and Cumulative Adrenocortical Activation: The 'First Straws' of Allostatic Load?," *Developmental Psychopathology* 23 (2011): 1089; A. Conejero et al., "Frontal Theta Activation Associated with Error Detection in Toddlers: Influence of Familial Socioeconomic Status," *Developmental Science* 21 (2018), doi:10.1111/desc.12494; S. Lu, R. Xu, and J. Cao, "The Left Dorsolateral Prefrontal Cortex Volume Is Reduced in Adults Reporting Childhood Trauma Independent of Depression Diagnosis," *Journal of Psychiatric Research* 12 (2019): 12; L. Betancourt, N. Brodsky, and H. Hurt, "Socioeconomic (SES) Differences in Language Are Evident in Female Infants at 7 Months of Age," *Early Human Development* 91 (2015): 719.

46. Y. Moriguchi and I. Shinohara, "Socioeconomic Disparity in Prefrontal Development during Early Childhood," *Science Reports* 9 (2019): 2585; M. Varnum and S. Kitayama, "The Neuroscience of Social Class," *Current Opinion in Psychology* 18 (2017): 147; K. Muscatell et al., "Social Status Modulates Neural Activity in the Mentalizing Network," *Neuroimage* 60 (2012): 1771; K. Sarsour et al., "Family Socioeconomic Status and Child Executive Functions: The Roles of Language, Home Environment, and Single Parenthood," *Journal of International Neuropsychology* 17 (2011): 120; M. Monninger, E. Kraaijenvanger, and T. Pollok, "The Long-Term Impact of Early Life Stress on Orbitofrontal Cortical Thickness," *Cerebral Cortex* 30 (2020): 1307; N. Hair et al., "Association of Child Poverty, Brain Development, and Academic Achievement," *JAMA Pediatrics* 169 (2015): 822.

47. L. Machlin, K. McLaughlin, and M. Sheridan, "Brain Structure Mediates the Association between Socioeconomic Status and Attention-Deficit/Hyperactivity Disorder," *Developmental Science* 23 (2020): e12844; K. Sarsour et al., "Family Socioeconomic Status and Child Executive Functions: The Roles of Language, Home Environment, and Single Parenthood," *Journal of the International Neuropsychological Society* 17 (2011): 120; M. Kim et al., "A Link between Childhood Adversity and Trait Anger Reflects Relative Activity of the Amygdala and Dorsolateral Prefrontal Cortex," *Biological Psychiatry Cognitive Neuroscience Neuroimaging* 3 (2019): 644; B. Hart and T. Risley, *Meaningful Differences in the Everyday Experience of Young American Children* (Brooke, 1995); E. Hoff, "How Social Contexts Support and Shape Language Development," *Developmental Review* 26 (2006): 55.

Footnote: J. Reed, E. D'Ambrosio, and S. Marenço, "Interaction of Childhood Urbanicity and Variation in Dopamine Genes Alters Adult Prefrontal Function as Measured by Functional Magnetic Resonance Imaging (fMRI)," *PLoS One* 13, no. 4 (2018): e0195189; B. Besteher et al., "Associations between Urban Upbringing and Cortical Thickness and Gyrification," *Journal of Psychiatry Research* 95 (2017): 114; J. Xu et al., "Global Urbanicity Is Associated with Brain and Behavior in Young People," *Nature Human Behaviour* 6 (2022): 279; V. Steinheuser et al., "Impact of Urban Upbringing on the (Re)activity of the Hypothalamus-Pituitary-Adrenal Axis," *Psychosomatic Medicine* 76 (2014): 678; F. Lederbogen, P. Kirsch, and L. Haddad, "City Living and Urban Upbringing Affect Neural Social Stress Processing in Humans," *Nature* 474 (2011): 498.

48. C. Franz et al., "Adult Cognitive Ability and Socioeconomic Status as Mediators of the Effects of Childhood Disadvantage on Salivary Cortisol in Aging Adults," *Psychoneuroendocrinology* 38 (2013): 2127; D. Barch et al., "Early Childhood Socioeconomic Status and Cognitive and Adaptive Outcomes at the Transition to Adulthood: The Mediating Role of Gray Matter

Development across 5 Scan Waves," *Biological Psychiatry: Cognitive Neuroscience and Neuroimaging* 7 (2021): 34; M. Farah, "Socioeconomic Status and the Brain: Prospects for Neuroscience-Informed Policy," *Nature Reviews Neuroscience* 19 (2018): 428.

49. J. Herzog and C. Schmahl, "Adverse Childhood Experiences and the Consequences on Neurobiological, Psychosocial, and Somatic Conditions across the Lifespan," *Frontiers of Psychiatry* 9 (2018): 420.

50. A variety of adverse neurobiological consequences of prenatal stress: Y. Lu, K. Kapse, and N. Andersen, "Association between Socioeconomic Status and In Utero Fetal Brain Development," *JAMA Network Open* 4 (2021): e213526.

Effects on risk of psychiatric disorders: A. Converse et al., "Prenatal Stress Induces Increased Striatal Dopamine Transporter Binding in Adult Nonhuman Primates," *Biological Psychiatry* 74 (2013): 502; C. Davies et al., "Prenatal and Perinatal Risk and Protective Factors for Psychosis: A Systematic Review and Meta-analysis," *Lancet Psychiatry* 7 (2010): 399; J. Markham and J. Koenig, "Prenatal Stress: Role in Psychotic and Depressive Diseases," *Psychopharmacology* 214 (2011): 89; B. Van den Bergh et al., "Prenatal Developmental Origins of Behavior and Mental Health: The Influence of Maternal Stress in Pregnancy," *Neuroscience and Biobehavioral Reviews* 117 (2020): 26.

How does maternal stress during pregnancy have these adverse effects on the fetal brain and the brain of that fetus as an adult? Elevated glucocorticoid levels getting from maternal to fetal circulation, elevated levels of damaging inflammatory messengers, decreased blood flow to the fetus. See: A. Kinnunen, J. Koenig, and G. Bilbe, "Repeated Variable Prenatal Stress Alters Pre- and Postsynaptic Gene Expression in the Rat Frontal Pole," *Journal of Neurochemistry* 86 (2003): 736; B. Van den Bergh, R. Dahnke, and M. Mennes, "Prenatal Stress and the Developing Brain: Risks for Neurodevelopmental Disorders," *Development and Psychopathology* 30 (2018): 743.

51. G. Winterer and D. Goldman, "Genetics of Human Prefrontal Function," *Brain Research Reviews* 43 (2003): 134.

52. A. Heinz et al., "Amygdala-Prefrontal Coupling Depends on a Genetic Variation of the Serotonin Transporter," *Nature Neuroscience* 8 (2005): 20; L. Passamonti et al., "Monoamine Oxidase-a Genetic Variations Influence Brain Activity Associated with Inhibitory Control: New Insight into the Neural Correlates of Impulsivity," *Biological Psychiatry* 59 (2006): 334; M. Nomura and Y. Nomura, "Psychological, Neuroimaging, and Biochemical Studies on Functional Association between Impulsive Behavior and the 5-HT2A Receptor Gene Polymorphism in Humans," *Annals of the New York Academy of Sciences* 1086 (2006): 134. The more gene variants of the "risk-taking" cluster, the smaller the dlPFC: G. Avdogan et al., "Genetic Underpinnings of Risky Behavior Relate to Altered Neuroanatomy," *Nature Human Behaviour* 5 (2021): 787.

53. K. Bruce et al., "Association of the Promoter Polymorphism -1438G/A of the 5-HT2A Receptor Gene with Behavioral Impulsiveness and Serotonin Function in Women with Bulimia Nervosa," *American Journal of Medical Genetics, Part B, Neuropsychiatric Genetics* 137B (2005): 40.

54. K. Honnegger and B. de Bivot, "Stoachasticity, Individuality and Behavior," *Current Biology* 28 (2018): R8; J. Ayroles et al., "Behavioral Idiosyncrasy Reveals Genetic Control of Phenotypic Variability," *Proceedings of the National Academy of Sciences of the United States of America* 112 (20150): 6706. Also see G. Linneweber et al., "A Neurodevelopmental Origin of Behavioral Individual in the *Drosophila* Visual System," *Science* 367 (2020): 1112.

55. J. Chiao et al., "Neural Basis of Individualistic and Collectivistic Views of Self," *Human Brain Mapping* 30 (2009): 2813.

56. S. Han and Y. Ma, "Cultural Differences in Human Brain Activity: A Quantitative Meta-analysis," *Neuroimage* 99 (2014): 293; Y. Ma et al., "Sociocultural Patterning of Neural Activity during Self-Reflection," *Social Cognitive and Affective Neuroscience* 9 (2014): 73; Lu, Kapse, and Andersen, "Association between Socioeconomic Status."

57. P. Chen et al., "Medial Prefrontal Cortex Differentiates Self from Mother in Chinese: Evidence from Self-Motivated Immigrants," *Culture and Brain* 1 (2013): 3.
58. General review: J. Sasaki and H. Kim, "Nature, Nurture, and Their Interplay: A Review of Cultural Neuroscience," *Journal of Cross-Cultural Psychology* 48 (2016): 4.

    Interactions between culture and genes: M. Palmatier, A. Kang and K. Kidd, "Global Variation in the Frequencies of Functionally Different Catechol-O-Methyltransferase Alleles," *Biological Psychiatry* 46 (1999): 557; Y. Chiao and K. Blizinsky, "Culture-Gene Coevolution of Individualism-Collectivism and the Serotonin Transporter Gene," *Proceedings of the Royal Society B: Biological Sciences* 277 (2010): 22; K. Ishii et al., "Culture Modulates Sensitivity to the Disappearance of Facial Expression Associated with Serotonin Transporter Polymorphism (5-HTTLPR)," *Culture and Brain* 2 (2014): 72; J. LeClair et al., "Gene-Culture Interaction: Influence of Culture and Oxytocin Receptor Gene (OXTR) Polymorphism on Loneliness," *Culture and Brain* 4 (2016): 21; S. Luo et al., "Interaction between Oxytocin Receptor Polymorphism and Interdependent Culture Values on Human Empathy," *Social Cognitive and Affective Neuroscience* 10 (2015): 1273.
59. K. Norton and M. Lilieholm, "The Rostrolateral Prefrontal Cortex Mediates a Preference for High-Agency Environments," *Journal of Neuroscience* 40 (2020): 4401. For a similar theme, also see J. Parvizi et al., "The Will to Persevere Induced by Electrical Stimulation of the Human Cingulate Gyrus," *Neuron* 80 (2013): 1359.

5. A PRIMER ON CHAOS

1. These concepts are discussed in A. Maar, "Kinds of Determinism in Science," *Principia* 23 (2019): 503.
2. E. Lorenz, "Deterministic Non-periodic Flow," *Journal of Atmospheric Sciences* 20 (1963): 130.
3. Laughable folklore: R. Bishop, "What Could Be Worse Than the Butterfly Effect?," *Canadian Journal of Philosophy* 38 (2008): 519. Strange attractors as repelling as well as attracting: J. Hobbs, "Chaos and Indeterminism," *Canadian Journal of Philosophy* 21 (1991): 141.
4. "A Sound of Thunder" can be found in R. Bradbury, *The Golden Apples of the Sun* (Doubleday, 1953).
5. Footnote: M. Mitchell, *Complexity: A Guided Tour* (Oxford University Press, 2009).
6. For a particularly clear discussion of these ideas, see: M. Bedau, "Weak Emergence," *Philosophical Perspectives* 11 (1997): 375.
7. C. Gu et al., "Three-Dimensional Cellular Automaton Simulation of Coupled Hydrogen Porosity and Microstructure during Solidification of Ternary Aluminum Alloys," *Scientific Reports* 9 (2019): 13099. YouTube has a number of videos showing 3D cellular automata, which are spectacular. For example: Softology, "3D Cellular Automata," December 5, 2017, YouTube video, 2:30, youtube.com/watch?v=dQJ5aEsP6Fs; Softology, "3D Accretor Cellular Automata," January 26, 2018, YouTube video, 4:45, youtube.com/watch?v=_W-n510Pca0.

    Footnote (p. 142): S. Wolfram, *A New Kind of Science* (Wolfram Media, 2002).

    Okay, I have a horrible confession to make. On p. 138, there is the picture of the wildly chaotic, thoroughly unpredictable complex cellular automata that can be generated with rule 22. Here's the confession: this isn't actually made with rule 22; instead, it's made with the closely related rule 90. The visual showing a crazily complex wonderful version of rule 22 was of terrible quality, I couldn't find anything better, made no headway in getting the Wolfram Empire to send a higher-resolution of the visual . . . and in a moment in the dark of night that tests one's soul, with the clock ticking, I decided to stick in a cool visual generated with rule 90 instead. It makes the same point—knowing the starting state and reproduction rule (90, in this case) gives you zero predictability as to what a complex version is going to look like. In fact, it makes a point about the chaoticism of cellular automata even more powerful—no one (hopefully? please) looking at it could tell that this complex pattern arose from application of rule 22 or rule 90. Now that's off my chest.

6. IS YOUR FREE WILL CHAOTIC?

1. Ideas discussed in: D. Porush, "Making Chaos: Two Views of a New Science," *New England Review and Bread Loaf Quarterly* 12 (1990): 439.
2. A sampling: M. Cutright, *Chaos Theory and Higher Education: Leadership, Planning, and Policy* (Peter Lang, 2001); S. Sule and S. Nilhan, *Chaos, Complexity and Leadership 2018: Explorations of Chaotic and Complexity Theory* (Springer, 2020); E. Peters, *Fractal Market Analysis: Applying Chaos Theory to Investment and Economics* (Wiley, 1994); R. Pryor, *The Chaos Theory of Careers* (Routledge, 2011); K. Yas et al., "From Natural to Artificial Selection: A Chaotic Reading of Shelagh Stephenson's *An Experiment with an Air Pump* (1998)," *International Journal of Applied Linguistics and English Literature* 7 (2018): 23; A. McLachlan, "Same but Different: Chaos and TV Drama Narratives" (doctoral thesis, Victoria University, Wellington, New Zealand, 2019), hdl.handle.net/10063/8046. Theological musings: D. Gray, *Toward a Theology of Chaos: The New Scientific Paradigm and Some Implications for Ministry* (Citeseer, 1997); D. Steenburg, "Chaos at the Marriage of Heaven and Hell," *Harvard Theological Review* 84 (1991): 447; J. Eigenauer, "The Humanities and Chaos Theory: A Response to Steenburg's 'Chaos at the Marriage of Heaven and Hell,'" *Harvard Theological Review* 86 (1993): 455; D. Steenburg, "A Response to John D. Eigenauer," *Harvard Theological Review* 86 (1993): 471.

   Footnote: J. Bassingthwaight, L. Liebovitch, and B. West, *Fractal Physiology* (American Physiological Society, 1994); N. Schweighofer et al., "Chaos May Enhance Information Transmission in the Inferior Olive," *Proceedings of the National Academy of Sciences of the United States of America* 101 (2004): 4655.
3. Simpsons Wiki, s.v. "Chaos Theory in Baseball Analysis," simpsons.fandom.com/wiki/Chaos_Theory_in_Baseball_Analysis; M. Farmer, *Chaos Theory, Nerds of Paradise book 2* (Amazon.com Services, 2017).
4. G. Eilenberger, "Freedom, Science, and Aesthetics," in *The Beauty of Fractals*, ed. H. Peitgen and P. Richter (Springer, 1986), p. 179.
5. K. Clancy, "Your Brain Is on the Brink of Chaos," *Nautilus*, Fall 2014, 144.
6. Farmer quoted in James Gleick, *Chaos: Making a New Science* (Viking, 1987), p. 251.
7. Steenburg, "Chaos at the Marriage."
8. Eilenberger, "Freedom, Science, and Aesthetics," p. 176.
9. A. Maar, "Kinds of Determinism in Science," *Principia* 23 (2019): 503. For a comparison of encompassing versus individual determinism, see: J. Doomen, "Cornering 'Free Will,'" *Journal of Mind and Behavior* 32 (2011): 165; H. Atmanspacher, "Determinism Is Ontic, Determinability Is Epistemic," in *Between Chance and Choice: Interdisciplinary Perspectives on Determinism*, ed. R. Bishop and H. Atmanspacher (Imprint Academic, 2002). To get further into the weeds with the likes of "partial determinism" and "adequate determinism," see: J. Earman, *A Primer on Determinism* (Reidel, 1986); S. Kellert, *In the Wake of Chaos: Unpredictable Order in Dynamic Systems* (University of Chicago Press, 1993).
10. S. Caprara and A. Vulpiani, "Chaos and Stochastic Models in Physics: Ontic and Epistemic Aspects," in *Models and Inferences in Science. Studies in Applied Philosophy, Epistemology and Rational Ethics*, vol. 25, ed. E. Ippoliti, F. Sterpetti, and T. Nickles (Springer, 2016), p. 133; G. Hunt, "Determinism, Predictability and Chaos," *Analysis* 47 (1987): 129; M. Stone, "Chaos, Prediction and Laplacean Determinism," *American Philosophical Quarterly* 26 (1989): 123; V. Batitsky and Z. Domotor, "When Good Theories Make Bad Predictions," *Synthese* 157 (2007): 79.
11. W. Seeley, "Behavioral Variant Frontotemporal Dementia," *Continuum* 25 (2019): 76; R. Dawkins, *The Blind Watchmaker* (Norton, 1986), p. 9.
12. W. Farnsworth and M. Grady, *Torts: Cases and Questions*, 3rd ed. (Wolters Kluwer, 2019).
13. R. Sapolsky, "Measures of Life," *The Sciences*, March/April 1994, p. 10.
14. M. Shandlen, "Comment on Adina Roskies," in *Moral Psychology*, vol. 4, *Free Will and Moral Responsibility*, ed. W. Sinnott-Armstrong (MIT Press, 2014), p. 139.

7. A PRIMER ON EMERGENT COMPLEXITY

1. Footnote: For more on this concept, see R. Carneiro, "The Transition from Quantity to Quality: A Neglected Causal Mechanism in Accounting for Social Evolution," *Proceedings of the National Academy of Sciences of the United States of America* 97 (2000): 12926.
2. For some impressive examples, see: W. Tschinkel, "The Architecture of Subterranean Ant Nests: Beauty and Mystery Underfoot," *Journal of Bioeconomics* 17 (2015): 271; M. Bollazzi and F. Roces, "The Thermoregulatory Function of Thatched Nests in the South American Grass-Cutting Ant, *Acromyrmex heyeri*," *Journal of Insect Science* 10 (2010): 137; I. Guimarães et al., "The Complex Nest Architecture of the Ponerinae Ant *Odontomachus chelifer*," *PLoS One* 13 (2018): e0189896; N. Mlot, C. Tovey, and D. Hu, "Diffusive Dynamics of Large Ant Rafts," *Communicative and Integrative Biology* 5 (2012): 590. For a theoretical approach to ant emergence, see D. Gordon, "Control without Hierarchy," *Nature* 446 (2007): 143. For a demonstration of how it's not all fun and games being an ant, see: N. Stroeymeyt et al., "Social Network Plasticity Decreases Disease Transmission in a Eusocial Insect," *Science* 363 (2018): 941 (the authors show that ant networks shift so that sick ants [experimentally infected with a fungus] are ostracized to limit infectivity).
3. P. Anderson, "More Is Different," *Science* 177 (1972): 393. Back to the fact that a single molecule of water cannot possess the property of "wetness"— it also cannot possess the property of surface tension (the emergent feature of water that allows basilisk Jesus lizards to run across the surface of a pond).
4. For an analysis of how people move in crowds in ways that resemble the fluid dynamics of waterfalls, see N. Bain and D. Bartolo, "Dynamic Response and Hydrodynamics of Polarized Crowds," *Science* 363 (2019): 46. For something similar in ants, see A. Dussutour et al., "Optimal Travel Organization in Ants under Crowded Conditions," *Nature* 428 (2003): 70.
5. Footnote: Discussed at length in P. Hiesenger, *The Self-Assembling Brain: How Neural Networks Grow Smarter* (Princeton University Press, 2021).
6. M. Bedau, "Is Weak Emergence Just in the Mind?," *Minds and Machines* 18 (2008): 443; J. Kim, "Making Sense of Emergence," *Philosophical Studies* 95 (1999): 3; O. Sartenaer, "Sixteen Years Later: Making Sense of Emergence (Again)," *Journal of General Philosophical Sciences* 47 (2016): 79.
7. E. Bonabeau and G. Theraulaz, "Swarm Smarts," *Scientific American* 282, no. 3 (2000): 72; M. Dorigo and T. Stutzle, *Ant Colony Optimization* (MIT Press, 2004); S. Garnier, J. Gautrais, and G. Theraulaz, "The Biological Principles of Swarm Intelligence," *Swarm Intelligence* 1 (2007): 3 (note that this topic was the first paper published in the history of this journal, which kind of makes sense, given its title).
8. L. Chen, D. Hall, and D. Chklovskii, "Wiring Optimization Can Relate Neuronal Structure and Function," *Proceedings of the National Academy of Sciences of the United States of America* 103 (2006): 4723; M. Rivera-Alba et al., "Wiring Economy and Volume Exclusion Determine Neuronal Placement in the *Drosophila* Brain," *Current Biology* 21 (2011): 2000; J. White et al., "The Structure of the Nervous System of the Nematode *Caenorhabditis elegans*," *Philosophical Transactions of the Royal Society B, Biological Sciences* 314 (1986): 1; V. Klyachko and C. Stevens, "Connectivity Optimization and the Positioning of Cortical Areas," *Proceedings of the National Academy of Sciences of the United States of America* 100 (2003): 7937; G. Mitchison, "Neuronal Branching Patterns and the Economy of Cortical Wiring," *Proceedings of the Royal Society B: Biological Sciences* 245 (1991): 151.
9. Y. Takeo et al., "GluD2- and Cbln1-Mediated Competitive Interactions Shape the Dendritic Arbor of Cerebellar Purkinje Cells," *Neuron* 109 (2020): 629.
10. S. Camazine and J. Sneyud, "A Model of Collective Nectar Source Selection by Honey Bees: Self-Organization through Simple Rules," *Journal of Theoretical Biology* 149 (1991): 547.

    Footnote (p. 160): K. von Frisch, *The Dancing Bees: An Account of the Life and Senses of the Honey Bee* (Harvest Books, 1953).

11. P. Visscher, "How Self-Organization Evolves," *Nature* 421 (2003): 799; M. Myerscough, "Dancing for a Decision: A Matrix Model for Net-Site Choice by Honey Bees," *Proceedings of the Royal Society of London B* 270 (2003): 577; D. Gordon, "The Rewards of Restraint in the Collective Regulation of Foraging by Harvester Ant Colonies," *Nature* 498 (2013): 91; D. Gordon, "The Ecology of Collective Behavior," *PLOS Biology* 12, no. 3 (2014): e1001805. For additional studies in this area see: J. Deneubourg and S. Goss, "Collective Patterns and Decision Making," *Ethology Ecology and Evolution* 1 (1989): 295; S. Edwards and S. Pratt, "Rationality in Collective Decision-Making by Ant Colonies," *Proceedings of the Royal Society B* 276 (2009): 3655; E. Bonabeau et al., "Self-Organization in Social Insects," *Trends in Ecology and Evolution* 12 (1997): 188.

    Footnote: G. Sherman and P. Visscher, "Honeybee Colonies Achieve Fitness through Dancing," *Nature* 419 (2002): 920.

    Second footnote: R. Goldstone, M. Roberts, and T. Gureckis, "Emergent Processes in Group Behavior," *Current Directions in the Psychological Sciences* 17 (2008): 10; C. Doctorow, "A Catalog of Ingenious Cheats Developed by Machine-Learning Systems," *BoingBoing*, November 12, 2018, boingboing.net/2018/11/12/local-optima-r-us.html.
12. C. Reid and M. Beekman, "Solving the Towers of Hanoi—How an Amoeboid Organism Efficiently Constructs Transport Networks," *Journal of Experimental Biology* 216 (2013): 1546; C. Reid and T. Latty, "Collective Behaviour and Swarm Intelligence in Slime Moulds," *FEMS Microbiology Reviews* 40 (2016): 798.
13. S. Tero et al., "Rules for Biologically Inspired Adaptive Network Design," *Science* 327 (2010): 439.
14. For another example of slime molds, see L. Tweedy et al., "Seeing around Corners: Cells Solve Mazes and Respond at a Distance Using Attractant Breakdown," *Science* 369 (2020): 1075.
15. Footnote: Hiesenger, *Self-Assembling Brain*. For an example of repulsion, see D. Pederick et al., "Reciprocal Repulsions Instruct the Precise Assembly of Parallel Hippocampal Networks," *Science* 372 (2021): 1058. See also: L. Luo, "Actin Cytoskeleton Regulation in Neuronal Morphogenesis and Structural Plasticity," *Annual Review of Cellular Developmental Biology* 18 (2002): 601; J. Raper and C. Mason, "Cellular Strategies of Axonal Pathfinding," *Cold Spring Harbor Perspectives in Biology* 2 (2010): a001933.

    How do neurons, when connecting up, also figure out which part of a target neuron to connect to (proximal or distal spines, cell body, axon in case of some neurotransmitters)? Neurons have the ability to control which branches of their axonal processes receive proteins needed for synapse construction: S. Falkner and P. Scheiffele, "Architects of Neuronal Wiring," *Science* 364 (2019): 437; O. Urwyler et al., "Branch-Restricted Localization of Phosphatase Prl-1 Specifies Axonal Synaptogenesis Domains," *Science* 364 (2019): 454; E. Favuzzi et al., "Distinct Molecular Programs Regulate Synapse Specificity in Cortical Inhibitory Circuits," *Science* 363 (2019): 413.
16. T. More, A. Buffo, and M. Gotz, "The Novel Roles of Glial Cells Revisited: The Contribution of Radial Glia and Astrocytes to Neurogenesis," *Current Topics in Developmental Biology* 69 (2005): 67; P. Malatesta, I. Appolloni, and F. Calzolari, "Radial Glia and Neural Stem Cells," *Cell and Tissue Research* 331 (2008): 165; P. Oberst et al., "Temporal Plasticity of Apical Progenitors in the Developing Mouse Neocortex," *Nature* 573 (2019): 370.
17. For an example of the molecular biology of the exquisite timing that goes into neuron interactions with radial glia, see K. Yoon et al., "Temporal Control of Mammalian Cortical Neurogenesis by m6A Methylation," *Cell* 171 (2017): 877.

    Footnote: N. Ozel et al., "Serial Synapse Formation through Filopodial Competition for Synaptic Seeding Factors," *Developmental Cell* 50 (2019): 447; M. Courgeon and C. Desplan, "Coordination between Stochastic and Deterministic Specification in the *Drosophila* Visual System," *Science* 366 (2019): 325.

18. T. Huxley, "On the Hypothesis That Animals Are Automata, and Its History," *Nature* 10 (1874): 362. Amid all these references to recent, cutting-edge science, it's kind of charming to reference a nineteenth-century scientific publication.
19. For more on this general topic, see the fantastic J. Gleick, *Chaos: Making a New Science* (Viking, 1987).
20. Forty-eight thousand miles: J. Castro, "11 Surprising Facts about the Circulatory System," *LiveScience*, August 8, 2022, livescience.com/39925-circulatory-system-facts-surprising.html.
21. The basis of this model: D. Iber and D. Menshykau, "The Control of Branching Morphogenesis," *Open Biology* 3 (2013): 130088130088; D. Menshykau, C. Kraemer, and D. Iber, "Branch Mode Selection during Early Lung Development," *PLOS Computational Biology* 8 (2012): e1002377. For an exploration of these issues at the laboratory bench, see R. Metzger et al., "The Branching Programme of Mouse Lung Development," *Nature* 453 (2008): 745.
22. A. Lindenmayer, "Developmental Algorithms for Multicellular Organisms: A Survey of L-Systems," *Journal of Theoretical Biology* 54 (1975): 3.
23. A. Ochoa-Espinosa and M. Affolter, "Branching Morphogenesis: From Cells to Organs and Back," *Cold Spring Harbor Perspectives in Biology* 4 (2004): a008243; P. Lu and Z. Werb, "Patterning Mechanisms of Branched Organs," *Science* 322 (2008): 1506–9.

    Footnote: A. Turing, "The Chemical Basis of Morphogenesis," *Philosophical Transactions of the Royal Society of London B* 237 (1952): 37.

    Second footnote: E. Azpeitia et al., "Cauliflower Fractal Forms Arise from Perturbations of Floral Gene Networks," *Science* 373 (2021): 192.
24. G. Vogel, "The Unexpected Brains behind Blood Vessel Growth," *Science* 307 (2005): 665; Metzger et al., "Branching Programme of Mouse Lung Development"; P. Carmeliet and M. Tessier-Lavigne, "Common Mechanisms of Nerve and Blood Vessel Wiring," *Nature* 436 (2005): 193. The second author, the superb neurobiologist and my departmental colleague Marc Tessier-Lavigne, expanded his portfolio a few years back and became president of Stanford University.
25. J. Bassingthwaighte, L. Liebovitch, and B. West, *Fractal Physiology, Methods in Physiology* (American Physiological Society, 1994).
26. "The World Religions Tree," 000024.org/religions_tree/religions_tree_8.html.
27. E. Favuzi et al., "Distinct Molecular Programs Regulate Synapse Specificity in Cortical Inhibitory Circuits," *Science* 363 (2019): 413; V. Hopker et al., "Growth-Cone Attraction to Netrin-1 Is Converted to Repulsion by Laminin-1," *Nature* 401 (1999): 69; J. Dorskind and A. Kolodkin, "Revisiting and Refining Roles of Neural Guidance Cues in Circuit Assembly," *Current Opinion in Neurobiology* 66 (2020): 10; S. McFarlane, "Attraction vs. Repulsion: The Growth Cones Decides," *Biochemistry and Cell Biology* 78 (2000): 563.
28. A. Bassem, A. Hassan, and P. R. Hiesinger, "Beyond Molecular Codes: Simple Rules to Wire Complex Brains," *Cell* 163 (2015): 285. For a two-rule system built around mechanical constraints explaining one aspect of human brain development, see E. Karzbrun et al., "Human Neural Tube Morphogenesis in Vitro by Geometric Constraints," *Nature* 599 (2021): 268.
29. D. Miller et al., "Full Genome Viral Sequences Inform Patterns of SARS-CoV-2 Spread into and within Israel," *Nature Communications* 11 (2020): 5518; D. Adam et al., "Clustering and Superspreading Potential of Severe Acute Respiratory Syndrome Coronavirus 2 (SARS-CoV-2) Infections in Hong Kong," *Nature Medicine* 26 (2020): 1714.
30. General reviews: A. Barabasi, "Scale-Free Networks: A Decade and Beyond," *Science* 325 (2009): 412; A. Barabasi and R. Albert, "Emergence of Scaling in Random Networks," *Science* 286 (1999): 509; C. Song, S. Havlin, and H. Makse, "Self-Similarity of Complex Networks," *Nature* 433 (2005): 392; P. Drew and L. Abbott, "Models and Properties of Power-Law Adaptation in Neural Systems," *Journal of Neurophysiology* 96 (2006): 826.

    Power law and related distributions in the brain: G. Buzsaki and A. Draguhn, "Neuronal Oscillations in Cortical Networks," *Science* 304 (2004): 1926; Power laws and the number of neurotransmitter vesicles released in response to an action potential: J. Lamanna et al., "A Pre-docking Source for the Power-Law Behavior of Spontaneous Quantal Release: Application to the Analysis of LTP," *Frontiers of Cellular Neuroscience* 9 (2015): 44.

Power law distributions and:

Spread of Covid: D. Miller et al., "Full Genome Viral Sequences Inform Patterns of SARS-CoV-2 Spread into and within Israel," *Nature Communications* 11 (2020): 5518; D. Adam et al., "Clustering and Superspreading Potential of Severe Acute Respiratory Syndrome Coronavirus 2 (SARS-CoV-2) Infections in Hong Kong," *Nature Medicine* 26 (2020): 1714.

Earthquakes: F. Meng, L. Wong, and H. Zhou, "Power Law Relations in Earthquakes from Microscopic to Macroscopic Scales," *Scientific Reports* 9 (2019): 10705.

Warfare and hate groups: N. Gilbert, "Modelers Claim Wars Are Predictable," *Nature* 462 (2009): 836; N. Johnson et al., "Hidden Resilience and Adaptive Dynamics of the Global Online Hate Ecology," *Nature* 573 (2019): 261; M. Schich et al., "Quantitative Social Science: A Network Framework of Cultural History," *Science* 345 (2014): 558.

Here's a topic that I barely understand, but I want to show off that I was able to force myself through a number of papers on the subject. A pattern can be highly structured, with repeating building blocks; its signal in a frequency spectrum is termed "white noise." This is akin to tight, uniform little interconnected clusters of neurons, isolated from each other. At the other extreme, a pattern that is random produces "brown noise" (named for Brownian motion, explained in chapter 9); these are connections between neurons of random distances, directions, and strengths. And as with porridge that is neither too hot nor too cold, there are patterns poised between the two extremes, termed "pink noise" (or 1/f noise). These are the networks of the brain balanced in scale-free ways between the robustness and efficiency of small, structured local networks and the creativity and evolvability of long-distance ones. The "critical brain" hypothesis posits that brains have evolved to be at this ideal spot and that this "criticality" optimizes all sorts of features of brain function. Moreover, in this model, the brain is able to correct itself as that perfect balancing point shifts with circumstances; this would be an example of the very trendy "self-organized criticality." This can be shown with some mathematically bruising analytical techniques, and a small subfield has grown examining brain criticality in normal and diseased circumstances. For example, there is a tilt toward white noise in epilepsy, reflecting the overly synchronized firing of clusters of epileptiform neurons (and, in fact, there is a remarkable similarity between the distribution of frequency and severity of seizures, and that of earthquakes). Similarly, autism spectrum disorder appears to have a different type of tilt toward white noise, reflecting the relatively isolated peninsulas of function in the cortex. And at the other end, Alzheimer's disease involves a tilt toward brown noise, as the death of neurons here and there begins to break down the patterning (and efficacy) of networks. See: J. Beggs and D. Plenz, "Neuronal Avalanches in Neocortical Circuits," *Journal of Neuroscience* 23 (2003): 11167; P. Bak, C. Tang, and K. Wiesenfeld, "Self-Organized Criticality: An Explanation of the 1/f Noise," *Physics Review Letters* 59 (1987): 381; L. Cocchi et al., "Criticality in the Brain: A Synthesis of Neurobiology, Models and Cognition," *Progress in Neurobiology* 158 (2017): 132; M. Gardner, "White and Brown Music, Fractal Curves and One-Over-f Fluctuation," *Scientific American*, April 1978; M. Belmonte et al., "Autism and Abnormal Development of Brain Connectivity," *Journal of Neuroscience* 24 (2004): 9228.

Footnote: A website celebrating the mathematics of Bacon numbers: coursehero.com/file/p12lp1kl/chosen-actors-can-be-linked-by-a-path-through-Kevin-Bacon-in-an-average-of-6/. For an excellent, accessible biography of Paul Erdös, see P. Hoffman, *The Man Who Loved Only Numbers: The Story of Paul Erdös and the Search for Mathematical Truth* (Hyperion, 1998).

31. For an example, see J. Couzin et al., "Effective Leadership and Decision-Making in Animal Groups on the Move," *Nature* 433 (2005): 7025.

Footnote: For example, see C. Candia et al., "The Universal Decay of Collective Memory and Attention," *Nature Human Behaviour* 3 (2018): 82. Also see V. Verbavatz and M. Barthelemy, "The Growth Equation of Cities," *Nature* 587 (2020): 397.

32. C. Song, S. Havlin, and H. Makse, "Self-Similarity of Complex Networks," *Nature* 433 (2005): 392.

Emergence in ecological contexts: M. Buchanan, "Ecological Modeling: The Mathematical Mirror to Animal Nature," *Nature* 453 (2008): 714; N. Humphries et al., "Environmental Context Explains Levy and Brownian Movement Patterns of Marine Predators," *Nature* 465 (2010): 1066; J. Banavar et al., "Scaling in Ecosystems and the Linkage of Macroecological Laws," *Physical Review Letters* 98 (2007): 068104; B. Houchmandzadeh and M. Vallade, "Clustering in Neutral Ecology," *Physical Reviews E* 68 (2003): 061912.

Emergence and behavior (including metaphorical behavior by white blood cells): "The Emergent Properties of a Dolphin Social Network," *Proceedings of the Royal Society of London B* 270, no supp. 2 (2003): S186; T. Harris et al., "Generalized Levy Walks and the Role of Chemokines in Migration of Effector CD8(+) T Cells," *Nature* 486 (2012): 545.

Emergence in neurons and neuronal circuits: D. Lusseau, S. Romano, and M. Eguia, "Characterization of Degree Frequency Distribution in Protein Interaction Networks," *Physical Reviews E* 71 (2005): 031901; D. Bray, "Molecular Networks: The Top-Down View," *Science* 301 (2003): 1864; B. Fulcher and A. Fornito, "A Transcriptional Signature of Hub Connectivity in the Mouse Connectome," *Proceedings of the National Academy of Sciences of the United States of America* 113 (2016): 1435.

33. Power laws and the evolutionary pressure to optimize wiring efficiency in the brain: S. Neubauer et al., "Evolution of Brain Lateralization: A Shared Hominid Pattern of Endocranial Asymmetry Is Much More Variable in Humans Than in Great Apes," *Science Advances* 6 (2020): eaax9935; I. Wang and T. Clandinin, "The Influence of Wiring Economy on Nervous System Evolution," *Current Biology* 26 (2016): R1101; T. Namba et al., "Metabolic Regulation of Neocortical Expansion in Development and Evolution," *Neuron* 109 (2021): 408; K. Zhang and T. Sejnowski, "A Universal Scaling Law between Gray Matter and White Matter of Cerebral Cortex," *Proceedings of the National Academy of Sciences of the United States of America* 97 (2000): 5621.

For an example of a debate in the field as to the degree to which clusters of highly interacting neurons contribute to brain function, see: J. Cohen and F. Tong, "The Face of Controversy," *Science* 293 (2001): 2405; P. Downing et al., "A Cortical Area Selective for Visual Processing of the Human Body," *Science* 293 (2001): 2470; J. Haxby et al., "Distributed and Overlapping Representations of Faces and Objects in Ventral Temporal Cortex," *Science* 293 (2001): 2425.

As a measure of just how much evolutionary pressure there has been to optimize spatial aspects of brain development, our brains contain approximately sixty thousand miles of projections among neurons: C. Filley, "White Matter and Human Behavior," *Science* 372 (2021): 1265.

Footnote: For an example, see A. Wissa, "Birds Trade Flight Stability for Manoeuvrability," *Nature* 603 (2022): 579.

Second footnote: Small-world networks: D. Bassett and E. Bullmore, "Small-World Brain Networks," *Neuroscientist* 12 (2006): 512; D. Bassett and E. Bullmore, "Small-World Brain Networks Revisited," *Neuroscientist* 23 (2017): 499; D. Watts and S. Strogatz, "Collective Dynamics of 'Small-World' Networks," *Nature* 393 (1998): 440. Two papers exploring just how important the sparse, long-distance projections are: J. Giles, "Making the Links," *Nature* 488 (2012): 448; M. Granovetter, "The Strength of Weak Ties," *American Journal of Sociology* 78 (1973): 1360.

34. Footnote: V. Zimmern, "Why Brain Criticality Is Clinically Relevant: A Scoping Review," *Frontiers in Neural Circuits* 26 (2020), doi.org/10.3389/fncir.2020.00054.

35. Footnote: Stigmergy: J. Korb, "Termite Mound Architecture, from Function to Construction," in *Biology of Termites: A Modern Synthesis*, ed. D. Bignell, Y. Roisin, and N. Lo (Springer,

2010), p. 349; J. Turner, "Termites as Models of Swarm Cognition," *Swarm Intelligence* 5 (2011): 19; E. Bonabeau et al., "Self-Organization in Social Insects," *Trends in Ecology and Evolution* 12 (1997): 188.

Applications to machine learning: J. Korb, "Robots Acting Locally and Building Globally," *Science* 343 (2014): 742.

Applications to the Wisdom of the Crowd phenomenon: A. Woolley et al., "Evidence for a Collective Intelligence Factor in the Performance of Human Groups," *Science* 330 (2010): 686; D. Wilson, J. Timmel, and R. Miller, "Cognitive Cooperation," *Human Nature* 15 (2004): 225. For a demonstration of how purely egalitarian wisdom-of-the-crowd phenomena aren't always the best, see P. Tetlock, B. Mellers, and J. Scoblic, "Bringing Probability Judgments into Policy Debates via Forecasting Tournaments," *Science* 355 (2017): 481.

Bottom-up curation systems: J. Giles, "Internet Encyclopedias Go Head to Head," *Nature* 438 (2005): 900; J. Beck, "Doctors' #1 Source for Healthcare Information: Wikipedia," *Atlantic*, March 5, 2014.

36. For a string of some of the dazzling findings in this field, see: M. Lancaster et al., "Cerebral Organoids Model Human Brain Development and Microcephaly," *Nature* 501 (2013): 373; J. Camp et al., "Human Cerebral Organoids Recapitulate Gene Expression Programs of Fetal Neocortex Development," *Proceedings of the National Academy of Science of the United States of America* 112 (2015): 15672; F. Birey, J. Andersen, and C. Makinson, "Assembly of Functionally Integrated Human Forebrain Spheroids," *Nature* 545 (2017): 54; S. Pasca, "The Rise of Three-Dimensional Human Brain Cultures," *Nature* 533 (2018): 437; S. Pasca, "Assembling Human Brain Organoids," *Science* 363 (2019): 126; C. Trujillo et al., "Complex Oscillatory Waves Emerging from Cortical Organoids Model Early Human Brain Network Development," *Cell Stem Cell* 25 (2019): 558; Frankfurt Radio Symphony, Manfred Honeck, conductor; L. Pelegrini et al., "Human CNS Barrier-Forming Organoids with Cerebrospinal Fluid Production," *Science* 369 (2020): 6500; I. Chiaradia and M. Lancaster, "Brain Organoids for the Study of Human Neurobiology at the Interface of in Vitro and in Vivo," *Nature Neuroscience* 23 (2020): 1496.

Footnote: For a thoroughly cool demonstration of the different types of brain organoids produced by different ape species, see: Z. Kronenberg et al., "High-Resolution Comparative Analysis of Great Ape Genomes," *Science* 360 (2018): 6393; C. Trujillo, E. Rice, and N. Schaefer, "Reintroduction of the Archaic Variant of *NOVA1* in Cortical Organoids Alters Neurodevelopment," *Science* 371 (2021): 6530; A. Gordon et al., "Long-Term Maturation of Human Cortical Organoids Matches Key Early Postnatal Transitions," *Nature Neuroscience* 24 (2021): 331.

Second footnote: S. Giandomenico et al., "Cerebral Organoids at the Air-Liquid Interface Generate Diverse Nerve Tracts with Functional Output," *Nature Neuroscience* 22 (2019): 669; V. Marx, "Reality Check for Organoids in Neuroscience," *Nature Methods* 17 (2020): 961; R. Menzel and M. Giurfa, "Cognitive Architecture of a Mini-Brain: The Honeybee," *Trends in Cognitive Sciences* 5 (2001): 62; S. Reardon, "Can Lab-Grown Brains Become Conscious?," *Nature* 586 (2020): 658; J. Koplin and J. Savulescu, "Moral Limits of Brain Organoid Research," *Journal of Law and Medical Ethics* 47 (2019): 760.

37. J. Werfel, K. Petersen, and R. Nagpal, "Designing Collective Behavior in a Termite-Inspired Robot Construction Team," *Science* 343 (2014): 754; W. Marwan, "Amoeba-Inspired Network Design," *Science* 327 (2019): 419; L. Shimin et al., "Slime Mould Algorithm: A New Method for Stochastic Optimization," *Future Generation Computer Systems* 111 (2020): 300; T. Umedachi et al., "Fully Decentralized Control of a Soft-Bodied Robot Inspired by True Slime Mold," *Biological Cybernetics* 102 (2010): 261. For a charming case of the student teaching the master, see: J. Halloy et al., "Social Integration of Robots into Groups of Cockroaches to Control Self-Organized Choices," *Science* 318 (2007): 5853.

Final footnote: S. Bazazi et al., "Collective Motion and Cannibalism in Locust Migratory Bands," *Current Biology* 18 (2008): 735. Just in case you thought locust cannibalism was dated science by now—as this book went to press, the presses were stopped for a May 5, 2023, paper detailing that locust have evolved pheromonal signaling mechanisms to decrease the odds of getting eaten by the locust just behind them. H. Chang et al., "A Chemical Defense Defers Cannibalism in Migratory Locusts," *Science* 380 (2023) 537.

8. DOES YOUR FREE WILL JUST EMERGE?

1. C. List, "The Naturalistic Case for Free Will: The Challenge," *Brains Blog*, August 12, 2019, https://philosophyofbrains.com/2019/08/12/1-the-naturalistic-case-for-free-will-the-challenge.aspx; R. Kane, "Rethinking Free Will: New Perspectives on an Ancient Problem," in *The Oxford Handbook of Free Will*, ed. R. Kane (Oxford University Press, 2002), 134.
2. List, "The Naturalistic Case for Free Will: The Challenge."
3. C. List and M. Pivato, "Emergent Chance," *Philosophical Review* 124 (2015): 119, quote from p. 122.
4. List and Pivato, "Emergent Chance," quote from p. 133. In addition to his book *Why Free Will Is Real* (Harvard University Press, 2019), List presents these ideas in: C. List, "Free Will, Determinism, and the Possibility of Doing Otherwise," *Noûs* 48 (2014): 156; C. List and P. Menzies, "My Brain Made Me Do It: The Exclusion Argument against Free Will, and What's Wrong with It," in *Making a Difference: Essays on the Philosophy of Causation*, ed. H. Beebee, C. Hitchcock, and H. Price (Oxford University Press, 2017).
5. W. Glannon, "Behavior Control, Meaning, and Neuroscience," in *Neuroexistentialism*, ed. G. Caruso and W. Flannagan (Oxford University Press, 2018). The Shadlen and Roskies quotes are both from: M. Shadlen and A. Roskies, "The Neurobiology of Decision-Making and Responsibility: Reconciling Mechanism and Mindedness," *Frontiers of Neuroscience* 6 (2012), doi.org/10.3389/fnins.2012.00056.
6. M. Bedau, "Weak Emergence," in *Philosophical Perspectives: Mind, Causation, and World*, ed. J. Tomberlin (Blackwell, 1997), p. 375, the two quotes are from pp. 376 and 397; D. Chalmers, "Strong and Weak Emergence," in *The Re-emergence of Emergence*, ed. P. Clayton and P. Davies (Oxford University Press, 2006); S. Carroll, *The Big Picture: On the Origins of Life, Meaning, and the Universe Itself* (Dutton, 2016); S. Carroll, personal communication, 5/22/2019.

   Footnote: G. Gomes, "Free Will, the Self, and the Brain," *Behavioral Sciences and the Law* 25 (2007): 221; quote is from p. 233.
7. G. Berns et al., "Neurobiological Correlates of Social Conformity and Independence during Mental Rotation," *Biology Psychiatry* 58 (2005): 245.

   Footnote: P. Rozin, "Social Psychology and Science: Some Lessons from Solomon Asch," *Personality and Social Psychology Review* 5 (2001): 2.
8. Footnote: H. Chua, J. Boland, and R. Nisbett, "Cultural Variation in Eye Movements during Scene Perception," *Proceedings of the National Academy of Sciences of the United States of America* 102 (2005): 12629.
9. M. Mascolo and E. Kallio, "Beyond Free Will: The Embodied Emergence of Conscious Agency," *Philosophical Psychology* 32 (2019): 437.
10. Mascolo and Kallio, "Beyond Free Will"; J. Bonilla, "Why Emergent Levels Will Not Save Free Will (1)," *Mapping Ignorance*, September 30, 2019, mappingignorance.org/2019/09/30/why-emergent-levels-will-not-save-free-will-1/.
11. Footnote: See, for example, C. Voyatzis, "'Even a Brick Wants to Be Something'—Louis Kahn," *Yatzer*, June 9, 2013, yatzer.com/even-brick-wants-be-something-louis-kahn.

9. A PRIMER ON QUANTUM INDETERMINACY

1. S. Janusonis et al., "Serotonergic Axons as Fractional Brownian Motion Paths: Insights into the Self-Organization of Regional Densities," *Frontiers in Computational Neuroscience* 14 (2020), doi.org/10.3389/fncom.2020.00056; H. Zhang and H. Peng, "Mechanism of Acetylcholine Receptor Cluster Formation Induced by DC Electric Field," *PLoS One* 6 (2011):

e26805; M. Vestergaard et al., "Detection of Alzheimer's Amyloid Beta Aggregation by Capturing Molecular Trails of Individual Assemblies," *Biochemistry and Biophysics Research Communications* 377 (2008): 725.

2. C. Finch and T. Kirkwood, *Chance, Development, and Aging* (Oxford University Press, 2000).
3. B. Brembs, "Towards a Scientific Concept of Free Will as a Biological Trait: Spontaneous Actions and Decision-Making in Invertebrates," *Proceedings of the Royal Society B: Biological Sciences* 278 (2011): 930; A. Nimmerjahn, F. Kirschhoff, and F. Helmchen, "Resting Microglial Cells Are Highly Dynamic Surveillants of Brain Parenchyma in Vivo," *Science* 308 (2005): 1314.
4. Footnote: M. Heisenberg, "The Origin of Freedom in Animal Behavior," in *Is Science Compatible with Free Will? Exploring Free Will and Consciousness in the Light of Quantum Physics and Neuroscience,* ed. A. Suarez and P. Adams (Springer, 2013).
5. T. Hellmuth Tet al., "Delayed-Choice Experiments in Quantum Interference," *Physics Reviews A* 35 (1987): 2532.
6. A. Ananthaswamy, *Through Two Doors at Once: The Elegant Experiment That Captures the Enigma of Our Quantum Reality* (Dutton, 2018); for an introduction to the many-world idea, see Y. Nomura, "The Quantum Multiverse," *Scientific American*, May 2017.
7. J. Yin et al., "Satellite-Based Entanglement Distribution over 1200 Kilometers," *Science* 356 (2017): 1140; J. Ren et al., "Ground-to-Satellite Quantum Teleportation," *Nature* 549 (2017): 70; G. Popkin, "China's Quantum Satellite Achieves 'Spooky Action' at Record Distance," *Science*, June 15, 2017.
8. Footnote: D. Simonton, *Creativity in Science: Chance, Logic, Genius, and Zeitgeist* (Cambridge University Press, 2004); R. Sapolsky, "Open Season," *New Yorker*, March 30, 1998.
9. C. Marletto et al., "Entanglement between Living Bacteria and Quantized Light Witnessed by Rabi Splitting," *Journal of Physics: Communications* 2 (2018): 101001; P. Jedlicka, "Revisiting the Quantum Brain Hypothesis: Toward Quantum (Neuro)biology?," *Frontiers in Molecular Neuroscience* 10 (2017): 366.

   Footnote: J. O'Callaghan, "'Schrödinger's Bacterium' Could Be a Quantum Biology Milestone," *Scientific American*, October 29, 2018.

10. IS YOUR FREE WILL RANDOM?

1. A selective tour of the quantum indeterminacy menagerie: R. Boni, *Quantum Christian Realism: How Quantum Mechanics Underwrites and Realizes Classical Christian Theism* (Wipf and Stock, 2019); D. O'Murchu, *Quantum Theology: Spiritual Implications of the New Physics* (Crossroads, 2004); I. Barbour, *Issues in Science and Religion* (Prentice Hall, 1966); quote by "New Age Physicist" from: Amit Goswami, as quoted in the film "*What the #$*! Do We Know*?!" and his website https://www.amitgoswami.org/2019/06/21/quantum-spirituality/; P. Fisher, "Quantum Cognition: The Possibility of Processing with Nuclear Spins in the Brain," *Annals of Physics* 362 (2015): 593; H. Hu and M. Wu, "Action Potential Modulation of Neural Spin Networks Suggests Possible Role of Spin," *NeuroQuantology* 2 (2004): 309; S. Tarlaci and M. Pregnolato, "Quantum Neurophysics: From Non-living Matter to Quantum Neurobiology and Psychopathology," *International Journal of Psychophysiology* 103 (2016): 161; E. Basar and B. Guntekin, "A Breakthrough in Neuroscience Needs a 'Nebulous Cartesian System' Oscillations, Quantum Dynamics and Chaos in the Brain and Vegetative System," *International Journal of Psychophysiology* 64 (2006): 108; M. Cocchi et al., "Major Depression and Bipolar Disorder: The Concept of Symmetry Breaking," *NeuroQuantology* 10 (2012): 676; P. Zizzi and M. Pregnolato, "Quantum Logic of the Unconscious and Schizophrenia," *NeuroQuantology* 10 (2012): 566. And naturally, there's a quantum diet: L. Fritz, *The Quantum Weight Loss Blueprint* (New Hope Health, 2020). Plus, don't miss: A. Amarasingam, "New Age Spirituality, Quantum Mysticism and Self-Psychology: Changing Ourselves from the Inside Out," *Mental Health, Religion & Culture* 12 (2009): 277.

Footnote: G. Pennycook et al., "On the Reception and Detection of Pseudo-profound Bullshit," *Judgment and Decision Making* 10 (2015): 549.

2. Goswami, amitgoswami.org.
3. Journal Citation Reports ranked *NeuroQuantology* 253rd out of 261 neuroscience journals in terms of impact on other scientists' work, making one curious as to what numbers 254–261 are like.

   Footnote: J. T. Ismael, *Why Physics Makes Us Free* (Oxford University Press, 2016).
4. Quote is from P. Kitcher, "The Mind Mystery," *New York Times*, February 4, 1990; for a similarly anguished review this time by a neuroscientist, see: J. Hobson, "Neuroscience and the Soul: The Dualism of John Carew Eccles," *Cerebrum: The Dana Forum on Brain Science* 6 (2004): 61.

   Footnote: J. Eccles, "Hypotheses Relating to the Brain-Mind Problem," *Nature* 168 (1951): 53.
5. G. Engel, T. Calhoun, and E. Read, "Evidence for Wavelike Energy Transfer through Quantum Coherence in Photosynthetic Systems," *Nature* 446 (2007): 782.
6. P. Tse, "Two Types of Libertarian Free Will Are Realized in the Human Brain," in *Neuroexistentialism*, ed. G. Caruso and O. Flanagan (Oxford University Press, 2018), p. 170.
7. J. Schwartz, H. Stapp, and M. Beauregard, "Quantum Physics in Neuroscience and Psychology: A Neurophysical Model of Mind-Brain Interaction," *Philosophical Transactions of the Royal Society London B, Biological Sciences* 360 (2005): 1309; Z. Ganim, A. Tokmako, and A. Vaziri, "Vibrational Excitons in Ionophores; Experimental Probes for Quantum Coherence–Assisted Ion Transport and Selectivity in Ion Channels," *New Journal of Physics* 13 (2011): 113030; A. Vaziri and M. Plenio, "Quantum Coherence in Ion Channels: Resonances, Transport and Verification," *New Journal of Physics* 12 (2010): 085001.
8. S. Hameroff, "How Quantum Biology Can Rescue Conscious Free Will," *Frontiers of Integrative Neuroscience* 6 (2012): 93; S. Hameroff and R. Penrose, "Orchestral Reduction of Quantum Coherence in Brain Microtubules: A Model for Consciousness," *Mathematical and Computational Simulation* 40 (1996): 453; E. Dent and P. Baas, "Microtubules in Neurons as Information Carriers," *Journal of Neurochemistry* 129 (2014): 235; R. Tas and L. Kapitein, "Exploring Cytoskeletal Diversity in Neurons," *Science* 361 (2018): 231.
9. M. Tegmark, "Why the Brain Is Probably Not a Quantum Computer," *Information Science* 128 (2000): 155; M. Tegmark, "Importance of Quantum Coherence in Brain Processes," *Physical Review E* 61 (2000): 4194; M. Kikkawa et al., "Direct Visualization of the Microtubule Lattice Seam Both in Vitro and in Vivo," *Journal of Cell Biology* 127 (1994): 1965; C. De Zeeuw, E. Hertzberg, and E. Mugnaini, "The Dendritic Lamellar Body: New Neuronal Organelle Putatively Associated with Dendrodendritic Gap Junctions," *Journal of Neuroscience* 15 (1995): 1587.
10. J. Tanaka et al., "Number and Density of AMPA Receptors in Single Synapses in Immature Cerebellum," *Journal of Neuroscience* 25 (2005): 799; M. West and H. Gundersen, "Unbiased Stereological Estimation of the Number of Neurons in the Human Hippocampus," *Comparative Neurology* 296 (1990): 1.
11. J. Hobbs, "Chaos and Indeterminism," *Canadian Journal of Philosophy* 21 (1991): 141; D. Lindley, *Where Does the Weirdness Go? Why Quantum Mechanics Is Strange, but Not as Strange as You Think* (Basic Books, 1996).
12. L. Amico et al., "Many-Body Entanglement," *Review of Modern Physics* 80 (2008): 517; Tarlaci and Pregnolato, "Quantum Neurophysics."
13. B. Katz, "On the Quantal Mechanism of Neural Transmitter Release" (Nobel Lecture, Stockholm, December 12, 1970), nobelprize.org/prizes/medicine/1970/katz/lecture/; Y. Wang et al., "Counting the Number of Glutamate Molecules in Single Synaptic Vesicles," *Journal of the American Chemical Society* 141 (2019): 17507.

    Footnote: J. Schwartz et al., "Quantum Physics in Neuroscience and Psychology: A Neurophysical Model of Mind-Brain Interactions," *Philosophical Transactions of the Royal Society B* 1360 (2005): 1309, the quote can be found on p. 1319.

14. C. Wasser and E. Kavalali, "Leaky Synapses: Regulation of Spontaneous Neurotransmission in Central Synapses," *Journal of Neuroscience* 158 (2008): 177; E. Kavalali, "The Mechanisms and Functions of Spontaneous Neurotransmitter Release," *Nature Reviews Neuroscience* 16 (2015): 5; C. Williams and S. Smith, "Calcium Dependence of Spontaneous Neurotransmitter Release," *Journal of Neuroscience Research* 96 (2018): 335.
15. Williams and Smith, "Calcium Dependence of Spontaneous Neurotransmitter Release"; K. Koga et al., "SCRAPPER Selectively Contributes to Spontaneous Release and Presynaptic Long-Term Potentiation in the Anterior Cingulate Cortex," *Journal of Neuroscience* 37 (2017): 3887; R. Schneggenburger and C. Rosenmund, "Molecular Mechanisms Governing Ca(2+) Regulation of Evoked and Spontaneous Release," *Nature Neuroscience* 18 (2015): 935; K. Hausknecht et al., "Prenatal Ethanol Exposure Persistently Alters Endocannabinoid Signaling and Endocannabinoid-Mediated Excitatory Synaptic Plasticity in Ventral Tegmental Area Dopamine Neurons," *Journal of Neuroscience* 37 (2017): 5798.
16. Determined indeterminacy under the control of:

    Hormones and stress: L. Liu et al., "Corticotropin-Releasing Factor and Urocortin I Modulate Excitatory Glutamatergic Synaptic Transmission," *Journal of Neuroscience* 24 (2004): 4020; H. Tan, P. Zhong, and Z. Yan, "Corticotropin-Releasing Factor and Acute Stress Prolongs Serotonergic Regulation of GABA Transmission in Prefrontal Cortical Pyramidal Neurons," *Journal of Neuroscience* 24 (2004): 5000.ohol.

    Alcohol: R. Renteria et al., "Selective Alterations of NMDAR Function and Plasticity in D1 and D2 Medium Spiny Neurons in the Nucleus Accumbens Shell Following Chronic Intermittent Ethanol Exposure," *Neuropharmacology* 112 (2017): 164; 1983: Technicolor, 116 minutes, starring Robert De Niro, Diane Keaton, and, in his film debut, the young Ryan Gosling as the sixth frontocortical neuron from the left; R. Shen, "Ethanol Withdrawal Reduces the Number of Spontaneously Active Ventral Tegmental Area Dopamine Neurons in Conscious Animals," *Journal of Pharmacology and Experimental Therapeutics* 307 (2003): 566.

    Other factors: J. Ribeiro, "Purinergic Inhibition of Neurotransmitter Release in the Central Nervous System," *Pharmacology and Toxicology* 77 (1995): 299; J. Li et al., "Regulation of Increased Glutamatergic Input to Spinal Dorsal Horn Neurons by mGluR5 in Diabetic Neuropathic Pain," *Journal of Neurochemistry* 112 (2010): 162; A. Goel et al., "Cross-Modal Regulation of Synaptic AMPA Receptors in Primary Sensory Cortices by Visual Experience," *Nature Neuroscience* 9 (2006): 1001.

    Just to hint at a whole additional world of determined indeterminacy in the brain, particular stretches of DNA will occasionally be copied, with the copy then randomly plunked into a different place in the genome (once pejoratively called "jumping genes" by skeptics, the reality of such "transposons" resulted in a Nobel Prize for their long-dismissed discoverer, Barbara McClintock, in 1983). Turns out that the brain can regulate when such randomness occurs in neurons (for example, during stress, by way of glucocorticoids). See R. Hunter et al., "Stress and the Dynamic Genome: Steroids, Epigenetics, and the Transposome," *Proceedings of the National Academy of Sciences of the United States of America* 112 (2014): 6828.
17. See the Kavalali papers in Reference #14; F. Varodayan et al., "CRF Modulates Glutamate Transmission in the Central Amygdala of Naïve and Ethanol-Dependent Rats," *Neuropharmacology* 125 (2017): 418; J. Earman, *A Primer on Determinism* (Reidel, 1986).
18. Spontaneous neurotransmitter release: D. Crawford et al., "Selective Molecular Impairment of Spontaneous Neurotransmission Modulates Synaptic Efficacy," *Nature Communications* 10 (2017): 14436; M. Garcia-Bereguiain et al., "Spontaneous Release Regulates Synaptic Scaling in the Embryonic Spinal Network in Vivo," *Journal of Neuroscience* 36 (2016): 7268; A. Blankenship and M. Feller, "Mechanisms Underlying Spontaneous Patterned Activity in Developing Neural Circuits," *Nature Reviews Neuroscience* 11 (2010): 18; C. O'Donnell and M. van Rossum, "Spontaneous Action Potentials and Neural Coding in Unmyelinated Axons," *Neural Computation* 27 (2015): 801; L. Andreae and J. Burrone, "The Role of Spontaneous Neurotransmission in Synapse and Circuit Development," *Journal of Neuroscience Research* 96 (2018): 354.

19. M. Raichle et al., "A Default Mode of Brain Function," *Proceedings of the National Academy of Sciences of the United States of America* 98 (2001): 676; M. Raichle and A. Snyder, "A Default Mode of Brain Function: A Brief History of an Evolving Idea," *NeuroImage* 37 (2007): 1083. For an interesting take on a circumstance where the brain actively works to make you daydream, see: V. Axelrod et al., "Increasing Propensity to Mind-Wander with Transcranial Direct Current Stimulation," *Proceedings of the National Academy of Sciences of the United States of America* 112 (2015): 3314. For additional relevant papers, see: R. Pena, M. Zaks, and A. Roque, "Dynamics of Spontaneous Activity in Random Networks with Multiple Neuron Subtypes and Synaptic Noise: Spontaneous Activity in Networks with Synaptic Noise," *Journal of Computational Neuroscience* 45 (2018): 1; A. Tozzi, M. Zare, and A. Benasich, "New Perspectives on Spontaneous Brain Activity: Dynamic Networks and Energy Matter," *Frontiers of Human Neuroscience* 10 (2016): 247.
20. J. Searle, "Free Will as a Problem in Neurobiology," *Philosophy* 76 (2001): 491; M. Shadlen and A. Roskies, "The Neurobiology of Decision-Making and Responsibility: Reconciling Mechanism and Mindedness," *Frontiers in Neuroscience* 6 (2021), doi.org/10.3389/fnins.2012.00056; S. Blackburn, *Think: A Compelling Introduction to Philosophy* (Oxford University Press, 1999), the quote is from p. 60. For a nice example of how individuality is built on consistency in the brain, and not randomness, see: T. Kurikawa et al., "Neuronal Stability in Medial Frontal Cortex Sets Individual Variability in Decision-Making," *Nature Neuroscience* 21 (2018): 1764.

    Footnote: M. Bakan, "Awareness and Possibility," *Review of Metaphysics* 14 (1960): 231.
21. D. Dennett, *Freedom Evolves* (Viking, 2003), p. 123; Tse: Reference #6, p. 123.
22. Z. Blount, R. Lenski, and J. Losos, "Contingency and Determinism in Evolution: Replaying Life's Tape," *Science* 362 (2018): 655; D. Noble, "The Role of Stochasticity in Biological Communication Processes," *Progress in Biophysics and Molecular Biology* 162 (2020): 122; R. Noble and D. Noble, "Harnessing Stochasticity: How Do Organisms Make Choices?," *Chaos* 28 (2018): 106309. The two authors of this last paper are Denis Noble of Oxford and Raymond Noble of University College London; after only a moderate amount of sleuthing on my part, I think they are father and son (Denis, father; Raymond, son), which is incredibly sweet; just to add to the charm, they both appear to be accomplished English troubadour singers—singing together and publishing papers on stochasticity together. As long as we're on that, and I feel confident that no one is reading this (Why are you reading this? Go for a walk outside somewhere nice), there's also C. McEwen and B. McEwen, "Social Structure, Adversity, Toxic Stress, and Intergenerational Poverty: An Early Childhood Model," *Annual Review of Sociology* 43 (2017): 445, by brothers Craig, sociologist at Bowdoin College, and Bruce, neurobiologist at Rockefeller University. This is interdisciplinary science and family relations at their finest. Bruce, an extraordinarily accomplished scientist, was my PhD adviser, mentor, and father figure for almost forty years. He died in 2020; I still feel his absence.
23. Dennett, *Brainstorms*, p. 295.
24. Shadlen and Roskies, "Neurobiology of Decision-Making and Responsibility."
25. Footnote: The record-holding monkey: Cited in D. Wershler-Henry, *Iron Whim: A Fragmented History of Typewriting* (McClelland and Stewart, 2005); R. Dawkins, *The Blind Watchmaker: Why the Evidence of Evolution Reveals a Universe without Design* (Norton, 1986); Borges's story appears in J. Borges, *Collected Fictions* (Viking, 1998). I suspect binge-reading the three of these without pause would make for a pretty interesting state of mind.
26. K. Mitchel. "Does Neuroscience Leave Room for Free Will?," *Trends in Neurosciences* 41 (2018): 573.
27. R. Kane, *The Significance of Free Will* (Oxford University Press, 1996), p. 130.
28. Tse: See Reference #6.
29. R. Kane, "Libertarianism" in *Four Views on Free Will*, ed. J. Fischer et al. (Wiley-Blackwell, 2007), p. 26.
30. The description/prescription distinction is explored in P. Cryle and E. Stephens, *Normality: A Critical Genealogy* (University of Chicago Press, 2017). For an interesting read running in

the opposite direction from this chapter, see J. Horgan, "Does Quantum Mechanics Rule Out Free Will?," *Scientific American*, March 2022. By the way, Horgan makes respectful reference to physicist Sabine Hossenfelder; I second that. Watch her YouTube lecture "You Don't Have Free Will, but Don't Worry" (youtube.com/watch?v=zpU_e3jh_FY). It's magnificent. In fact, watch it, rather than reading this book . . .

10.5. INTERLUDE

1. H. Sarkissian et al., "Is Belief in Free Will a Cultural Universal?," *Mind & Language* 25 (2010): 346.

   Footnote: W. Phillips et al., "'Unwilling' versus 'Unable': Capuchin Monkeys' (*Cebus apella*) Understanding of Human Intentional Action," *Developmental Science* 12 (2009): 938; J. Call et al., "'Unwilling' versus 'Unable': Chimpanzees' Understanding of Human Intentional Action," *Developmental Science* 7 (2004): 488; E. Furlong and L. Santos, "Evolutionary Insights into the Nature of Choice: Evidence from Nonhuman Primates," in *Moral Psychology*, vol. 4, *Free Will and Moral Responsibility*, ed. W. Sinnott-Armstrong (MIT Press, 2014), p. 347.
2. Footnote: Loeb was knifed to death in prison by an inmate who said Loeb had aggressively propositioned him. Several pundits (which in the original isn't clear) noted that, surprisingly for someone who was educated enough to presumably know his grammar, Loeb had "ended his sentence with a proposition" (e.g., Mark Hellinger, *Syracuse Journal*, February 19, 1936). His killer was exonerated.
3. Society for Neuroscience, "Timeline," n.d., sfn.org/about/history-of-sfn/1969-2019/timeline. Philosopher Thomas Nadelhoffer writes explicitly about this "threat of shrinking agency." T. Nadelhoffer, "The Threat of Shrinking Agency and Free Will Disillusionism," in *Conscious Will and Responsibility: A Tribute to Benjamin Libet*, ed. L. Nadel and W. Sinnott-Armstrong (Oxford University Press, 2011).

11. WILL WE RUN AMOK?

1. D. Walker, *Rights in Conflict: The Walker Report* (Bantam Books, 1968); N. Steinberg, "The Whole World Watched: 50 Years after the 1968 Chicago Convention," *Chicago Sun Times*, August 17, 2018; J. Schultz, *No One Was Killed: The Democratic National Convention, August 1968* (University of Chicago Press, 1969); H. Johnson, "1968 Democratic Convention: The Bosses Strike Back," *Smithsonian*, August 2008. For a review of anonymity and enhanced violence in traditional cultures, see chapter 11 of R. Sapolsky, *Behave: The Biology of Humans at Our Best and Worst* (Penguin Press, 2017).
2. M. L. Saint Martin, "Running Amok: A Modern Perspective on a Culture-Bound Syndrome," *Primary Care Companion for the Journal of Clinical Psychiatry* 1 (1999): 66.
3. Francis Crick, *The Astonishing Hypothesis: The Scientific Search for the Soul* (Scribner, 1994), p. 1.
4. Footnote: For variations, see E. Seto and J. Hicks, "Disassociating the Agent from the Self: Undermining Belief in Free Will Diminishes True Self-Knowledge," *Social Psychological and Personality* 7 (2016): 726.
5. D. Rigoni et al., "Inducing Disbelief in Free Will Alters Brain Correlates of Preconscious Motor Preparation Whether We Believe in Free Will or Not," *Psychological Science* 22 (2011): 613.
6. D. Rigoni, G. Pourtois, and M. Brass, "'Why Should I Care?' Challenging Free Will Attenuates Neural Reaction to Errors," *Social Cognitive and Affective Neuroscience* 10 (2015): 262; D. Rigoni et al., "When Errors Do Not Matter: Weakening Belief in Intentional Control Impairs Cognitive Reaction to Errors," *Cognition* 127 (2013): 264.
7. K. Vohs and J. Schooler, "The Value of Believing in Free Will," *Psychological Science* 19 (2002) 49; A. Shariff and K. Vohs, "The World without Free Will," *Scientific American*, June 2014; M. MacKenzie, K. Vohs, and R. Baumeister, "You Didn't Have to Do That: Belief in Free Will Promotes Gratitude," *Personality and Social Psychology Bulletin* 40 (2014): 14223;

B. Moynihan, E. Igou, and A. Wijnand, "Free, Connected, and Meaningful: Free Will Beliefs Promote Meaningfulness through Belongingness," *Personality and Individual Differences* 107 (2017): 54. Also see: Seto and Hicks, "Disassociating the Agent from the Self"; R. Baumeister, E. Masicampo, and C. DeWall, "Prosocial Benefits of Feeling Free: Disbelief in Free Will Increases Aggression and Reduces Helpfulness," *Personality and Social Psychology Bulletin* 35 (2009): 260.

8. M. Lynn et al., "Priming Determinist Beliefs Diminishes Implicit (but Not Explicit) Components of Self-Agency," *Frontiers in Psychology* 5 (2014), doi.org/10.3389/fpsyg.2014.01483.

   Footnote: S. Obhi and P. Hall, "Sense of Agency in Joint Action: Influence of Human and Computer Co-actors," *Experimental Brain Research* 211 (2011): 663–70.

9. A. Vonash et al., "Ordinary People Associate Addiction with Loss of Free Will," *Addictive Behavior Reports* 5 (2017): 56; K. Vohs and R. Baumeiser, "Addiction and Free Will," *Addiction Research and Theory* 17 (2009): 231; G. Heyman, "Do Addicts Have Free Will? An Empirical Approach to a Vexing Question," *Addictive Behavior Reports* 5 (2018): 85; E. Racine, S. Sattler, and A. Escande, "Free Will and the Brain Disease Model of Addiction: The Not So Seductive Allure of Neuroscience and Its Modest Impact on the Attribution of Free Will to People with an Addiction," *Frontiers in Psychology* 8 (2017): 1850.

10. T. Nadelhoffer et al., "Does Encouraging a Belief in Determinism Increase Cheating? Reconsidering the Value of Believing in Free Will," *Cognition* 203 (2020): 104342; A. Monroe, G. Brady, and B. Malle, "This Isn't the Free Will Worth Looking For: General Free Will Beliefs Do Not Influence Moral Judgments, Agent-Specific Choice Ascriptions Do," *Social Psychological and Personality Science* 8 (2017): 191; D. Wisniewski et al., "Relating Free Will Beliefs and Attitudes," *Royal Society Open Science* 9 (2022): 202018.

11. See: Nadelhoffer et al., "Does Encouraging a Belief in Determinism"; Monroe, Brady, and Malle, "This Isn't the Free Will"; J. Harms et al., "Free to Help? An Experiment on Free Will Belief and Altruism," *PLoS One* 12 (2017): e0173193; L. Crone and N. Levy, "Are Free Will Believers Nicer People? (Four Studies Suggest Not)," *Social Psychological and Personality Science* 10 (2019): 612; E. Caspar et al., "The Influence of (Dis)belief in Free Will on Immoral Behaviour," *Frontiers in Psychology* 8 (2017): 20. Meta-analysis: O. Genschow, E. Cracco, and J. Schneider, "Manipulating Belief in Free Will and Its Downstream Consequences: A Meta-analysis," *Personality and Social Psychology Review* 27 (2022): 52 B. Nosek, "Estimating the Reproducibility of Psychological Inference," *Science* 349 (2015), DOI:10.1126/science.aac4716.

    Footnote: O. Genschow et al., "Professional Judges' Disbelief in Free Will Does Not Decrease Punishment," *Social Psychological and Personality Science* 12 (2020): 357.

12. A. Norenzayan, *Big Gods: How Religion Transformed Cooperation and Conflict* (Princeton University Press, 2013). For an interesting review of the book by complexity scientist Peter Turchin, see P. Turchin, "From Big Gods to the Big Brother," *Cliodynamica* (blog), September 4, 2015, peterturchin.com/cliodynamica/from-big-gods-to-the-big-brother/.

13. P. Edgell et al., "Atheists and Other Cultural Outsiders: Moral Boundaries and the Nonreligious in the United States," *Social Forces* 95 (2016): 607; E. Volokh, "Parent-Child Speech and Child Custody Speed Restrictions," *New York University Law Review* 81 (2006): 631; A. Furnham, N. Meader, and A. McCelland. "Factors Affecting Nonmedical Participants' Allocation of Scarce Medical Resources," *Journal of Social Behavior and Personality* 12 (1996): 735; J. Hunter, "The Williamsburg Charter Survey: Methodology and Findings," *Journal of Law and Religion* 8 (1990): 257; M. Miller and B. Bornstein, "The Use of Religion in Death Penalty Sentencing Trials," *Law and Human Behavior* 30 (2006): 675. For a demonstration of prescience, see J. Joyner, "Black President More Likely Than Mormon or Atheist," *Outside the Beltway*, February 20, 2007, outsidethebeltway.com/archives/black_president_more_likely_than_mormon_or_atheist_/.

    Footnote: S. Weber et al., "Psychological Distress among Religious Nonbelievers: A Systematic Review," *Journal of Religion and Health* 51 (2012): 72.

14. W. Gervais and M. Najle, "Nonreligious People in Religious Societies," in *The Oxford Handbook of Secularism*, ed. P. Zuckerman and J. Shook (Oxford University Press, 2017); "USPS Discrimination against Atheism?," https://atheist.shoes/pages/usps-study. Also, some chilling news: R. Evans, "Atheists Face Death in 13 Countries, Global Discrimination: Study," Reuters, December 9, 2013, reuters.com/article/us-religion-atheists-idUSBRE9B900G20131210; International Humanist and Ethical Union, "You Can Be Put to Death for Atheism in 13 Countries around the World," October 12, 2013, iheu.org/you-can-be-put-death-atheism-13-countries-around-world/; Human Rights Watch, "Saudi Arabia: New Terrorism Regulations Assault Rights," March 20, 2014, hrw.org/news/2014/03/20/saudi-arabia-new-terrorism-regulations-assault-rights.
15. C. Tamir et al., "The Global God Divide," Pew Research Center, July 20, 2020; S. Weber et al., "Psychological Distress among Religious Nonbelievers," *Journal of Religion and Health* 51 (2012): 72; M. Gervais, "Everything Is Permitted? People Intuitively Judge Immorality as Representative of Atheists," *PLoS One* 9, no. 4 (2014): e92302; R. Ritter and J. Preston, "Representations of Religious Words: Insights for Religious Priming Research," *Journal for the Scientific Study of Religion* 52 (2013): 494; W. Gervais et al., "Global Evidence of Extreme Intuitive Moral Prejudice against Atheists," *Nature Human Behaviour* 1 (2017): 0151.

    Footnote: B. Rutjens and S. Heine, "The Immoral Landscape? Scientists Are Associated with Violations of Morality," *PLoS One* 11 (2016): e0152798.
16. See Weber et al., "Psychological Distress among Religious Nonbelievers."
17. Footnote: A. Norenzayan and W. Gervais, "The Origins of Religious Disbelief," *Trends in Cognitive Sciences* 17 (2013): 20; G. Pennycook et al., "On the Reception and Detection of Pseudo-profound Bullshit," *Judgment and Decision Making* 10 (2015): 549; A. Shenhav, D. Rand, and J. Greene, "Divine Intuition: Cognitive Style Influences Belief in God," *Journal of Experimental Psychology: General* 141 (2011): 423; W. Gervais and A. Norenzayan, "Analytic Thinking Promotes Religious Disbelief," *Science* 336 (2012): 493; A. Jack et al., "Why Do You Believe in God? Relationships between Religious Belief, Analytic Thinking, Mentalizing and Moral Concern," *PLoS One* 11 (2016): e0149989; Pew Forum on Religion and Public Life, "2008 U.S. Religious Landscape Survey: Religious Affiliation: Diverse and Dynamic," religions.pewforum.org/pdf/report-religious-landscape-study-full.pdf.
18. Self-reporting: B. Pelham and S. Crabtree, "Worldwide, Highly Religious More Likely to Help Others," Gallup, October 8, 2008, news.gallup.com/poll/111013/worldwide-highly-religious-more-likely-help-others.aspx; M. Donahue and M. Nielsen, "Religion, Attitudes, and Social Behavior," in *Handbook of the Psychology of Religion and Spirituality*, ed. R. Paloutzian and C. Park (Guilford, 2005); I. Pichon and V. Saroglou, "Religion and Helping: Impact of Target, Thinking Styles and Just-World Beliefs," *Archive for the Psychology of Religion* 31 (2009): 215. Caring about giving a good impression: L. Galen, "Does Religious Belief Promote Prosociality? A Critical Examination," *Psychological Bulletin* 138 (2012): 876; R. Putnam and R. Campbell, *American Grace: How Religion Divides and Unites Us* (Simon & Schuster, 2010).
19. Religiosity and pro-sociality: V. Saroglou, "Religion's Role in Prosocial Behavior: Myth or Reality?," *Psychology of Religion Newsletter* 31 (2006): 1; V. Saroglou et al., "Prosocial Behavior and Religion: New Evidence Based on Projective Measures and Peer Ratings," *Journal for the Scientific Study of Religion* 44 (2005): 323; L. Anderson and J. Mellor, "Religion and Cooperation in a Public Goods Experiment," *Economics Letters* 105 (2009): 58; C. Ellison, "Are Religious People Nice People? Evidence from the National Survey of Black Americans," *Social Forces* 71 (1992): 411.

    Religiosity and self-enhancement: K. Eriksson and A. Funcke, "Humble Self-Enhancement: Religiosity and the Better-Than-Average Affect," *Social Psychological and Personality Science* 5 (2014): 76; C. Sedikides and J. Gebauer, "Religiosity as Self-Enhancement: A Meta-analysis of the Relation between Socially Desirable Responding and Religiosity," *Personality and Social Psychology Review* 14 (2010): 17; P. Brenner, "Identity Importance and the Over-Reporting of Religious Service Attendance: Multiple Imputation of Religious Attendance Using the American Time Use Study and the General Social Survey," *Journal for the Scientific Study of Religion* 50 (2011): 103; P. Brenner, "Exceptional Behavior or Exceptional Identity? Over-Reporting of Church Attendance in the U.S.," *Public Opinion Quarterly* 75 (2011): 19.

Religiosity and life satisfaction: E. Diener, L. Tay, and D. Myers, "The Religion Paradox: If Religion Makes People Happy, Why Are So Many Dropping Out?" *Journal of Personality and Social Psychology* 101 (2011): 1278; C. Sabatier et al., "Religiosity, Family Orientation, and Life Satisfaction of Adolescents in Four Countries," *Journal of Cross-Cultural Psychology* 42 (2011): 1375.

20. Charitability: R. Gillum and K. Master, "Religiousness and Blood Donation: Findings from a National Survey," *Journal of Health Psychology* 15 (2010): 163; P. Grossman and M. Parrett, "Religion and Prosocial Behaviour: A Field Test," *Applied Economics Letters* 18 (2011): 523; McCullough and Worthington, "Religion and the Forgiving Personality"; G. Pruckner and R. Sausgruber, "Honesty on the Streets: A Field Experiment on Newspaper Purchasing," *Journal of the European Economic Association* 11 (2008): 661; A. Tsang, A. Schulwitz, and R. Carlisle, "An Experimental Test of the Relationship between Religion and Gratitude," *Psychology of Religion and Spirituality* 4 (2011): 40.

    Religiosity and aggression: J. Blogowska, C. Lambert, and V. Saroglou, "Religious Prosociality and Aggression: It's Real," *Journal for the Scientific Study of Religion* 52 (2013): 524. Being retributive: T. Greer et al., "We Are a Religious People; We Are a Vengeful People," *Journal for the Scientific Study of Religion* 44 (2005): 45; M. Leach, M. Berman, and L. Eubanks, "Religious Activities, Religious Orientation, and Aggressive Behavior," *Journal for the Scientific Study of Religion* 47 (2008): 311.

21. L. Galen and J. Kloet, "Personality and Social Integration Factors Distinguishing Nonreligious from Religious Groups: The Importance of Controlling for Attendance and Demographics," *Archive for the Psychology of Religion* 33 (2011): 205; L. Galen, M. Sharp, and A. McNulty, "The Role of Nonreligious Group Factors versus Religious Belief in the Prediction of Prosociality," *Social Indicators Research* 122 (2015): 411; R. Stark, "Physiology and Faith: Addressing the 'Universal' Gender Difference in Religious Commitment," *Journal for the Scientific Study of Religion* 41 (2002): 495; M. Argyle, *Psychology and Religion: An Introduction* (Routledge, 2000); G. Lenski, "Social Correlates of Religious Interest," *American Sociological Review* 18 (1953): 533; A. Miller and J. Hoffmann, "Risk and Religion: An Explanation of Gender Differences in Religiosity," *Journal for the Scientific Study of Religion* 34 (1995): 63.

22. Putnam and Campbell, *American Grace*; T. Smith, M. McCullough, and J. Poll, "Religiousness and Depression: Evidence for a Main Effect and the Moderating Influence of Stressful Life Events," *Psychological Bulletin* 129 (2003): 614; L. Galen and J. Kloet, "Mental Well-Being in the Religious and the Non-religious: Evidence for a Curvilinear Relationship," *Mental Health, Religion & Culture* 14 (2011): 673; M. McCullough and T. Smith, "Religion and Depression: Evidence for a Main Effect and the Moderating Influence of Stress Life Events," *Psychological Bulletin* 129 (2003): 614; L. Manning, "Gender and Religious Differences Associated with Volunteering in Later Life," *Journal of Women and Aging* 22 (2010): 125.

23. Pichon and Saroglou, "Religion and Helping"; N. Mazar, O. Ami, and D. Ariely, "The Dishonesty of Honest People: A Theory of Self-Concept Maintenance," *Journal of Marketing Research* 45 (2008): 633; M. Lang et al., "Moralizing Gods, Impartiality and Religious Parochialism across 15 Societies," *Proceedings of the Royal Society B: Biological Sciences* 286 (2019): 20190202; A. Shariff et al., "Religious Priming: A Meta-analysis with a Focus on Prosociality," *Personality and Social Psychology Review* 20 (2016): 27.

24. Pichon and Saroglou, "Religion and Helping"; A. Shariff and A. Norenzayan, "God Is Watching You: Priming God Concepts Increases Prosocial Behavior in an Anonymous Economic Game," *Psychological Science* 18 (2007): 803; K. Laurin, A. Kay, and G. Fitzsimons, "Divergent

Effects of Activating Thoughts of God on Self-Regulation," *Journal of Personality and Social Psychology* 102 (2012): 4; K. Rounding et al., "Religion Replenishes Self-Control," *Psychological Science* 23 (2012): 635; J. Saleam and A. Moustafa, "The Influence of Divine Rewards and Punishments on Religious Prosociality," *Frontiers in Psychology* 7 (2016): 1149.

25. Shariff and Norenzayan, "God Is Watching You"; B. Randolph-Seng and M. Nielsen, "Honesty: One Effect of Primed Religious Representations," *International Journal of Psychology and Religion* 17 (2007): 303.

    Footnote: M. Quirin, J. Klackl, and E. Jonas, "Existential Neuroscience: A Review and Brain Model of Coping with Death Awareness," in *Handbook of Terror Management Theory*, ed. C. Routledge and M. Vess (Elsevier, 2019).

26. J. Haidt, *The Righteous Mind: Why Good People Are Divided by Politics and Religion* (Pantheon, 2012); J. Weedon and R. Kurzban, "What Predicts Religiosity? A Multinational Analysis of Reproductive and Cooperative Morals," *Evolution and Human Behavior* 34 (2012): 440; P. Zuckerman, *Society without God* (New York University Press, 2008).

27. M. Regnerus, C. Smith, and D. Sikkink, "Who Gives to the Poor? The Influence of Religious Tradition and Political Location on Personal Generosity of Americans toward the Poor," *Journal for the Scientific Study of Religion* 37 (1998): 481; J. Jost and M. Krochik, "Ideological Differences in Epistemic Motivation: Implications for Attitude Structure, Depth of Information Processing, Susceptibility to Persuasion, and Stereotyping," *Advances in Motivation Science* 1 (2014): 181; F. Grupp and W. Newman, "Political Ideology and Religious Preference: The John Birch Society and Americans for Democratic Action," *Journal for the Scientific Study of Religion* 12 (1974): 401.

28. Center for Global Development, "Commitment to Development Index 2021," cgdev.org/section/initiatives/_active/cdi/; Center for Global Development, "Ranking the Rich," *Foreign Policy* 142 (2004): 46; Center for Global Development, "Ranking the Rich," *Foreign Policy* 150 (2005): 76; Zuckerman, *Society without God*; P. Norris and R. Inglehart, *Sacred and Secular: Religion and Politics Worldwide* (Cambridge University Press, 2004); S. Bruce, *Politics and Religion* (Polity, 2003).

    Footnote: Center for Global Development, "Ranking the Rich," *Foreign Policy* 150 (2005): 76.

29. P. Zuckerman, "Atheism, Secularity, and Well-Being: How the Findings of Social Science Counter Negative Stereotypes and Assumptions," *Sociology Compass* 3 (2009): 949; B. Beit-Hallahmi, "Atheists: A Psychological Profile," in *The Cambridge Companion to Atheism*, ed. M. Martin, Cambridge Companions to Philosophy (Cambridge University Press, 2007); S. Crabtree and B. Pelham, "More Religious Countries, More Perceived Ethnic Intolerance," Gallup, April 7, 2009, gallup.com/poll/117337/Religious-Countries Perceived-Ethnic-Intolerance.aspx; J. Lyne, "Who's No. 1? Finland, Japan and Korea, Says OECD Education Study," *Site Selection*, December 10, 2001, siteselection.com/ssinsider/snapshot/sf011210.htm; United Nations Office on Drugs and Crime, "UNODC Statistics Online."

30. Inglehart and Norris, *Sacred and Secular*.

31. H. Tan and C. Vogel, "Religion and Trust: An Experimental Study," *Journal of Economic Psychology* 29 (2008): 332; J. Preston and R. Ritter, "Different Effects of Religion and God on Prosociality with the Ingroup and Outgroup," *Personality and Social Psychology Bulletin* 39 (2013): 1471; A. Ahmed, "Are Religious People More Prosocial? A Quasi-experimental Study with Madrasah Pupils in a Rural Community in India," *Journal for the Scientific Study of Religion* 48 (2009): 368; A. Ben-Ner et al., "Identity and In-group/Out-group Differentiation in Work and Giving Behaviors: Experimental Evidence," *Journal of Economic Behavior & Organization* 72 (2009): 153; C. Fershtman, U. Gneezy, and F. Verboven, "Discrimination and Nepotism: The Efficiency of the Anonymity Rule," *Journal of Legal Studies* 34 (2005): 371; R. Reich, *Just Giving: Why Philanthropy Is Failing Democracy and How It Can Do Better* (Princeton University Press, 2018).

32. Lang et al., "Moralizing Gods, Impartiality and Religious Parochialism."

33. J. Blogowska and V. Saroglou, "Religious Fundamentalism and Limited Prosociality as a Function of the Target," *Journal for the Scientific Study of Religion* 50 (2011): 44; M. Johnson et al., "A Mediational Analysis of the Role of Right-Wing Authoritarianism and Religious Fundamentalism in the Religiosity-Prejudice Link," *Personality and Individual Differences* 50 (2011): 851.
34. D. Gay and C. Ellison, "Religious Subcultures and Political Tolerance: Do Denominations Still Matter?," *Review of Religious Research* 34 (1993): 311; T. Vilaythong, N. Lindner, and B. Nosek, "'Do unto Others': Effects of Priming the Golden Rule on Buddhists' and Christians' Attitudes toward Gay People," *Journal for the Scientific Study of Religion* 49 (2010): 494; J. LaBouff et al., "Differences in Attitudes towards Outgroups in a Religious or Non-religious Context in a Multi-national Sample: A Situational Context Priming Study," *International Journal for the Psychology of Religion* 22 (2012): 1; M. Johnson, W. Rowatt, and J. LaBouff, "Priming Christian Religious Concepts Increases Racial Prejudice," *Social Psychological and Personality Science* 1 (2010): 119; Pichon and Saroglou, "Religion and Helping"; R. McKay et al., "Wrath of God: Religious Primes and Punishment," *Proceedings of the Royal Society B: Biological Sciences* 278 (2011): 1858; G. Tamarin, "The Influence of Ethnic and Religious Prejudice on Moral Judgment," *New Outlook* 9 (1996): 49; J. Ginges, I. Hansen, and A. Norenzayan, "Religion and Support for Suicide Attacks," *Psychological Science* 20 (2009): 224. See: Leach, Berman, and Eubanks, "Religious Activities, Religious Orientation"; H. Ledford, "Scriptural Violence Can Foster Aggression," *Nature* 446 (2007): 114; B. Bushman et al., "When God Sanctions Killing: Effect of Scriptural Violence on Aggression," *Psychological Science* 18 (2007): 204.
35. Crone and Levy, "Are Free Will Believers Nicer People?"
36. C. Ma-Kellams and J. Blascovich, "Does 'Science' Make You Moral? The Effects of Priming Science on Moral Judgments and Behavior," *PLoS One* 8 (2013): e57989.
37. See the two Brenner papers, reference 19; A. Keysar, "Who Are America's Atheists and Agnostics?," in *Secularism and Secularity: Contemporary International Perspectives*, ed. B. Kosmin and A. Keysar (Institute for the Study of Secularism in Society and Culture, 2007).
38. Galen, Sharp, and McNulty, "The Role of Nonreligious Group Factors"; A. Jorm and H. Christensen, "Religiosity and Personality: Evidence for Non-linear Associations," *Personality and Individual Differences* 36 (2004): 1433; D. Bock and N. Warren, "Religious Belief as a Factor in Obedience to Destructive Demands," *Review of Religious Research* 13 (1972): 185; F. Curlin et al., "Do Religious Physicians Disproportionately Care for the Underserved?," *Annals of Family Medicine* 5 (2007): 353; S. Oliner and P. Oliner. *The Altruistic Personality: Rescuers of Jews in Nazi Europe* (Free Press, 1988).

12. THE ANCIENT GEARS WITHIN US: HOW DOES CHANGE HAPPEN?

1. For Eric Kandel's magisterial (the only appropriate word) review of his life's work, see this written version of his 2000 Nobel Prize lecture: E. Kandel, "The Molecular Biology of Memory Storage: A Dialogue between Genes and Synapses," *Science* 294 (2001): 1030.
2. E. Alnajjar and K. Murase, "A Simple *Aplysia*-Like Spiking Neural Network to Generate Adaptive Behavior in Autonomous Robots," *Adaptive Behavior* 16 (2008): 306.
3. Footnote: H. Boele et al., "Axonal Sprouting and Formation of Terminals in the Adult Cerebellum during Associative Motor Learning," *Journal of Neuroscience* 33 (2013): 17897.
4. Footnote: M. Srivastava et al., "The *Amphimedon queenslandica* Genome and the Evolution of Animal Complexity," *Nature* 466 (2010): 720.
5. J. Medina et al., "Parallels between Cerebellum- and Amygdala-Dependent Conditioning," *Nature Reviews Neuroscience* 3 (2002): 122.
6. M. Kalinichev et al., "Long-Lasting Changes in Stress-Induced Corticosterone Response and Anxiety-Like Behaviors as a Consequence of Neonatal Maternal Separation in Long-Evans

Rats," *Pharmacology Biochemistry and Behavior* 73 (2002): 13; B. Aisa et al., "Cognitive Impairment Associated to HPA Axis Hyperactivity after Maternal Separation in Rats," *Psychoneuroendocrinology* 32 (2007): 256; B. Aisa et al., "Effects of Maternal Separation on Hypothalamic-Pituitary-Adrenal Responses, Cognition and Vulnerability to Stress in Adult Female Rats," *Neuroscience* 154 (2008): 1218; M. Moffett et al., "Maternal Separation Alters Drug Intake Patterns in Adulthood in Rats," *Biochemical Pharmacology* 73 (2007): 321. Interestingly, the effects of transient maternal separation on the development of the offspring's brain and behavior are heavily due to changes in Mom's behavior when she is returned: R. Alves et al., "Maternal Separation Effects on Mother Rodents' Behaviour: A Systematic Review," *Neuroscience and Biobehavioral Reviews* 117 (2019): 98.

7. A. Wilber, G. Lin, and C. Wellman, "Glucocorticoid Receptor Blockade in the Posterior Interpositus Nucleus Reverses Maternal Separation–Induced Deficits in Adult Eyeblink Conditioning," *Neurobiology of Learning and Memory* 94 (2010): 263; A. Wilber et al., "Neonatal Maternal Separation Alters Adult Eyeblink Conditioning and Glucocorticoid Receptor Expression in the Interpositus Nucleus of the Cerebellum," *Developmental Neurobiology* 67 (2011): 751.
8. J. LeDoux, "Evolution of Human Emotion," *Progress in Brain Research* 195 (2012): 431; also see any of LeDoux's various excellent books on the broad subject, such as J. LeDoux, *The Deep History of Ourselves: The Four-Billion-Year Story of How We Got Conscious Brains* (Viking, 2019); L. Johnson et al., "A Recurrent Network in the Lateral Amygdala: A Mechanism for Coincidence Detection," *Frontiers in Neural Circuits* 2 (2008): 3; W. Haubensak et al., "Genetic Dissection of an Amygdala Microcircuit That Gates Conditioned Fear," *Nature* 468 (2010): 270.
9. P. Zhu and D. Lovinger, "Retrograde Endocannabinoid Signaling in a Postsynaptic Neuron/Synaptic Bouton Preparation from Basolateral Amygdala," *Journal of Neuroscience* 25 (2005): 6199; M. Monsey et al., "Chronic Corticosterone Exposure Persistently Elevates the Expression of Memory-Related Genes in the Lateral Amygdala and Enhances the Consolidation of a Pavlovian Fear Memory," *PLoS One* 9 (2014): e91530; R. Sobota et al., "Oxytocin Reduces Amygdala Activity, Increases Social Interactions, and Reduces Anxiety-Like Behavior Irrespective of NMDAR Antagonism," *Behavioral Neuroscience* 129 (2015): 389; O. Kozanian et al., "Long-Lasting Effects of Prenatal Ethanol Exposure on Fear Learning and Development of the Amygdala," *Frontiers in Behavioral Neuroscience* 12 (2018): 200; E. Pérez-Villegas et al., "Mutation of the HERC 1 Ubiquitin Ligase Impairs Associative Learning in the Lateral Amygdala," *Molecular Neurobiology* 55 (2018): 1157.
10. Footnote: T. Moffitt et al., "Deep-Seated Psychological Histories of COVID-19 Vaccine Hesitance and Resistance," *PNAS Nexus* 1 (2022): pgac034.
11. A. Baddeley, "Working Memory: Looking Back and Looking Forward," *Nature Reviews Neuroscience* 4 (2003): 829; J. Jonides et al., "The Mind and Brain of Short-Term Memory," *Annual Review of Psychology* 59 (2008): 193.
12. For examples of how the actual circuitry in the brain has these properties, see: D. Zeithamova, A. Dominick, and A. Preston, "Hippocampal and Ventral Medial Prefrontal Activation during Retrieval-Mediated Learning Supports Novel Inference," *Neuron* 75 (2012): 168; D. Cai et al., "A Shared Neural Ensemble Links Distinct Contextual Memories Encoded Close in Time," *Nature* 534 (2016): 115.

Footnote (p. 294): J. Alvarez, *In the Time of the Butterflies* (Algonquin Books, 2010).

Footnote (p. 295): J. Harris, "Anorexia Nervosa and Anorexia Miracles: Miss K. R— and St. Catherine of Siena," *JAMA Psychiatry* 71 (2014): 12; F. Forcen, "Anorexia Mirabilis: The Practice of Fasting by Saint Catherine of Siena in the Late Middle Ages," *American Journal of Psychiatry* 170 (2013): 370; F. Galassi, N. Bender, and M. Habicht, "St. Catherine of Siena (1347–1380 AD): One of the Earliest Historic Cases of Altered Gustatory Perception in Anorexia Mirabilis," *Neurological Sciences* 39 (2018): 939.

13. The picture on the right is of Private Donald Brown, who, at age twenty-four, was killed when his Sherman tank was destroyed by Nazi fire in France in 1944. The unidentified remains of his tank crew were recovered in 1947, and it was not until 2018 that Brown was identified by DNA analysis. The news release about his identification gives no indication of how many family members went to their graves over those seventy-four years never knowing what happened to him. I thank the Defense MIA/POW Accounting Agency for permission to use his image for this expository purpose. More than seventy-two thousand American soldiers remain missing in action from World War II. (You can read the news release at dpaa.mil/News-Stories/News-Releases/PressReleaseArticleView/Article/1647847/funeral-announcement-for-soldier-killed-during-world-war-ii-brown-d/.) For an anthropological analysis of the desire we have to know what has happened to the dead, as well as a personal story of a twenty-seven-year wait for such information, see R. Sapolsky, "Why We Want Their Bodies Back," *Discover*, January 31, 2002, reprinted in R. Sapolsky, *Monkeyluv and Other Essays on Our Lives as Animals* (Simon & Schuster/Scribner, 2005).

13. WE REALLY HAVE DONE THIS BEFORE

1. E. Magiorkinis et al., "Highlights in the History of Epilepsy: The Last 200 Years," *Epilepsy Research and Treatment* 2014 (2014): 582039.
2. J. Rho and H. White, "Brief History of Anti-seizure Drug Development," *Epilepsia Open* 3 (2018): 114.

   Footnote (p. 307): J. Russell, *Witchcraft in the Middle Ages* (Cornell University Press, 1972), p. 234.
3. See, for example: R. Sapolsky and G. Steinberg, "Gene Therapy for Acute Neurological Insults," *Neurology* 10 (1999): 1922.
4. A. Walker, "Murder or Epilepsy?," *Journal of Nervous and Mental Disease* 133 (1961): 430; J. Livingston, "Epilepsy and Murder," *Journal of the American Medical Association* 188 (1964): 172; M. Ito et al., "Subacute Postictal Aggression in Patients with Epilepsy," *Epilepsy & Behavior* 10 (2007): 611; J. Gunn, "Epileptic Homicide: A Case Report," *British Journal of Psychiatry* 132 (1978): 510; C. Hindler, "Epilepsy and Violence," *British Journal of Psychiatry* 155 (1989): 246; N. Pandya et al., "Epilepsy and Homicide," *Neurology* 57 (2001): 1780.
5. S. Fazel et al., "Risk of Violent Crime in Individuals with Epilepsy and Traumatic Brain Injury: A 35-Year Swedish Population Study," *PLoS Medicine* 8 (2011): e1001150; C. Älstrom, *Study of Epilepsy and Its Clinical, Social and Genetic Aspects* (Monksgaard, 1950); J. Kim et al., "Characteristics of Epilepsy Patients Who Committed Violent Crimes: Report from the National Forensic Hospital," *Journal of Epilepsy Research* 1 (2011): 13; D. Treiman, "Epilepsy and Violence: Medical and Legal Issues," *Epilepsia* 27 (1986): S77; D. Hill and D. Pond, "Reflections on One Hundred Capital Cases Submitted to Electroencephalography," *Journal of Mental Science* 98 (1952): 23; E. Rodin, "Psychomotor Epilepsy and Aggressive Behavior," *Archives of General Psychiatry* 28 (1973): 210.
6. J. Falret, "De l'etat mental des epileptiques," *Archives generales de médicine* 16 (1860): 661.

   Footnote: P. Pichot, "Circular Insanity, 150 Years On," *Bulletin de l'académie nationale de médecine* 188 (2004): 275.
7. S. Fernandes et al., "Epilepsy Stigma Perception in an Urban Area of a Limited-Resource Country," *Epilepsy & Behavior* 11 (2007): 25; A. Jacoby, "Epilepsy and Stigma: An Update and Critical Review," *Current Neurology and Neuroscience Reports* 8 (2008): 339; G. Baker et al., "Perceived Impact of Epilepsy in Teenagers and Young Adults: An International Survey," *Epilepsy and Behavior* 12 (2008): 395; R. Kale, "Bringing Epilepsy Out of the Shadows," *British Medical Journal* 315 (1997): 2.
8. G. Krauss, L. Ampaw, and A. Krumholz, "Individual State Driving Restrictions for People with Epilepsy in the US," *Neurology* 57 (2001): 1780.

9. C. Bonanos, "What New York Should Learn from the Park Slope Crash That Killed Two Children," *Intelligencer, New York*, March 30, 2018.
10. T. Moore and K. Sheehy, "Driver in Crash That Killed Two Kids Suffers from MS, Seizures," *New York Post*, March 6, 2018.
11. C. Moynihan, "Driver Charged with Manslaughter in Deaths of 2 Children," *New York Times*, May 3, 2018; A. Winston, "Driver Who Killed Two Children in Brooklyn Is Found Dead," *New York Times*, November 7, 2018.
12. L. Italiano, "Judge Gives Trash-Haul Killer Life," *New York Post*, November 19, 2009; B. Aaron, "Driver Who Killed 3 People on Bronx Sidewalk Charged with Manslaughter," *StreetsBlog NYC*, September 20, 2016; B. Aaron, "Cab Driver Pleads to Homicide for Killing 2 on Bronx Sidewalk While Off Epilepsy Meds," *StreetsBlog NYC*, November 13, 2017.

    Footnote: S. Billakota, O. Devinsky, and K. Kim, "Why We Urgently Need Improved Epilepsy Therapies for Adult Patients," *Neuropharmacology* 170 (2019): 107855; K. Meador et al., "Neuropsychological and Neurophysiologic Effects of Carbamazepine and Levetiracetam," *Neurology* 69 (2007): 2076; D. Buck et al., "Factors Influencing Compliance with Antiepileptic Drug Regimes," *Seizure* 6 (1997): 87.
13. Italiano, "Judge Gives Trash-Haul Killer Life."
14. Second footnote: A. Weil, *The Natural Mind: An Investigation of Drugs and the Higher Consciousness* (Houghton Mifflin, 1998), p. 211.

    Third footnote: D. Rosenhan, "On Being Sane in Insane Places," *Science* 179 (1973): 250; S. Cahalan, *The Great Pretender* (Canongate Trade, 2019); also see: A. Abbott, "On the Troubling Trail of Psychiatry's Pseudopatients Stunt," *Nature* 574 (2019): 622.
15. P. Maki et al., "Predictors of Schizophrenia—a Review," *British Medical Bulletin* 73 (2005): 1; S. Stilo, M. Di Forti, and R. Murray, "Environmental Risk Factors for Schizophrenia: Implications for Prevention," *Neuropsychiatry* 1 (2011): 457; E. Walker and R. Lewine, "Prediction of Adult-Onset Schizophrenia from Childhood Home Movies of the Patients," *American Journal of Psychiatry* 147 (1990): 1052.
16. S. Bo et al., "Risk Factors for Violence among Patients with Schizophrenia," *Clinical Psychology Reviews* 31 (2014): 711; B. Rund, "A Review of Factors Associated with Severe Violence in Schizophrenia," *Nordic Journal of Psychiatry* 72 (2018): 561.
17. J. Lieberman and O. Ogas, *Shrinks: The Untold Story of Psychiatry* (Little Brown, 2015); E. Torrey, *Freudian Fraud: The Malignant Effect of Freud's Theory on American Thought and Culture* (HarperCollins, 1992).
18. A. Harrington, *Mind Fixers: Psychiatry's Troubled Search for the Biology of Mental Illness* (Norton, 2019); see Torrey, *Freudian Fraud.*
19. The quote regarding the conceptual advance of blaming schizophrenogenic families, rather than solely women, comes from P. Bart, "Sexiam and Social Science: From the Gilded Cage to the Iron Cage, or, the Perils of Pauline," *Journal of Marriage and the Family* (November 1971), 741.
20. Stilo, Di Forti, and Murray, "Environmental Risk Factors for Schizophrenia"; Maki et al., "Predictors of Schizophrenia."
21. R. Gentry, D. Schuweiler, and M. Roesch, "Dopamine Signals Related to Appetitive and Aversive Events in Paradigms That Manipulate Reward and Avoidability," *Brain Research* 1713 (2019): 80; P. Glimcher, "Understanding Dopamine and Reinforcement Learning: The Dopamine Reward Prediction Error Hypothesis," *Proceedings of the National Academy of Sciences of the United States of America* 108, supp. 3 (2011): 15647; M. Happel, "Dopaminergic Impact on Local and Global Cortical Circuit Processing during Learning," *Behavioral Brain Research* 299 (2016): 32.
22. A. Boyd et al., "Dopamine, Cognitive Biases and Assessment of Certainty: A Neurocognitive Model of Delusions," *Clinical Psychology Review* 54 (2017): 96; C. Chun, P. Brugger, and T. Kwapil, "Aberrant Salience across Levels of Processing in Positive and Negative

Schizotypy," *Frontiers of Psychology* 10 (2019): 2073; T. Winton-Brown et al., "Dopaminergic Basis of Salience Dysregulation in Psychosis," *Trends in Neurosciences* 37 (2014): 85.

23. P. Mallikarjun et al., "Aberrant Salience Network Functional Connectivity in Auditory Verbal Hallucinations: A First Episode Psychosis Sample," *Translational Psychiatry* 8 (2018): 69; K. Schonauer et al., "Hallucinatory Modalities in Prelingually Deaf Schizophrenic Patients: A Retrospective Analysis of 67 Cases," *Acta Psychiatrica Scandinavica* 98 (1998): 377; J. Atkinson, "The Perceptual Characteristics of Voice-Hallucinations in Deaf People: Insights into the Nature of Subvocal Thought and Sensory Feedback Loops," *Schizophrenia Bulletin* 32 (2006): 701; E. Anglemyer and C. Crespi, "Misinterpretation of Psychiatric Illness in Deaf Patients: Two Case Reports," *Case Reports in Psychiatry* 2018 (2018): 3285153; B. Engmann, "Peculiarities of Schizophrenic Diseases in Prelingually Deaf Persons," *MMW Fortschritte der Medizin* 153 supp. 1 (2011): 10. It is generally assumed (including by me) that everyone has an inner voice in their heads; this turns out to be wrong: D. Coffey, "Does Everyone Have an Inner Monologue?," *Livescience*, June 12, 2021. I thank Hilary Roberts for sending me to this source.
24. S. Lawrie et al., "Brain Structure and Function Changes during the Development of Schizophrenia: The Evidence from Studies of Subjects at Increased Genetic Risk," *Schizophrenia Bulletin* 34 (2008): 330; C. Pantelis et al., "Neuroanatomical Abnormalities Before and After Onset of Psychosis: A Cross-Sectional and Longitudinal MRI Comparison," *Lancet* 361 (2003): 281.
25. J. Harris et al., "Abnormal Cortical Folding in High-Risk Individuals: A Predictor of the Development of Schizophrenia?," *Biological Psychiatry* 56 (2004): 182; R. Birnbaum and D. Weinberger, "Functional Neuroimaging and Schizophrenia: A View towards Effective Connectivity Modeling and Polygenic Risk," *Dialogues in Clinical Neuroscience* 15 (2022): 279.
26. D. Eisenberg and K. Berman, "Executive Function, Neural Circuitry, and Genetic Mechanisms in Schizophrenia," *Neuropsychopharmacology* 35, no. 1 (2010): 258.
27. B. Birur et al., "Brain Structure, Function, and Neurochemistry in Schizophrenia and Bipolar Disorder—a Systematic Review of the Magnetic Resonance Neuroimaging Literature," *NPJ Schizophrenia* 3 (2017): 15; J. Fitzsimmons, M. Kubicki, and M. Shenton, "Review of Functional and Anatomical Brain Connectivity Findings in Schizophrenia," *Current Opinions in Psychiatry* 26 (2013): 172; K. Karlsgodt, D. Sun, and T. Cannon, "Structural and Functional Brain Abnormalities in Schizophrenia," *Current Directions in Psychological Sciences* 19 (2010): 226.
28. Footnote: For an exploration of some of the complexities of making sense of both schizophrenia *and* Parkinson's at the same time, see: J. Waddington, "Psychosis in Parkinson's Disease and Parkinsonism in Antipsychotic-Naive Schizophrenia Spectrum Psychosis: Clinical, Nosological and Pathobiological Challenges," *Acta Pharmacologica Sinica* 41 (2020): 464.
29. Footnote: K. Terkelsen, "Schizophrenia and the Family: II. Adverse Effects of Family Therapy," *Family Processes* 22 (1983): 191.
30. A. McLean, "Contradictions in the Social Production of Clinical Knowledge: The Case of Schizophrenia," *Social Science and Medicine* 30 (1990): 969. For a moving, horrendous history of a family as afflicted by schizophrenia as any in America, see: R. Kolker, *Hidden Valley Road: Inside the Mind of an American Family* (Doubleday, 2020).
31. T. McGlashan, "The Chestnut Lodge Follow-up Study. II. Long-Term Outcome of Schizophrenia and the Affective Disorders," *Archives of General Psychiatry* 41 (1984): 586. Torrey's satire: E. Fuller Torrey, "A Fantasy Trial about a Real Issue," *Psychology Today* (March 1977), 22.

Interviews with Eleanor Owen, Laurie Flynn, and Ron Honberg were conducted on 7/23/2019, 7/24/2019, and 7/25/2019, respectively.

Footnote: M. Sheridan et al., "The Impact of Social Disparity on Prefrontal Function in Childhood," *PLoS One* 7 (2012): e35744; J. L. Hanson et al., "Structural Variations in Prefrontal Cortex Mediate the Relationship between Early Childhood Stress and Spatial Working

Memory," *Journal of Neuroscience* 32 (2012): 7917; R. Sapolsky, "Glucocorticoids and Hippocampal Atrophy in Neuropsychiatric Disorders," *Archives of General Psychiatry* 57 (2000): 925.
32. Bruno Bettelheim, *Surviving—and Other Essays* (Knopf, 1979), p. 110.
33. M. Finn, "In the Case of Bruno Bettelheim," *First Things*, June 1997; R. Pollak, *The Creation of Dr. B: A Biography of Bruno Bettelheim* (Simon & Schuster, 1997).
34. D. Kaufer et al., "Acute Stress Facilitates Long-Lasting Changes in Cholinergic Gene Expression," *Nature* 393 (1998): 373; A. Friedman et al., "Pyridostigmine Brain Penetration under Stress Enhances Neuronal Excitability and Induces Early Immediate Transcriptional Response," *Nature Medicine* 2 (1996): 1382; R. Sapolsky, "The Stress of Gulf War Syndrome," *Nature* 393 (1998): 308; C. Amourette et al., "Gulf War Illness: Effects of Repeated Stress and Pyridostigmine Treatment on Blood-Brain Barrier Permeability and Cholinesterase Activity in Rat Brain," *Behavioral Brain Research* 203 (2009): 207; P. Landrigan, "Illness in Gulf War Veterans: Causes and Consequences," *Journal of the American Medical Association* 277 (1997): 259.
35. E. Klingler et al., "Mapping the Molecular and Cellular Complexity of Cortical Malformations," *Science* 371 (2021): 361; S. Mueller et al., "The Neuroanatomy of Transgender Identity: Mega-analytic Findings from the ENIGMA Transgender Persons Working Group," *Journal of Sexual Medicine* 18 (2021): 1122.

14. THE JOY OF PUNISHMENT

1. B. Tuchman, *A Distant Mirror* (Random House, 1994); P. Shipman, "The Bright Side of the Black Death," *American Science* 102 (2014): 410.
2. "In the Middle Ages There Was No Such Thing as Childhood," *Economist*, January 3, 2019; J. Robb et al., "The Greatest Health Problem of the Middle Ages? Estimating the Burden of Disease in Medieval England," *International Journal of Paleopathology* 34 (2021): 101; M. Shirk, "Violence and the Plague in Aragón, 1348–1351," *Quidditas* 5 (1984): article 5.
3. The Lepers' Plot: S. Tibble, "Medieval Strategy? The Great 'Leper Conspiracy' of 1321," Yale University Books, September 11, 2020, yalebooks.yale.edu/2020/09/11/medieval-strategy-the-great-leper-conspiracy-of-1321/; D. Nirenberg, *Communities of Violence: Persecution of Minorities in the Middle Ages* (Princeton University Press, 1996); I. Ritzmann, "The Black Death as a Cause of the Massacres of Jews: A Myth of Medical History?," *Medizin, Gesellschaft und Geschichte* 17 (1998): 101 [in German]; M. Barber, "Lepers, Jews and Moslems: The Plot to Overthrow Christendom in 1321," *History* 66 (1989): 1; T. Barzilay, "Early Accusations of Well Poisoning against Jews: Medieval Reality or Historiographical Fiction?," *Medieval Encounters* 22 (2016): 517.
4. Weyer: V. Hoorens, "The Link between Witches and Psychiatry: Johan Weyer," KU Lueven News, September 9, 2011, nieuws.kuleuven.be/en/content/2011/jan_wier.html; Encyclopedia.com, s.v. "Weyer, Johan," encyclopedia.com/science/encyclopedias-almanacs-transcripts-and-maps/weyer-johan-also-known-john-wier-or-wierus-1515-1588.
5. Execution of Robert Damiens, Château de Versailles: "Assassination Attempt on King Louis XV by Damiens, 1757," n.d., en.chateauversailles.fr/discover/history/key-dates/assassination-attempt-king-louis-xv-damiens-1757; "Letter from a Gentleman in Paris to His Friend in London," in Anonymous, *A Particular and Authentic Narration of the Life, Examination, Torture, and Execution of Robert Francis Damien* [*sic*], trans. Thomas Jones (London, 1757), also available at revolution.chnm.org/d/238. Renting out the box seats to the wealthy: "The Truly Horrific Execution of Robert-François Damiens," Unfortunate Ends, June 25, 2021, YouTube video, 14:40, youtube.com/watch?v=K7q8VSEBOMI; executedtoday.com/2008/03/28/1757-robert-francois-damiens-discipline-and-punish/.
6. A. Lollini, *Constitutionalism and Transitional Justice in South Africa* (*Human Rights in Context*, vol. 5) (Berghahn Books, 2011). For an exploration of the psychological weight of truth and

reconciliation, see P. Gobodo-Madikizela, *A Human Being Died That Night* (Houghton-Mifflin, 2003).

7. Positive reviews of restorative justice: V. Camp and J. Wemmers, "Victim Satisfaction with Restorative Justice: More Than Simply Procedural Justice," *International Review of Victimology* 19 (2013): 117; L. Walgrave, "Investigating the Potentials of Restorative Justice Practice," *Washington University Journal of Law and Policy* 36 (2011): 91.
8. F. Marineli et al., "Mary Mallon (1869–1938) and the History of Typhoid Fever," *Annals of Gastroenterology* 26 (2013): 132; J. Leavitt, *Typhoid Mary: Captive to the Public's Health* (Putnam, 1996).
9. For recent and forceful advocacy of quarantine approaches to criminality, see: D. Pereboom, *Wrongdoing and the Moral Emotions* (Oxford University Press, 2021); G. Caruso and D. Pereboom, *Moral Responsibility Reconsidered* (Cambridge University Press, 2022); G. Caruso, *Rejecting Retributivism* (Cambridge University Press, 2021); G. Caruso, "Free Will Skepticism and Criminal Justice: The Public Health–Quarantine Model," in *Oxford Handbook of Moral Responsibility*, ed. D. Nelkin and D. Pereboom (Oxford University Press, 2022).
10. M. Powers and R. Faden, *Social Justice: The Moral Foundations of Public Health and Health Policy* (Oxford University Press, 2006).
11. S. Smilansky, "Hard Determinism and Punishment: A Practical *Reductio*,' *Law and Philosophy* 30 (2011): 353. The worry that quarantine models will lead to indefinite detention: M. Corrado, "Fichte and the Psychopath: Criminal Justice Turned Upside Down," in *Free Will Skepticism in Law and Society*, ed. E. Shaw, D. Pereboom, and G. Caruso (Cambridge University Press, 2019).
12. Footnote: D. Zweig, "They Were the Last Couple in Paradise. Now They're Stranded," *New York Times*, April 5, 2020, nytimes.com/2020/04/05/style/coronavirus-honeymoon-stranded.html. Tracking down their resort island on Google Earth reveals that their Devil's Island with mai tais was approximately one thousand by six hundred feet.
13. R. Dundon, "Photos: Less Than a Century Ago, 20,000 People Traveled to Kentucky to See a White Woman Hang a Black Man," Timeline, *Medium*, February 22, 2018, timeline.com/rainy-bethea-last-public-execution-in-america-lischia-edwards-6f035f61c229; "Denies Owning Ring Found in Widow's Room," *Messenger-Inquirer* (Owensboro, KY), June 11, 1936; "Negro's Second Confession Bares Hiding Place," *Owensboro (KY) Messenger*, June 13, 1936; "10,000 See Hanging of Kentucky Negro; Woman Sheriff Avoids Public Appearance as Ex-policeman Springs Trap. CROWD JEERS AT CULPRIT Some Grab Pieces of Hood for Souvenirs as Doctors Pronounce Condemned Man Dead," *New York Times*, August 15, 1936; "Souvenir Hunters at Hanging Tear Hood Face," *Evening Star* (Washington, DC), August 14, 1936; C. Pitzulo, "The Skirted Sheriff: Florence Thompson and the Nation's Last Public Execution," *Register of the Kentucky Historical Society* 115 (2017): 377.
14. P. Kropotkin, *Mutual Aid: A Factor of Evolution* (1902; Graphic Editions, 2020). For a nice biography of him, see G. Woodcock, *Peter Kropotkin: From Prince to Rebel* (Black Rose Books, 1990).
15. K. Foster et al., "Pleiotropy as a Mechanism to Stability Cooperation," *Nature* 431 (2004): 693.

    Footnote: The merging of eukaryotic cells and mitochondria is one of the most important events in the history of life on earth, and its occurrence was first proposed by the visionary evolutionary biologist Lynn Margulis; naturally, it was roundly rejected and ridiculed for years by most in the field, until modern molecular techniques fully vindicated her. Her seminal paper is (published as Lynn Sagan, reflecting her marriage at the time to astronomer Carl Sagan): L. Sagan, "On the Origin of Mitosing Cells," *Journal of Theoretical Biology* 14 (1967): 255.

    W. Eberhard, "Evolutionary Consequences of Intracellular Organelle Competition," *Quarterly Review of Biology* 55 (1980): 231; J. Agren and S. Wright, "Co-evolution between

Transposable Elements and Their Hosts: A Major Factor in Genome Size," *Chromosome Research* 19 (2011): 777. Selfish mitochondria: J. Havird, "Selfish Mitonuclear Conflict," *Current Biology* 29 (2019): PR496.

16. Chimps: F. de Waal, *Chimpanzee Politics* (Allen & Unwin, 1982). Wrens: R. Mulder and N. Langmore, "Dominant Males Punish Helpers for Temporary Defection in Superb Fairy-Wrens," *Animal Behavior* 45 (1993): 830. Naked mole rats: H. Reeve, "Queen Activation of Lazy Workers in Colonies of the Eusocial Naked Mole-Rat," *Nature* 358 (1992): 147. Reefer/wrasse fish: R. Bshary and A. Grutter, "Punishment and Partner Switching Cause Cooperative Behaviour in a Cleaning Mutualism," *Biology Letters* 1 (2005): 396. Social bacteria: Foster et al., "Pleiotropy as a Mechanism to Stability Cooperation." Transposon hegemony: E. Kelleher, D. Barbash, and J. Blumenstiel, "Taming the Turmoil Within: New Insights on the Containment of Transposable Elements," *Trends in Genetics* 36 (2020): 474; J. Agren, N. Davies, and K. Foster, "Enforcement Is Central to the Evolution of Cooperation," *Nature Ecology and Evolution* 3 (2019): 1018. Transposon exploitation: E. Kelleher, "Reexamining the P-Element Invasion of *Drosophila melanogaster* through the Lens of piRNA Silencing," *Genetics* 203 (2016): 1513.
17. R. Boyd, H. Gintis, and S. Bowles, "Coordinated Punishment of Defectors Sustains Cooperation and Can Proliferate When Rare," *Science* 328 (2010): 617.
18. R. Axelrod and W. D. Hamilton, "The Evolution of Cooperation," *Science* 211 (1981): 1390. Also see J. Henrich and M. Muthukrishna, "The Origins and Psychology of Human Cooperation," *Annual Review of Psychology* 72 (2021): 207. Much of this game-theory literature is based on the assumed social equality of the players. For an analysis of how cooperation goes down the tubes once it is unequals who are playing, see O. Hauser et al., "Social Dilemmas among Unequals," *Nature* 572 (2019): 524.
19. G. Aydogan et al., "Oxytocin Promotes Altruistic Punishment," *Social Cognitive and Affective Neuroscience* 12 (2017): 1740; T. Yamagishi et al., "Behavioural Differences and Neural Substrates of Altruistic and Spiteful Punishment," *Science Reports* 7 (2017): 14654; T. Baumgartner et al., "Who Initiates Punishment, Who Joins Punishment? Disentangling Types of Third-Party Punishers by Neural Traits," *Human Brain Mapping* 42 (2021): 5703; O. Klimeck, P. Vuilleumier, and D. Sander, "The Impact of Emotions and Empathy-Related Traits on Punishment Behavior: Introduction and Validation of the Inequality Game," *PLoS One* 11 (2016): e0151028.

    Costly punishment during child development: Y. Kanakogi et al., "Third-Party Punishment by Preverbal Infants," *Nature Human Behaviour* 6 (2022): 1234; G. D. Salali, M. Juda, and J. Henrich, "Transmission and Development of Costly Punishment in Children," *Evolution and Human Behavior* 36, no. 2 (2015): 86–94;

    Just us: K. Riedl et al., "No Third-Party Punishment in Chimpanzees," *PNAS* 109, no. 37 (2012): 14824–29.
20. B. Herrmann, C. Thöni, and S. Gächter, "Antisocial Punishment across Societies," *Science* 319 (2008): 1362; J. Henrich and N. Henrich, "Fairness without Punishment: Behavioral Experiments in the Yasawa Island, Fiji," in *Experimenting with Social Norms: Fairness and Punishment in Cross-Cultural Perspective*, ed. J. Ensminger and J. Henrich (Russell Sage Foundation, 2014); J. Engelmann, E. Herrmann, and M. Tomasello, "Five-Year Olds, but Not Chimpanzees, Attempt to Manage Their Reputations," *PLoS One* 7 (2012): e48433; R. O'Gorman, J. Henrich, and M. Van Vugt, "Constraining Free Riding in Public Goods Games: Designated Solitary Punishers Can Sustain Human Cooperation," *Proceedings of the Royal Society B: Biological Sciences* 276 (2009): 323.
21. A. Norenzayan, *Big Gods: How Religion Transformed Cooperation and Conflict* (Princeton University Press, 2013); M. Lang et al., "Moralizing Gods, Impartiality and Religious Parochialism across 15 Societies," *Proceedings of the Royal Society B: Biological Sciences* 286 (2019): 1898. Also see: J. Henrich et al., "Market, Religion, Community Size and the Evolution of Fairness and Punishment," *Science* 327 (2010): 1480.

Footnote: Herrmann, Thöni, and Gächter, "Antisocial Punishment across Societies"; M. Cinyabuguma, T. Page, and L. Putterman, "Can Second-Order Punishment Deter Perverse Punishment?," *Experimental Economics* 9 (2006): 265.

22. J. Jordan et al., "Third-Party Punishment as a Costly Signal of Trustworthiness," *Nature* 530 (2016): 473.

Footnote: Henrich and Henrich, "Fairness without Punishment."

23. The costs and benefits of being a third party punisher: Jordan et al., "Third-Party Punishment as a Costly Signal"; N. Nikiforakis and D. Engelmann, "Altruistic Punishment and the Threat of Feuds," *Journal of Economic Behavior and Organization* 78 (2011): 319; D. Gordon, J. Madden, and S. Lea, "Both Loved and Feared: Third Party Punishers Are Viewed as Formidable and Likeable, but Those Reputational Benefits May Only Be Open to Dominant Individuals," *PLoS One* 27 (2014): e110045; M. Milinski, "Reputation, a Universal Currency for Human Social Interactions," *Philosophical Transactions of the Royal Society London B: Biological Sciences* 371 (2016): 20150100.

The emergence of third party punishment: K. Panchanathan and R. Boyd, "Indirect Reciprocity Can Stabilize Cooperation without the Second-Order Free Rider Problem," *Nature* 432 (2004): 499.

Features of third party punishment among hunter-gatherers: C. Boehm, *Hierarchy in the Forest: The Evolution of Egalitarian Behavior* (Harvard University Press, 1999).

24. T. Kuntz, "Tightening the Nuts and Bolts of Death by Electric Chair," *New York Times*, August 3, 1997.

25. For sober coverage of Bundy's life, trial, and execution, see: J. Nordheimer, "All-American Boy on Trial," *New York Times*, December 10, 1978; J. Nordheimer, "Bundy Is Put to Death in Florida after Admitting Trail of Killings," *New York Times*, January 25, 1989. Also see: B. Bearak, "Bundy Electrocuted after Night of Weeping, Praying: 500 Cheer Death of Murderer," *Los Angeles Times*, January 24, 1989; G. Bruney, "Here's What Happened to Ted Bundy after the Story Portrayed in *Extremely Wicked* Ended," *Esquire*, May 4, 2019, esquire.com/entertainment/a27363554/ted-bundy-extremely-wicked-execution/. For additional photos of celebrations related to his execution, visit gettyimages.com/detail/news-photo/sign-at-music-instrument-store-announcing-sale-on-electric-news-photo/72431549?adppopup=true and gettyimages.ie/detail/news-photo/sign-of-naked-lady-saloon-celebrating-the-execution-of-news-photo/72431550?adppopup=true and see M. Hodge, "THE DAY A MONSTER FRIED: How Ted Bundy's Electric Chair Execution Was Celebrated by Hundreds Shouting 'Burn, Bundy, Burn' Outside Serial Killer's Death Chamber," *Sun* (UK), January 16, 2019, thesun.co.uk/news/8202022/ted-bundy-execution-electric-chair-netflix-conversation-with-a-killer/.

For a memoir of someone who worked beside him throughout his murderous period, see A. Rule, *The Stranger Beside Me* (Norton, 1980). For an analysis by a pioneer in the psychological study of psychopathy, see R. Hare, *Without Conscience: The Disturbing World of the Psychopath among Us* (Guildford Press, 1999). For a truly insightful analysis of our obsession with serial murderers, see S. Marshall, "Violent Delights," *Believer*, December 22, 2022.

26. N. Mendes et al., "Preschool Children and Chimpanzees Incur Costs to Watch Punishment of Antisocial Others," *Nature Human Behaviour* 2 (2018): 45. Also see: M. Cant et al., "Policing of Reproduction by Hidden Threats in a Cooperative Mammal," *Proceedings of the National Academy of Sciences of the United States of America* 111 (2014): 325; T. Clutton-Brock and G. Parker, "Punishment in Animal Societies," *Nature* 373 (1995): 209.

27. Footnote: R. Deaner, A. Khera, and M. Platt, "Monkeys Pay per View: Adaptive Valuation of Social Images by Rhesus Macaques," *Current Biology* 15 (2005): 543; K. Watson et al., "Visual Preferences for Sex and Status in Female Rhesus Macaques," *Animal Cognition* 15 (2012): 401; A. Lacreuse et al., "Effects of the Menstrual Cycle on Looking Preferences for Faces in Female Rhesus Monkeys," *Animal Cognition* 10 (2007): 105.

28. Y. Wu et al., "Neural Correlates of Decision Making after Unfair Treatment," *Frontiers of Human Neuroscience* 9 (2015): 123; E. Du and S. Chang, "Neural Components of Altruistic Punishment," *Frontiers of Neuroscience* 9 (2015): 26; A. Sanfey et al., "Neuroeconomics: Cross-Currents in Research on Decision-Making," *Trends in Cognitive Sciences* 10 (2006): 108; M. Haruno and C. Frith, "Activity in the Amygdala Elicited by Unfair Divisions Predicts Social Value Orientation," *Nature Neuroscience* 13 (2010): 160; T. Burnham, "High-Testosterone Men Reject Low Ultimatum Game Offers," *Proceedings of the Royal Society B: Biological Sciences* 274 (2007): 2327.
29. G. Bellucci et al., "The Emerging Neuroscience of Social Punishment: Meta-analytic Evidence," *Neuroscience and Biobehavioral Reviews* 113 (2020): 426; H. Ouyang et al., "Empathy-Based Tolerance towards Poor Norm Violators in Third-Party Punishment," *Experimental Brain Research* 239 (2021): 2171.
30. D. de Quervain et al., "The Neural Basis of Altruistic Punishment," *Science* 305 (2004): 1254; B. Knutson, "Behavior. Sweet Revenge?," *Science* 305 (2004): 1246; D. Chester and C. DeWall, "The Pleasure of Revenge: Retaliatory Aggression Arises from a Neural Imbalance towards Reward," *Social Cognitive and Affective Neuroscience* 11 (2016): 1173; Y. Hu, S. Strang, and B. Weber, "Helping or Punishing Strangers: Neural Correlates of Altruistic Decisions as Third-Party and of Its Relation to Empathic Concern," *Frontiers of Behavioral Neuroscience* 9 (2015): 24; Baumgartner et al., "Who Initiates Punishment, Who Joins Punishment?"; G. Holstege et al., "Brain Activation during Human Male Ejaculation," *Journal of Neuroscience* 23 (2003): 9185.

    A belief in free will can even be motivated by the desire to cite it as a justification for righteous punishment: C. Clark et al., "Free to Punish: A Motivated Account of Free Will Belief," *Journal of Personality and Social Psychology* 106 (2014): 541. For an opposing view, see A. Monroe and D. Ysidron, "Not So Motivated after All? Three Replication Attempts and a Theoretical Challenge to a Morally Motivated Belief in Free Will," *Journal of Experimental Psychology: General* 150 (2021): e1.
31. T. Hu et al., "Helping Others, Warming Yourself: Altruistic Behaviors Increase Warmth Feelings of the Ambient Environment," *Frontiers of Psychology* 7 (2016): 1359; Y. Wang et al., "Altruistic Behaviors Relieve Physical Pain," *Proceedings of the National Academy of Sciences of the United States of America* 117 (2020): 950.
32. For an overview of where McVeigh was coming from, see: N. McCarthy, "The Evolution of Anti-government Extremist Groups in the U.S.," *Forbes*, January 18, 2021.

    Some of the statistics regarding the bombing: Office for Victims of Crime, *Responding to Terrorism Victims: Oklahoma City and Beyond* (U.S. Department of Justice, 2000), "Chapter I: Bombing of the Alfred P. Murrah Federal Building," ovc.ojp.gov/sites/g/files/xyckuh226/files/publications/infores/respterrorism/chap1.html.

    For contemporary reports about Timothy McVeigh's terrorist act, trial, and eventual execution, see: "Eyewitness Accounts of McVeigh's Execution," ABC News, June 11, 2001, abc-news.go.com/US/story?id=90542&page=1; "Eyewitness Describes Execution," *Wired*, June 11, 2001, wired.com/2001/06/eyewitness-describes-execution/; P. Carlson, "Witnesses for the Execution," *Washington Post*, April 11, 2001, washingtonpost.com/archive/lifestyle/2001/04/11/witnesses-for-the-execution/5b3083a2-364c-47bf-9696-1547269a6490/; J. Borger, "A Glance, a Nod, Silence and Death," *Guardian*, June 11, 2001, theguardian.com/world/2001/jun/12/mcveigh.usa . . .

    For an interesting take regarding the witnesses to executions, see A. Freinkel, C. Koopman, and D. Spiegel, "Dissociative Symptoms in Media Eyewitnesses of an Execution," *American Journal of Psychiatry* 151 (1994): 1335.

    "Invictus" is available at poetryfoundation.org/poems/51642/invictus.
33. A. Linders, *The Execution Spectacle and State Legitimacy: The Changing Nature of the American Execution Audience, 1833–1937* (Law and Society Association, 2002); R. Bennett, *Capital Punishment and the Criminal Corpse in Scotland, 1740–1834* (Palgrave Macmillan, 2017).

Footnote: M. Foucault, *Discipline and Punish: The Birth of the Prison* (Vintage, 1995); C. Alford, "What Would It Matter if Everything Foucault Said about Prison Were Wrong? *Discipline and Punish* after Twenty Years," *Theory and Society* 29 (2000): 125.

34. S. Bandes, "Closure in the Criminal Courtroom: The Birth and Strange Career of an Emotion," in *Research Handbook on Law and Emotion*, ed. S. Bandes et al. (Edward Elgar, 2021); M. Armour and M. Umbreit, "Assessing the Impact of the Ultimate Penal Sanction on Homicide Survivors: A Two State Comparison," *Marquette Law Review* 96 (2012), scholarship .law.marquette.edu/mulr/vol96/iss1/3/. Also see J. Madeira, "Capital Punishment, Closure, and Media," 2016, in *Oxford Research Encyclopedia of Criminology and Criminal Justice*, doi.org /10.1093/acrefore/9780190264079.013.20.
35. "The Death Penalty and the Myth of Closure," Death Penalty Information Center, January 19, 2021, deathpenaltyinfo.org/news/the-death-penalty-and-the-myth-of-closure.
36. A definitive history of Anders Breivik's life, his terrorist actions, and the aftermath can be found in A. Seierstad, *One of Us: The Story of Anders Breivik and the Massacre in Norway* (Farrar, Straus and Giroux, 2015); sharpening the horror of what he had done, the book also contained mini biographies of some of his numerous victims. Just to air a parental response to them, these were all great kids—humane, progressive, intent on doing good with their lives, and very likely to achieve exactly that.

While reading Seierstad's book, I was also reading Masha Gessen's *The Brothers: The Road to an American Tragedy* (Riverhead Books, 2015), an account of the Boston Marathon bombing carried out by the Tsarnaev brothers. The older brother, Tamerlan, was clearly the dominating force and catalyst of the two; Gessen's profile of him paints someone who seems astonishingly similar to Breivik; as opposite of ideologies as can be found, but the same mediocrities stewing with a sense of being entitled to glory and domination, and externalizing fault when they fall far short—pointless, empty vessels waiting to be filled with some sort of poison that would finally make them someone who could not be ignored. This same point was explored by Tom Nichols in "The Narcissism of the Angry Young Men," *Atlantic*, January 29, 2023: "They are man-boys who maintain a teenager's sharp sense of self-absorbed grievance long after adolescence; they exhibit a combination of childish insecurity and lethally bold arrogance; they are sexually and socially insecure. Perhaps most dangerous, they go almost unnoticed until they explode." The German writer Hans Magnus Enzensberger aptly call these young men "radical losers."

An interesting analysis of the trial: B. de Graaf et al., "The Anders Breivik Trial: Performing Justice, Defending Democracy," *Terrorism and Counter-Terrorism Studies* 4, no. 6 (2013), doi:10.19165/2013.1.06. Statement by Jens Stoltenberg: D. Rickman, "Norway's Prime Minister Jens Stoltenberg: We Are Crying with You after Terror Attacks," *Huffington Post*, July 24, 2011, huffingtonpost.co.uk/2011/07/24/norways-prime-minister-je_n_907937.html.

Mocking of Breivik's uniform: G. Toldnes, L. K. Lundervold, and A. Meland, "Slik skaffet han seg sin enmannshær" (in Norwegian), *Dagbladet Nyheter*, July 30, 2011. For more of the Norwegian response to the tragedy, see N. Jakobsson and S. Blom, "Did the 2011 Terror Attacks in Norway Change Citizens' Attitudes towards Immigrants?," *International Journal of Public Opinion Research* 26 (2014): 475. Statement of the rector of the university: "Anders Breivik accepted at Norway's University of Oslo," BBC, July 17, 2015, bbc.com/news/world -europe-33571929. Breivik getting to socialize with retired police officers: "Breivik saksøkte staten" (in Norwegian), NRK, October 23, 2015.

Casting light on every page of this book, Breivik's father, Jens, apparently self-published a book entitled *My Fault?*

37. The Emanuel African Methodist Episcopal Church massacre: M. Schiavenza, "Hatred and Forgiveness in Charleston," *Atlantic*, June 20, 2015; "Dylann Roof Told by Charleston Shooting Survivor 'the Devil Has Come Back to Claim' Him," CBS News, January 11, 2017, cbsnews.com/news/dylann-roof-charleston-shooting-survivor-devil-come-back-claim-him/; "Families of Charleston Shooting Victims to Dylann Roof: We Forgive You," *Yahoo!*

News, June 19, 2015, yahoo.com/news/familes-of-charleston-church-shooting-victims-to-dylann-roof--we--forgive-you-185833509.html?.

The Tree of Life Synagogue massacre: K. Davis, "Not Guilty Plea Entered for Alleged Synagogue Shooter on 109 Federal Charges" *San Diego Tribune*, May 14, 2019, sandiegouniontribune.com/news/courts/story/2019-05-14/alleged-synagogue-shooter-pleads-not-guilty-to-109-federal-charges. The efforts made by Jewish staffers at the hospital to which the shooter was taken: D. Andone, "Jewish Hospital Staff Treated Synagogue Shooting Suspect as He Spewed Hate, Administrator Says," CNN, November 1, 2018, cnn.com/2018/11/01/health/robert-bowers-jewish-hospital-staff/index.html. Quote by Jeff Cohen: E. Rosenberg, "'I'm Dr. Cohen': The Powerful Humanity of the Jewish Hospital Staff That Treated Robert Bowers," *Washington Post*, October 30, 2018, washingtonpost.com/health/2018/10/30/im-dr-cohen-powerful-humanity-jewish-hospital-staff-that-treated-robert-bowers/.

Deputy mayor of Oslo: H. Mauno, "Fikk brev fra Breivik: 'Da jeg leste navnet ditt, fikk jeg frysninger nedover ryggen'" (in Norwegian), *Dagsavisen*, April 8, 2021.

38. There was some outrage in the U.S. and UK about how lax the Norwegians were being in their punishment: S. Cottee, "Norway Doesn't Understand Evil," *UnHerd*, February 8, 2022, unherd.com/2022/02/norway-doesnt-understand-evil/; K. Weill, "All the Fun Things Anders Breivik Can Do in His 'Inhumane' Prison," *Daily Beast*, April 13, 2017, thedailybeast.com/all-the-fun-things-anders-breivik-can-do-in-his-inhumane-prison; J. Kirchick, "Mocking Justice in Norway: The Breivik Trial Targets Contrarian Intellectuals," *World Affairs* 175 (2012): 75; H. Gass, "Anders Breivik: Can Norway Be Too Humane to a Terrorist?," *Christian Science Monitor*, April 20, 2016. For an additional, obviously relevant analysis, and the quote from "one commentator," see S. Lucas, "Free Will and the Anders Breivik Trial," *Humanist*, August 13, 2012.

The U.S.'s 9/11 tragedy and the Breivik rampage were similar, insofar as the two terrorist acts killed roughly the same percentage of the population of the country; in both cases, the head of state addressed the mourning nation in the days afterward, both giving talks roughly five minutes long. Which is where the differences are stark. Bush cited God three times, evil four; for Stoltenberg, there was one mention of evil and none of God. Bush used the words *despicable*, *anger*, and *enemy*. In contrast, Stoltenberg used the words *compassion*, *dignity*, and *love*. Bush stated that this act of terror "cannot dent the steel of American resolve." Stoltenberg addressed the loved ones of victims, saying, "We are weeping with you."

Though the likes of a beheading or a public hanging are things of the past in the West, they're not all that far past—you could have taken the London subway to attend the last public hanging in the UK, in 1868; while the last guillotining was being carried out in France, you could spend your evening watching a *Star Wars* movie, dancing to the Bee Gees in a disco, or feeding (or not) your pet rock—1977.

For superb overviews of the focus of this entire chapter, see M. Hoffman, *The Punisher's Brain: The Evolution of Judge and Jury* (Cambridge University Press, 2014), as well as P. Alces, *Trialectic: The Confluence of Law, Neuroscience, and Morality* (University of Chicago Press, 2023).

15. IF YOU DIE POOR

1. To get a sense of the world of fifty-year-olds, in effect, dying of old age, see: A. Case and A. Deaton, *Deaths of Despair and the Future of Capitalism* (Princeton University Press, 2020).
2. M. Shermer, *Heavens on Earth: The Scientific Search for the Afterlife, Immortality, and Utopia* (Henry Holt, 2018); M. Quirin, J. Klackl, and E. Jonas, "Existential Neuroscience: A Review and Brain Model of Coping with Death Awareness," in *Handbook of Terror Management Theory*, ed. C. Routledge and M. Vess (Elsevier, 2019).
3. L. Alloy and L. Abramson, "Judgment of Contingency in Depressed and Nondepressed Students: Sadder but Wiser?," *Journal of Experimental Psychology* 108 (1979): 441. For a general

overview, see chapter 13, "Why Is Psychological Stress Stressful?," in R. Sapolsky, *Why Zebras Don't Get Ulcers: A Guide to Stress, Stress-Related Disease and Coping*, 3rd ed. (Holt, 2004).

4. R. Trivers, *Deceit and Self-Deception: Fooling Yourself to Better Fool Others* (Allen Lane, 2011).
5. Gomes quote: G. Gomes, "The Timing of Conscious Experience: A Critical Review and Reinterpretation of Libet's Research," *Consciousness and Cognition* 7 (1998): 559. Gazzaniga quote: M. Gazzaniga, "On Determinism and Human Responsibility," in *Neuroexistentialism: Meanings, Morals and Purpose in the Age of Neuroscience*, ed. G. Caruso (Oxford University Press, 2017), p. 232. Dennett quote: G. Caruso and D. Dennett, "Just Deserts," *Aeon*, https://aeon.co/essays/on-free-will-daniel-dennett-and-gregg-caruso-go-head-to-head.
6. For Dennett's charges of neuroscientists skeptical of free will being nefarious and irresponsible, see D. Dennett, "Daniel Dennett: Stop Telling People They Don't Have Free Will," n.d., Big Think video, 5:33, bigthink.com/videos/daniel-dennett-on-the-nefarious-neurosurgeon/.

   Footnote: Jin Park, "Harvard Orator Jin Park | Harvard Class Day 2018," Harvard University, May 23, 2018, YouTube video, 10:23, youtube.com/watch?v=TlWgdLzTPbc.
7. T. Chiang, "What's Expected of Us," *Nature* 436 (2005): 150.
8. R. Lake, "The Limits of a Pragmatic Justification of Praise and Blame," *Journal of Cognition and Neuroethics* 3 (2015): 229; P. Tse, "Two Types of Libertarian Free Will Are Realized in the Human Brain," in Caruso, *Neuroexistentialism*; R. Bishop, "Contemporary Views on Compatibilism and Incompatibilism: Dennett and Kane," *Mind and Matter* 7 (2009): 91.
9. Footnote: A. Sokal, "Transgressing the Boundaries: Toward a Transformative Hermeneutics of Quantum Gravity," *Social Text* 46/47 (1996): 217.
10. "Free will" as irrelevant to the biology, psychology and sociology of obesity:

    Genetic aspects: S. Alsters et al., "Truncating Homozygous Mutation of Carboxypeptidase E in a Morbidly Obese Female with Type 2 Diabetes Mellitus, Intellectual Disability and Hypogonadotrophic Hypogonadism," *PLoS One* 10 (2015): e0131417; G. Paz-Filho et al., "Whole Exam Sequencing of Extreme Morbid Obesity Patients: Translational Implications for Obesity and Related Disorders," *Genes* 5 (2014): 709; R. Singh, P. Kumar, and K. Mahalingam, "Molecular Genetics of Human Obesity: A Comprehensive Review," *Comptes rendus biologies* 340 (2017): 87; H. Reddon, J. Gueant, and D. Meyre, "The Importance of Gene-Environment Interactions in Human Obesity," *Clinical Sciences* (London) 130 (2016): 1571; D. Albuquerque et al., "The Contribution of Genetics and Environment to Obesity," *British Medical Bulletin* 123 (2017): 159.

    Evolutionary aspects: Z. Hochberg, "An Evolutionary Perspective on the Obesity Epidemic," *Trends in Endocrinology and Metabolism* 29 (2018): 819.

    Contribution of low social status to obesity: R. Wilkinson and K. Pickett, *The Spirit Level: Why More Equal Societies Almost Always Do Better* (Allen Lane, 2009); E. Goodman et al., "Impact of Objective and Subjective Social Status on Obesity in a Biracial Cohort of Adolescents," *Obesity Research* 11, no. 8 (2003): 1018–26;

    Footnote: Dutch Hunger Winter: B. Heijmans et al., "Persistent Epigenetic Differences Associated with Prenatal Exposure to Famine in Humans," *Proceedings of the National Academy of Sciences of the United States of America* 105 (2008): 17046. A recent, remarkable paper shows a similar phenomenon. In it, researchers examined individuals who were fetuses during the time that their parents' community was in its most dire economic state during the Great Depression; such individuals, many decades later, had an epigenetic profile associated with accelerated aging. L. Schmitz and V. Duque, "In Utero Exposure to the Great Depression Is Reflected in Late-Life Epigenetic Aging Signatures—Accelerated Epigenetic Markers of Aging," *Proceedings of the National Academy of Sciences of the United States of America* 119 (2022): e2208530119.
11. T. Charlesworth and M. Banaji, "Patterns of Implicit and Explicit Attitudes: I. Long-Term Changes and Stability from 2007 to 2016," *Psychological Sciences* 30 (2019): 174; S. Phelan et al., "Implicit and Explicit Weight Bias in a National Sample of 4,732 Medical Students: The

Medical Student CHANGES Study," *Obesity* 22 (2014): 1201; R. Carels et al., "Internalized Weight Stigma and Its Ideological Correlates among Weight Loss Treatment Seeking Adults," *Eating and Weight Disorders* 14 (2019): e92; M. Vadiveloo and J. Mattei, "Perceived Weight Discrimination and 10-Year Risk of Allostatic Load among US Adults," *Annals of Behavioral Medicine* 51 (2017): 94; R. Puhl and C. Heuer, "Obesity Stigma: Important Considerations for Public Health," *American Journal of Public Health* 100 (2010): 1019; L. Vogel, "Fat Shaming Is Making People Sicker and Heavier," *CMAJ* 191 (2019): E649.

12. Quote from Sam: S. Finch, "9 Affirmations You Deserve to Receive if You Have a Mental Illness," *Let's Queer Things Up!*, August 29, 2015, letsqueerthingsup.com/2015/08/29/9-affirmations-you-deserve-to-receive-if-you-have-a-mental-illness/. Quote from Arielle: D. Lavelle, "'I Assumed It Was All My Fault': The Adults Dealing with Undiagnosed ADHD," *Guardian*, September 5, 2017, theguardian.com/society/2017/sep/05/i-assumed-it-was-all-my-fault-the-adults-dealing-with-undiagnosed-adhd?scrlybrkr=74e99dd8. Quote from Marianne: M. Eloise, "I'm Autistic. I Didn't Know Until I Was 27," *New York Times*, December 5, 2020, nytimes.com/2020/12/05/opinion/autism-adult-diagnosis-women.html?action=click&module=Opinion&pgtype=Homepage
13. Quote from unnamed individual: QuartetQuarter, "Is it my fault I'm short?," Reddit, February 4, 2020, reddit.com/r/short/comments/ez3tcy/is_it_my_fault_im_short/. Quote from Manas: https://www.quora.com/How-do-I-get-past-the-fact-that-my-dad-blamed-me-for-being-short-and-not-pretty-I-shouldve-exercised-a-lot-more-and-eaten-a-lot-of-protein-while-growing-up-but-my-parents-are-short-too-Whose-fault-is-it.
14. Quotes from Kat and Erin: sane.org/information-stories/the-sane-blog/wellbeing/how-has-diagnosis-affected-your-sense-of-self. Quote from Michelle: Lavelle, "'I Assumed It Was All My Fault.'" Quote from Marianne: Eloise, "I'm Autistic. I Didn't Know." Quote from Sam: S. Finch, "4 Ways People with Mental Illness Are 'Gaslit' into Self-Blame," *Healthline*, July 30, 2019, healthline.com/health/mental-health/gaslighting-mental-illness-self-blame?scrlybrkr=74e99dd8.
15. LeVay's landmark study: S. LeVay, "A Difference in Hypothalamic Structure between Heterosexual and Homosexual Men," *Science* 253 (1991): 1034. Quote from father concerning his son's summer camp: sane.org/information-stories/the-sane-blog/wellbeing/how-has-diagnosis-affected-your-sense-of-self. The execrable preacher, Fred Phelps of the notorious Westboro Baptist Church, is discussed in: "Active U.S. Hate Groups (Kansas)," Southern Poverty Law Center. For a statement by the American Psychiatric Association condemning conversion therapy as pseudoscience, see American Psychiatric Association, "APA Maintains Reparative Therapy Not Effective," January 15, 1999, psychiatricnews.org/pnews/99-01-15/therapy.html.
16. For a review of how the effects of stress on reproductive physiology are far from simple, see J. Wingfield and R. Sapolsky, "Reproduction and Resistance to Stress: When and How," *Journal of Neuroendocrinology* 15 (2003): 711.

    The psychological impact of infertility: R. Clay, "Battling the Self-Blame of Infertility," *APA Monitor* 37 (2006): 44; A. Stanton et al., "Psychosocial Aspects of Selected Issues in Women's Reproductive Health: Current Status and Future Directions," *Journal of Consulting and Clinical Psychology* 70 (2002): 751; A. Domar, P. Zuttermeister, and R. Friedman, "The Psychological Impact of Infertility: A Comparison with Patients with Other Medical Conditions," *Journal of Psychosomatic Obstetrics and Gynecology* 14 (1993): 45.
17. S. James, "John Henryism and the Health of African-Americans," *Culture, Medicine and Psychiatry* 18 (1994): 163.
18. M. Sandel, *The Tyranny of Merit: What's Become of the Common Good?* (Farrar, Straus and Giroux, 2020); E. Anderson, "It's Not Your Fault if You Are Born Poor, but It's Your Fault if You Die Poor," *Medium*, January 21, 2022, medium.com/illumination-curated/its-not-your-fault-if-you-are-born-poor-but-it-s-your-fault-if-you-die-poor-36cf3d56da3f.

19. The Yiddish lyrics and English translation are available at genius.com/Daniel-kahn-and-the-painted-bird-mayn-rue-plats-where-i-rest-lyrics. A video of a performance is available at youtube.com/watch?v=lNRaU7zUGRo.

Science as being about statistical properties of populations, unable to predict enough about the individual: for a thoughtful discussion of this, see D. Faigman et al., "Group to Individual (G2i) Inferences in Scientific Expert Testimony," *University of Chicago Law Review* 81 (2014): 417.

Footnote: D. Von Drehle, "No, History Was Not Unfair to the Triangle Shirtwaist Factory Owners," December 20, 2018, washingtonpost.com/opinions/no-history-was-not-unfair-to-the-triangle-shirtwaist-factory-owners/2018/12/20/10fb050e-046a-11e9-9122-82e98f91ee6f_story.html. Collapse of Rana Plaza: Wikipedia, s.v. "2013 Rana Plaza Factory Collapse," wikipedia.org/wiki/2013_Rana_Plaza_factory_collapse?scrlybrkr=74e99dd8.

Illustration Credits

p. xii BookyBuggy/Shutterstock.com

p. 20 EEG illustration used with permission of Mayo Foundation for Medical Education and Research, all rights reserved; Harrison image North Wind Picture Archives/Alamy

p. 66 Robert Wood Johnson Foundation/Centers for Disease Control and Prevention

p. 67 Steve Lawrence, Oxford University RAE profile 2004/13/Wikimedia Commons

p. 131 Courtesy James Gleick, *Chaos: Making a New Science* (1987)

p. 134 Courtesy American Association for the Advancement of Science (AAAS), 139th Annual Meeting address, December 29, 1972

p. 137 (bottom) Beojan Stanislaus/Wikimedia Commons

p. 138 (top) Eouw0o83hf/Wikimedia Commons

p. 139 Courtesy Ebrahim Patel/The London Interdisciplinary School

p. 146 Courtesy E. Dameron-Hill, M. Farmer/*Chaos Theory: Nerds of Paradise*, Book 2 (2017)

p. 164 (top) Yamaoyaji/Shutterstock.com

p. 164 (bottom) Courtesy Nakagaki, T., et al., "Maze-solving by an amoeboid organism," *Nature* 407 (2000): 470

p. 165 Courtesy Tero, A., et al., "Rules for Biologically Inspired Adaptive Network Design," *Science* 327 (2010): 439

p. 167 Santiago Ramón y Cajal/Wikimedia Commons

p. 168 Central Historic Books/Alamy Stock Photo

p. 169 Alejandro Miranda/Alamy Stock Vector

p. 170 (top) © Alejandro Miranda/Dreamstime.com

p. 171 (top) Robert Brook/Science Photo Library/Alamy Stock Photo

p. 171 (bottom) Storman/istock.com

p. 172 Santiago Ramón y Cajal/Wikimedia

pp. 177, 178 Courtesy Hares Youssef/GAIIA Foundation, https://gaiia.foundation

p. 180 Courtesy Mohsen Afshar/Penney Gilbert Lab, University of Toronto

p. 190 (left) Courtesy Momoko Watanabe, Lab University of California, Irvine/Ben Novitch Lab, University of California, Los Angeles

p. 190 (right) Courtesy Arnold Krigstein, University of California, San Francisco

p. 194 Courtesy Christian List

pp. 270, 273, 277 Part of the Nobel Prize lecture of Erik Kandel, copyright © The Nobel Foundation 2000

p. 279 (left) © Seadan/Dreamstime.com

p. 279 (right) IrinaK/Shutterstock.com

p. 294 (below, left) Wikimedia Commons

p. 294 (below center) DeMarsico, Dick, photographer. Dr. Martin Luther King, Jr., half-length portrait, facing front/*World Telegram & Sun* photo by Dick DeMarsico, 1964. Photograph. https://www.loc.gov/item/00651714/

p. 294 (below right) Prachaya Roekdeethaweesab/Portrait from Dominican Republic 200 Pesos 2007 Banknotes/Shutterstock.com OR Diegobib/Dreamstime.com

p. 296 (top left) Guy Corbishley/Alamy Stock Photo

p. 296 (top right) Courtesy of DPAA Public Affairs

p. 296 (bottom left) Wikimedia Commons

p. 296 (bottom center) Ilbusca/Jesus on the Cross by Michelangelo/istock.com

p. 296 (bottom right) Moviestore Collection Ltd /Alamy Stock Photo

p. 298 (bottom left) Michael Flippo /Alamy Stock Photo

p. 298 (bottom right) FlamingoImages/iStock.com

p. 334 Courtesy Fuller Torrey, The Stanley Medical Research Institute

p. 343 Follower of the Virgil Master/Chapter 7, "Philip V"/British Library

p. 346 Chronicle/Alamy Stock Photo

p. 357 Bettmann/Getty Images

p. 358 Everett Collection Historical/Alamy Stock Photo

p. 367 Ken Hawkings/ Alamy Stock Photo

p. 368 Mark Foley/Associated Press

p. 373 GL Archive/Alamy Stock Photo

p. 381 AP Photo/via Scanpix

p. 409 Santiago Ramón y Cajal/Wikimedia

Index

Page numbers in *italics* refer to illustrations.